Compendium of Organic Synthetic Methods

Compendium of Organic Synthetic Methods

Volume 9

MICHAEL B. SMITH

DEPARTMENT OF CHEMISTRY
THE UNIVERSITY OF CONNECTICUT
STORRS, CONNECTICUT

A Wiley-Interscience Publication

JOHN WILEY & SONS, INC.
New York • Chichester • Weinheim • Brisbane • Singapore • Toronto

Cover illustration was adapted from "Disconnect By the Numbers: A Beginner's Guide to Synthesis" by M. B. Smith. *Journal of Chemical Education*, **1990**, 67, 848–856.

This book is printed on acid-free paper. ♾

Published simultaneously in Canada.

For ordering and customer service, call 1-800-CALL-WILEY.

Library of Congress Cataloging Card Number: 71-162800

ISBN 0-471-14579-3

Printed in the United States of America

10 9 8 7 6 5 4 3 2 1

CONTENTS

PREFACE

Since the original volume in this series by Ian and Shuyen Harrison, the goal of the *Compendium of Organic Synthetic Methods* was to facilitate the search for functional group transformations in the original literrature of Organic Chemistry. In Volume 2, difunctional compounds were added and this compilation was continued by Louis Hegedus and Leroy Wade for Voume 3 of the series. Wade became the author for Volume 4 and continued with Volume 5. I began editing the series with Volume 6, where I introduced an author index for the first time and added a new chapter (Chapter 15, Oxides). Volume 7 introduced Sections 378 (Oxides–Alkynes) through Section 390 (Oxides–Oxides). The *Compendium* is a handy desktop reference that will remain a valuable tool to the working Organic chemist, allowing a "quick check" of the literature. It also allows one to "browse" for new reactions and transformations that may be of interest. The body of Organic literature is very large and the *Compendium* is a focused and highly representative review of the literature and is offered in that context.

Compendium of Organic Synthetic Methods, Volume 9 contains both functional group transformations and carbon-carbon bond forming reactions from the literature appearing in the years 1993, 1994 and 1995. The classification schemes used for volumes 6–8 have been continued. Difunctional compounds appear in Chapter 16. The experienced user of the *Compendium* will require no special insructions for the use of Volume 9. Author citations and the Author Index have been continued as in Volumes 6–8.

Every effort has been made to keep the manuscript error free. Where there are errors I take full responsibility. If there are questions or comments, the reader is encouraged to contact me directly at the address, phone, fax, or Email addresses given below.

As I have througout my writing career, I thank my wife Sarah and my son Steven who have shown unfailing patience and devotion during this work. I also thank Darla Henderson, the editor of this volume.

Michael B. Smith

Department of Chemistry
University of Connecticut
55 N. Eagleville Road
Storrs, Connecticut 06269-3060

Voice phone: (860)-486-2881
Fax: (860)-486-2981
Email: smith@nucleus.chem.unconn.edu

ABBREVIATIONS

Ac	Acetyl	$C(=O)CH_3$
acac	Acetylacetonate	
AIBN	*azo-bis*-isobutyronitrile	
aq.	Aqueous	
	9-Borabicyclo[3.3.1]nonylboryl	
9-BBN	9-Borabicyclo[3.3.1]nonane	
BER	Borohydride exchange resin	
BINAP	*2R,3S*-2,2′-*bis*-(diphenylphosphino)-1,1′-binapthyl	
Bn	benzyl	
Bz	benzoyl	
BOC	*t*-Butoxycarbonyl	C(=O)O*t*-Bu
bpy (Bipy)	2,2′-Bipyridyl	
Bu	*n*-Butyl	$-CH_2CH_2CH_2CH_3$
CAM	Carboxamidomethyl	
CAN	Ceric ammonium nitrate	$(NH)_2Ce(NO_3)_6$
c-	cyclo-	
cat.	Catalytic	
Cbz	Carbobenzyloxy	$C(=O)OCH_2Ph$
Chirald	2S,3R-(+)-4-dimethylamino-1,2-diphenyl-3-methylbutan-2-ol	
COD	1,5-Cyclooctadienyl	
COT	1,3,5-cyclooctatrienyl	
Cp	Cyclopentadienyl	
CSA	Camphorsulfonic acid	
CTAB	cetyltrimethylammonium bromide	$C_{16}H_{33}NMe_3{+}Br^-$
Cy (c-C_6H_{11})	Cyclohexyl	
°C	Temperature in Degrees Centigrade	
DABCO	1,4-Diazobicyclo[2.2.2]octane	
dba	dibenzylidene acetone	
DBE	1,2-Dibromoethane	$BrCH_2CH_2Br$
DBN	1,8-Diazabicyclo[5.4.0]undec-7-ene	
DBU	1,5-Diazabicyclo[4.3.0]non-5-ene	
DCC	1,3-Dicyclohexylcarbodiimide	c-C_6H_{13}-N=C=N-c-C_6H_{13}

DCE	1,2-Dichloroethane	$ClCH_2CH_2Cl$
DDQ	2, 3-Dichloro-5,6-dicyano-1,4-benzoquinone	
% de	% Diasteromeric excess	
DEA	Diethylamine	$HN(CH_2CH_3)_2$
DEAD	Diethylazodicarboxylate	$EtO_2C\text{-}N{=}NCO_2Et$
Dibal-H	Diisobutylaluminum hydride	$(Me_2CHCH_2)_2AlH$
Diphos (dppe)	1,2-*bis*-(Diphenylphosphino)ethane	$Ph_2PCH_2CH_2PPh_2$
Diphos-4 (dppb)	1,4-*bis*-(Diphenylphosphino)butane	$Ph_2P(CH_2)_4PPh_2$
DMAP	4-Dimethylaminopyridine	
DMA	Dimethylacetamide	
DME	Dimethoxyethane	$MeOCH_2CH_2OMe$
DMF	*N,N′*-Dimethylformamide	$HC(=O)N(CH_3)_2$
dmp	*bis*-[1,3-Di(*p*-methoxyphenyl)-1,3-propanedionato]	
dpm	dipivaloylmethanato	
dppb	1,4-*bis*-(Diphenylphosphino)butane	$Ph_2P(CH_2)_4PPh_2$
dppe	1,2-*bis*-(Diphenylphosphino)ethane	$Ph_2PCH_2CH_2PPh_2$
dppf	*bis*-(Diphenylphosphino)ferrocene	
dppp	1,3-*bis*-(Diphenylphosphino)propane	$Ph_2P(CH_2)_3PPh_2$
dvb	Divinylbenzene	
e^-	Electrolysis	
% ee	% Enantiomeric excess	
EE	1-Ethoxyethyl	EtO(Me)HCO-
Et	Ethyl	$\text{-}CH_2CH_3$
EDA	Ethylenediamine	$H_2NCH_2CH_2NH_2$
EDTA	Ethylenediaminetetraacetic acid	
FMN	Flavin mononucleotide	
fod	*tris*-(6,6,7,7,8,8,8)-Heptafluoro-2,2-dimethyl-3,5-octanedionate	
Fp	Cyclopentadienyl-*bis*-carbonyl iron	
FVP	Flash Vacuum Pyrolysis	
h	hour (hours)	
hν	Irradiation with light	
1,5-HD	1,5-Hexadienyl	
HMPA	Hexamethylphosphoramide	$(Me_3N)_3P{=}O$
HMPT	Hexamethylphosphorous triamide	$(Me_3N)_3P$
iPr	Isopropyl	$\text{-}CH(CH_3)_2$
LICA (LIPCA)	Lithium cyclohexylisopropylamide	
LDA	Lithium diisopropylamide	$LiN(iPr)_2$
LHMDS	Lithium hexamethyl disilazide	$LiN(SiMe_3)_2$
LTMP	Lithium 2,2,6,6-tetramethylpiperidide	
MABR	Methylaluminum *bis*-(4-bromo-2,6-di-*tert*-butylphenoxide)	
MAD	*bis*-(2,6-di-*t*-butyl-4-methylphenoxy)methyl aluminum	
mCPBA	*meta*-Chloroperoxybenzoic acid	
Me	Methyl	$\text{-}CH_3$
MEM	β-Methoxyethoxymethyl	$MeOCH_2CH_2OCH_2\text{-}$
Mes	Mesityl	$2,4,6\text{-tri-Me-}C_6H_2$

MOM	Methoxymethyl	$MeOCH_2$-
Ms	Methanesulfonyl	CH_3SO_2-
MS	Molecular Sieves (3Å or 4Å)	
MTM	Methylthiomethyl	CH_3SCH_2-
NAD	Nicotinamide adenine dinucleotide	
NADP	Sodium triphosphopyridine nucleotide	
Napth	Napthyl ($C_{10}H_8$)	
NBD	Norbornadiene	
NBS	*N*-Bromosuccinimide	
NCS	*N*-Chlorosuccinimide	
NIS	*N*-Iodosuccinimide	
Ni(R)	Raney nickel	
NMP	N-Methyl-2-pyrrolidinone	
Oxone	2 $KHSO_5$•$KHSO_4$•K_2SO_4	
Ⓟ	Polymeric backbone	
PCC	Pyridinium chlorochromate	
PDC	Pyridinium dichromate	
PEG	Polyethylene glycol	
Ph	Phenyl	[structure]
PhH	Benzene	
PhMe	Toluene	
Phth	Phthaloyl	
pic	2-Pyridinecarboxylate	
Pip	Piperidine	H—N [structure]
PMP	4-methoxyphenyl	
Pr	*n*-Propyl	$-CH_2CH_2CH_3$
Py	Pyridine	N [structure]
quant.	Quantitative yield	
Red-Al	$[(MeOCH_2CH_2O)_2AlH_2]Na$	
sBu	*sec*-Butyl	$CH_3CH_2CH(CH_3)$
sBuLi	*sec*-Butyllithium	$CH_3CH_2CH(Li)CH_3$
Siamyl	Diisoamyl	$(CH_3)_2CHCH(CH_3)$-
TADDOL	α,α,α′,α′-tetraaryl-4,5-dimethoxy-1,3-dioxolane	
TASF	*tris*-(Diethylamino)sulfonium difluorotrimethyl silicate	
TBAF	Tetrabutylammonium fluoride	n-Bu_4N^+F
TBDMS	*t*-Butyldimethylsilyl	t-$BuMe_2Si$
TBHP (*t*-BuOOH)	*t*-Butylhydroperoxide	Me_3COOH
t-Bu	*tert*-Butyl	$-C(CH_3)_3$
TEBA	Triethylbenzylammonium	$Bn(CH_3)_3N^+$
TEMPO	Tetramethylpiperdinyloxy free radical	

TFA	Trifluoroacetic acid	CF_3COOH
TFAA	Trifluoroacetic anhydride	$(CF_3CO)_2O$
Tf (OTf)	Triflate	$-SO_2CF_3(-OSO_2CF_3)$
THF	Tetrahydrofuran	
THP	Tetrahydropyran	
TMEDA	Tetramethylethylenediamine	$Me_2NCH_2CH_2NMe_2$
TMG	1,1,3,3-Tetramethylguanidine	
TMS	Trimethylsilyl	$-Si(CH_3)_3$
TMP	2,2,6,6-Tetramethylpiperidine	
TPAP	tetra-*n*-Propylammonium perruthenate	
Tol	Tolyl	$4-C_6H_4CH_3$
Tr	Trityl	$-CPh_3$
TRIS	Triisopropylphenylsulfonyl	
Ts(Tos)	Tosyl = *p*-Toluenesulfonyl	$4-MeC_6H_4$
))))))	Sonication	
X_c	Chiral auxiliary	

INDEX, MONOFUNCTIONAL COMPOUNDS

Sections—**heavy type**
Pages—light type

Blanks in the table correspond to sections for which no additional examples were found in the literature

PROTECTION

	Sect.	Pg.
Carboxylic acids	30A	9
Alcohols, phenols	45A	42
Aldehydes	60A	55
Amides	90A	115
Amines	105A	137
Ketones	180A	206

PREPARATION OF →

FROM ↓

FROM	Alkynes	Carboxylic acid derivatives	Alcohols, phenols	Aldehydes	Alkyls, methylenes, aryls	Amides	Amines	Esters	Ethers, epxoides	Halides, sulfonates	Hydrides (RH)	Ketones	Nitriles	Alkenes	Miscellaneous
Alkynes	**1** 1									**10** 2		**12** 3		**14** 3	**15** 3
Carboxylic acid derivatives		**17** 5	**18** 6	**19** 6				**23** 7	**24** 7	**25** 7		**27** 8		**29** 8	**30** 9
Alcohols, phenols	**31** 11	**32** 11	**33** 12	**34** 13		**36** 25		**38** 26	**39** 26	**40** 29	**41** 30	**42** 30	**43** 39	**44** 39	**45** 41
Aldehydes		**47** 48	**48** 48	**49** 49	**50** 50		**52** 51	**53** 51	**54** 52	**55** 52		**57** 53	**58** 53	**59** 54	**70** 55
Alkyls, methylenes, aryls	**61** 58		**63** 60	**64** 60	**65** 61			**68** 64	**69** 65	**70** 65	**71** 70	**72** 71		**74** 72	**75** 99
Amides	**76** 100	**77** 100	**78** 101		**80** 102	**81** 102	**82** 108	**83** 110		**85** 111		**87** 112	**88** 113	**89** 113	**90** 113
Amines			**93** 117	**94** 118		**96** 119	**97** 120	**98** 129	**99** 129	**100** 130	**101** 131	**102** 131	**103** 132	**104** 132	**105** 133
Esters	**106** 140	**107** 140	**108** 142	**109** 144		**111** 145		**113** 146	**114** 148	**115** 149	**116** 150	**117** 151	**118** 153	**119** 154	**120** 154
Ethers, epxoides	**121** 156		**123** 156	**124** 158				**128** 159	**129** 160	**130** 161	**131** 161	**132** 161		**134** 162	**135** 169
Halides, sulfonates		**137** 171	**138** 171		**140** 172		**142** 172			**145** 173	**146** 173	**147** 174		**149** 175	**150** 175
Hydrides (RH)			**153** 176				**157** 177	**158** 178	**159** 179	**160** 179		**162** 181	**163** 182		**165** 182
Ketones	**166** 183	**167** 184	**168** 186	**169** 189	**170** 191	**171** 191	**172** 192	**173** 193	**174** 193	**175** 195	**176** 196	**177** 198		**179** 202	**180** 204
Nitriles			**183** 208	**184** 208		**186** 209		**188** 209		**190** 209		**192** 210			**195** 211
Alkenes	**196** 212	**197** 216	**198** 217	**199** 217	**200** 219	**201** 220	**202** 220	**203** 220	**204** 221	**205** 221		**207** 223		**209** 223	**210** 225
Miscellaneous	**211** 227	**212** 227	**213** 227				**217** 228	**218** 228	**219** 228	**220** 232	**221** 232	**222** 232	**223** 233	**234** 233	**225** 233

INDEX, DIFUNCTIONAL COMPOUNDS

Sections—**heavy type**
Pages—light type

Blanks in the table correspond to sections for which no additional examples were found in the literature

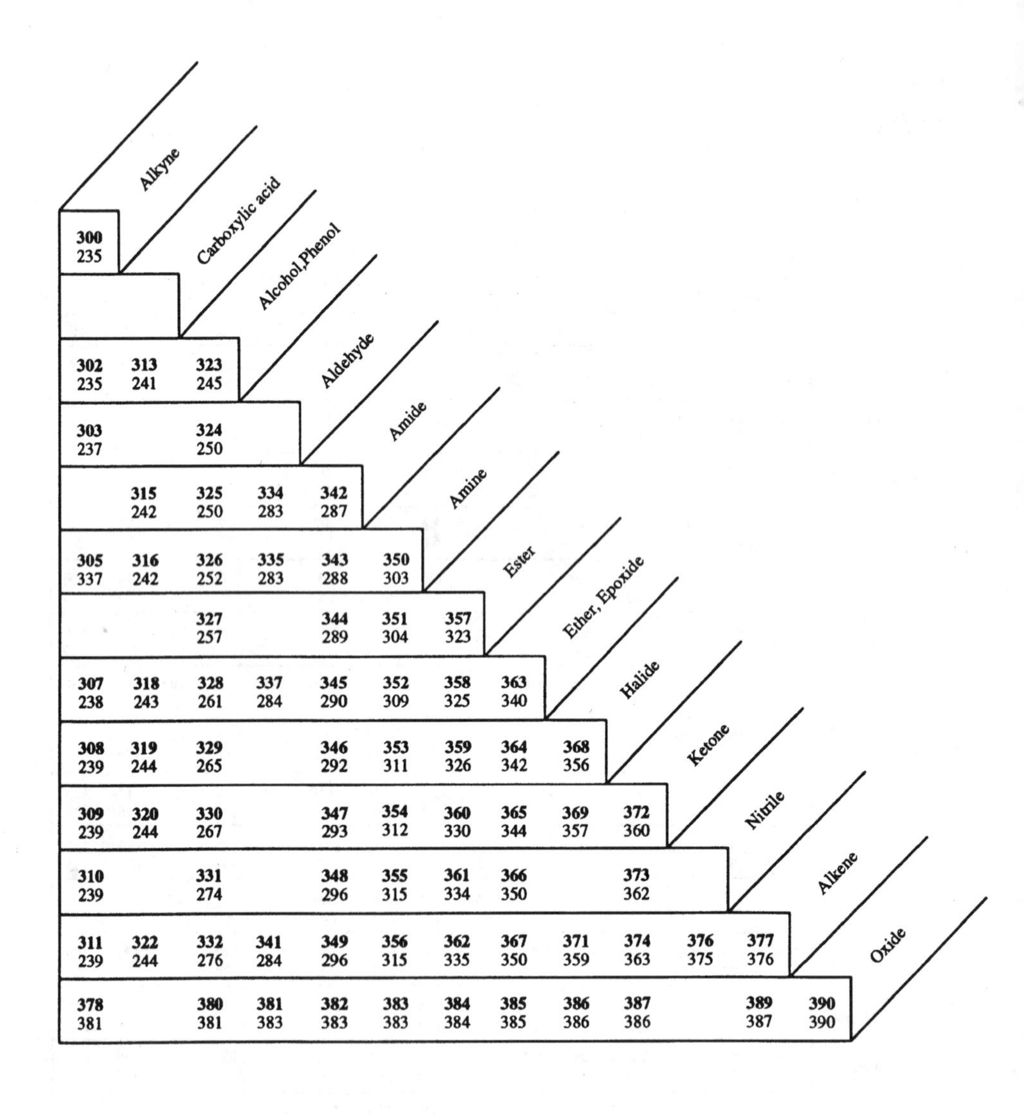

Alkyne	Carboxylic acid	Alcohol,Phenol	Aldehyde	Amide	Amine	Ester	Ether, Epoxide	Halide	Ketone	Nitrile	Alkene	Oxide
300 235												
302 235	**313** 241	**323** 245										
303 237		**324** 250										
	315 242	**325** 250	**334** 283	**342** 287								
305 337	**316** 242	**326** 252	**335** 283	**343** 288	**350** 303							
		327 257		**344** 289	**351** 304	**357** 323						
307 238	**318** 243	**328** 261	**337** 284	**345** 290	**352** 309	**358** 325	**363** 340					
308 239	**319** 244	**329** 265		**346** 292	**353** 311	**359** 326	**364** 342	**368** 356				
309 239	**320** 244	**330** 267		**347** 293	**354** 312	**360** 330	**365** 344	**369** 357	**372** 360			
310 239		**331** 274		**348** 296	**355** 315	**361** 334	**366** 350		**373** 362			
311 239	**322** 244	**332** 276	**341** 284	**349** 296	**356** 315	**362** 335	**367** 350	**371** 359	**374** 363	**376** 375	**377** 376	
378 381		**380** 381	**381** 383	**382** 383	**383** 383	**384** 384	**385** 385	**386** 386	**387** 386		**389** 387	**390** 390

INTRODUCTION

Relationship between Volume 9 and Previous Volumes. *Compendium of Organic Synthetic Methods, Volume 9* presents about 1200 examples of published reactions for the preparation of monofunctional compounds, updating the 10650 in Volumes 1–8. Volume 9 contains about 800 examples of reactions which prepare of difunctional compounds with various functional groups. Reviews have long been a feature of this series and Volume 9 adds almost 90 pertinent reviews in the various sections. Volume 9 contains approximately 1000 fewer entries than Volume 8 for an identical three-year period, primarily for difunctional compunds. Interestingly, there are about 500 fewer citations from the most cited journal (*Tetrahedron Letters*) than in the previous edition. Whether this represents a trend in the literature or an inadvertent selectivity on my part is unknown, but there has been a clear increase in biochemical and total synthesis papers which may account for this.

Chapters 1–14 continue as in Volumes 1–8, as does Chapter 15, introduced in Volume 6. Difunctional compounds appear in Chapter 16, as in Volumes 6 and 7. The sections on oxides as part of difunctional compounds, introduced in Volume 7, continues in Chapter 16 of Volumes 8 and 9 with Sections 378 (Oxides–Alkynes) through Section 390 (Oxides–Oxides).

Following Chapter 16 is a complete alphabetical listing of all authors (last name, initials). The authors for each citation appear below the reaction. The principle author is indicated by underlining (i.e., Kwon, T.W.; Smith, M. B.), as in Volumes 7 and 8.

Classification and Organization of Reactions Forming Monofunctional Compounds. Chemical transformations are classified according to the reacting functional group of the starting material and the functional group formed. Those reactions that give products with the same functional group form a chapter. The reactions in each chapter are further classified into sections on the basis of the functional group of the starting material. Within each section, reactions are loosely arranged in ascending order of year cited (1993–1995), although an effort has been made to put similar reactions together when possible. Review articles are collected at the end of each appropriate section.

The classification is unaffected by allylic, vinylic, or acetylenic unsaturation appearing in both starting material and product, or by increases or decreases in the length of carbon chains; for example, the reactions *t*-BuOH → *t*-BuCOOH, $PhCH_2OH$ → PhCOOH, and $PhCH=CHCH_2OH$ → PhCH=CHCOOH would all be considered as preparations of carboxylic acids from alcohols. Conjugate reduction and alkylation of unsaturated ketones, aldehydes, esters, acids, and nitriles have been placed in Sections 74D and 74E (Alkyls from Alkenes), respectively.

The terms hydrides, alkyls, and aryls classify compounds containing reacting hydrogens, alkyl groups, and aryl groups, respectively; for example, RCH_2-H → RCH_2COOH (carboxylic acids from hydrides), RMe → RCOOH (carboxylic acids from alkyls), RPh → RCOOH (carboxylic acids from aryls). Note the distinction between R_2CO → R_2CH_2 (methylenes from ketones) and RCOR′ → RH (hydrides from ketones). Alkylations involving additions across double bonds are found in Section 74 (alkyls, methylenes, and aryls from alkenes).

The following examples illustrate the classfication of some potentially confusing cases:

Reaction		Classification
RCH=CHCOOH →	$RCH=CH_2$	Hydrides from carboxylic acids
$RCH=CH_2$ →	RCH=CHCOOH	Carboxylic acids from hydrides
ArH →	ArCOOH	Carboxylic acids from hydrides
ArH →	ArOAc	Esters from hydrides
RCHO →	RH	Hydrides from aldehydes
RCH=CHCHO →	$RCH=CH_2$	Hydrides from aldehydes
RCHO →	RCH_3	Alkyls from aldehydes
R_2CH_2 →	R_2CO	Ketones from methylenes
RCH_2COR →	R_2CHCOR	Ketones from ketones
$RCH=CH_2$ →	RCH_2CH_3	Alkyls from alkenes (Hydrogenation of Alkenes)
RBr + HC≡CH →	RCH≡CR	Alkynes from halides; also alkynes from alkynes
ROH + RCOOH →	RCOOR	Esters from alcohols; also esters from carboxylic acids
RCH=CHCHO →	RCH_2CH_2CHO	Alkyls from alkenes (Conjugate Reduction)
RCH=CHCN →	RCH_2CH_2CN	Alkyls from alkenes (Conjugate Rduction)

How to Use the Book to Locate Examples of the Preparation of Protection of Monofunctional Compounds. Examples of the preparation of one functional group from another are found in the monofunctional index on p. x, which lists the corresponding section and page. Sections that contain examples of the reactions of a functional group are found in the horizontal rows of this index. Section 1 gives examples of the reactions of alkynes that form new alkynes; Section 16 gives reactions of alkynes that form carboxylic acids; and Section 31 gives reactions of alkynes that form alcohols.

Examples of alkylation, dealkylation, homologation, isomerization, and transposition are found in Sections 1, 17, 33, and so on, lying close to a diagonal of the index. These sections correspond to such topics as the preparation of alkynes from alkynes; carboxylic acids from carboxylic acids; and alcohols, thiols, and phenols from alcohols, thiols, and phenols. Alkylations that involve conjugate additions across a double bond are found in Section 74E (Alkyls, Methylenes, and Aryls from Alkenes).

Examples of name reactions can be found by first considering the nature of the starting material and product. The Wittig reaction, for instance, is in Section 199 (Alkenes from Aldehydes) and Section 207 (Alkenes frorm Ketones). The aldol condensation can be found in the chapters on difunctional compounds in Section 324 (Alcohol, Thiol-Aldehyde) and in Section 330 (Alcohol, Thiol-Ketone).

Examples of the protection of alkynes, carboxylic acids, alcohols, phenols, aldehydes, amides, amines, esters, ketones, and alkenes are also indexed on p. xvii. Section (designated with an A: 15A, 30A, etc.) with "protecting group: reactions are located at the end of pertinent chapters.

Some pairs of functional groups such as alcohol, ester; carboxylic acid, ester; amine, amide; and carboxylic acid, amide can be interconverted by simple reactions. When a member of these groups is the desired product or starting material, the other member should also be consulted in the text.

The original literature must be used to determine the generality of reactions, although this is occasionally stated in the citation. This is only done in cases where such generality is stated clearly in the original citation. A reaction given in this book for a primary aliphatic substrate may also be applicable to tertiary or aromatic compounds. This book provides very limited experimental conditions or precautions and the reader is referred to the original literature before attempting a reaction. **In no instance should a citation in this book be taken as a complete experimental procedure. Failure to refer to the original literature prior to beginning laboratory work could be hazardous.** The original papers usually yield a further set of references to previous work. Papers that appear after those publications can usually be found by consulting *Chemical Abstracts* and the *Science Citation Index*.

Classification and Organization of Reactions Forming Difunctional Compounds. This chapter considers all possible difunctional compounds formed from the groups acetylene, carboxylic acid, alcohol, thiol, aldehyde, amide, amine, ester, ether, epoxide, thioether, halide, ketone, nitrile, and alkene. Reactions that form difunctional compounds are classified into sections on the basis of two functional grups in the product that are pertinent to the reaction. The relative positions of the groups do not affect the classification. Thus preparations of 1,2-amino-alcohols, 1,3-amino-alcohols, and 1,4-amino-alcohols are included in a single section (Section 326, Alcohol-Amine). Difunctional compounds that have an oxide as the second group are found in the appropriate section (Sections 278–290). The nitroketone product of oxidation of a nitroalcohol is found in Section 386 (Ketone-Oxide). Conversion of an oxide to another functional grup is generally found in the "Miscellaneous" section of the sections concerning monofunctional compounds. Conversion of a nitroalkane to an amine, for example is found in Section 105 (Amines from Miscellaneous Compounds). The following examples illustrate applications of this classification system:

Difunctional Product	*Section Title*
RC≡C—C≡CR	Alkyne-Alkyne
RCH(OH)COOH	Carboxylic acid-Alcohol
RCH=CHOMe	Ether-Alkene
$RCHF_2$	Halide-Halide
$RCH(Br)CH_2F$	Halide-Halide
$RCH(OAc)CH_2OH$	Alcohol-Ester
$RCH(OH)CO_2Me$	Alcohol-Ester
$RCH=CHCH_2CO_2Me$	Ester-Alkene
RCH=CHOAc	Ester-Alkene
$RCH(OMe)CH_2SO_2CH_2CH_2OH$	Alcohol-Ether
$RSO_2CH_2CH_2OH$	Alcohol-Oxide

How to Use the Book to Locate Examples of the Preparation of Difunctional Compounds. The difunctional index on p. xi gives the section and page corresponding to each difunctional product. Thus Section 327 (Alcohol, Thiol-Ester) contains examples of the preparation of hydroxyesters; Section 323 (Alcohol, Thiol-Alcohol, Thiol) contains examples of the preparation of diols.

Some preparations of alkene and acetylenic compounds from alkene and acetylenic starting materials can, in principle, be classified in either the monofunctional or difunctional sections; for example, the transformation RCH=CHBr → RCH=CHCOOH could be considered as preparing carboxylic acids from halides (Section 25, monofunctional compounds) or preparing a carboxylic acid-alkene (Section 322, difunctional compounds). The choice usually depends on the focus of the particular paper where this reaction was found. In such cases both sections should be consulted.

Reactions applicable to both aldehyde and ketone starting materials are in many cases illustrated by an example that uses only one of them. Likewise, many citations for reactions found in the Aldehyde-X sections, will include examples that could be placed in the Ketone-X section. Again the choice is dictated by the paper where the reaction was found.

Many literature preparations of difunctional compounds are extensions of the methods applicable to monofunctional compounds. As an example, the reaction RCl → ROH might be used for the preparation of diols from an appropriate dichloro compund. Such methods are difficult to categorize and may be found in either the monofunctional or difunctional sections, depending on the focus of the original paper.

The user should bear in mind that the pairs of functional groups alcohol, ester; carboxylic acids, ester; amine, amide; and carboxylic acid, amide can be interconverted by simple reactions. Compounds of the type

$RCH(OAc)CH_2OAc$ (ester-ester) would thus be of interest to anyone preparing the diol $RCH(OH)CH_2OH$ (alcohol-alcohol).

Sources of Literature Citations. I thought it would be useful for a reader of this *Compendium* to see the distribution of citations used to this book (i.e., which journals have the most new synthetic methodology). As seen in the accompanying graph, *Tetrahedron Letters* and *Journal of Organic Chemistry* account for roughly 60% of all the citations in Volume 9. This book was not edited to favor one journal, category or type of article over another. Undoubtedly, my own personal preferences are part of the selection but I believe that this compilation is an accurate represention of new synthetic methods that appear in the literature for this period. Therefore, I believe the accompanying graph reflects those journals where new synthetic methodology is located. I should point out that the category "18 other journals" includes: *Accts. Chem. Res.; Acta Chem. Scand.; Angew. Chem. Int. Ed. Engl.; Bull. Chim. Soc. Belg.; Bull. Chim. Soc. Fr.; Can. J. Chem.; Chem. Ber.; Gazz. Chim. Ital.; Heterocycles; J. Chem. Soc.; J. Het. Chem.; J. Indian Chem. Soc; Liebigs Ann. Chem.; Org. Prep. Proceed Int.; Recl. Trav. Chim., Pays-Bas;* and *Tetrahedron Asymmetry*. In addition, nine more journals were examined but no references were recorded.

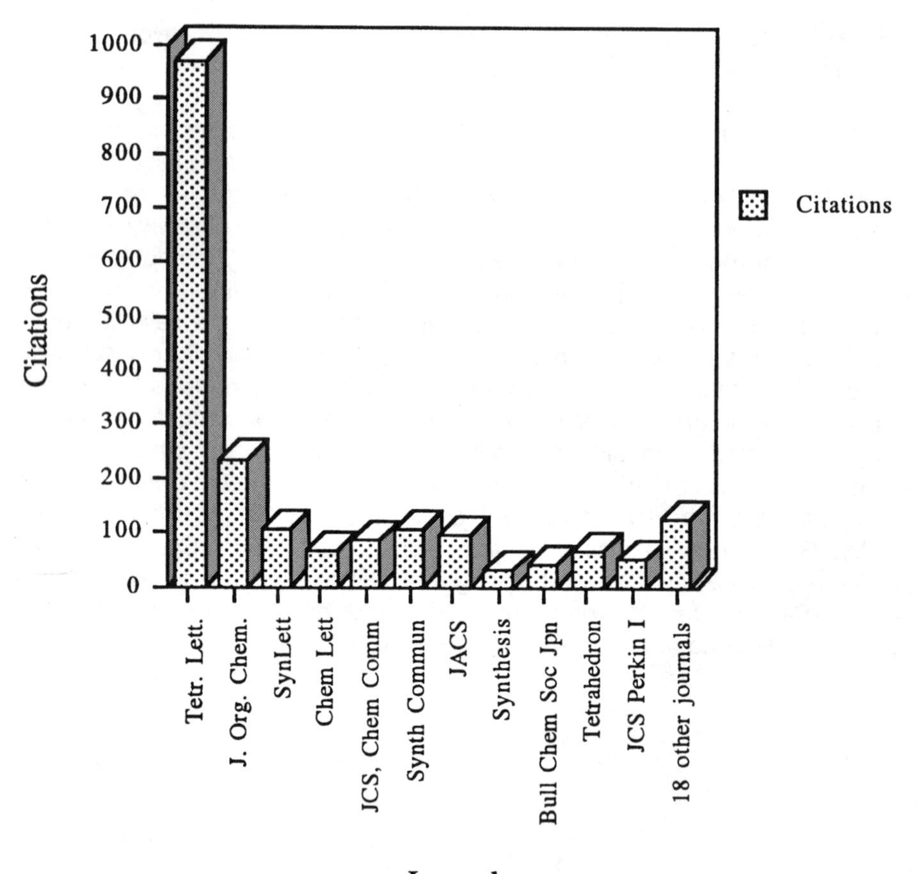

1000
900
800
700
600
500
400
300
200
100
0
Citations
Citations
Tetr. Lett.
J. Org. Chem.
SynLett
Chem Lett
JCS, Chem Comm
Synth Commun
JACS
Synthesis
Bull Chem Soc Jpn
Tetrahedron
JCS Perkin I
18 other journals
Journal

Compendium of Organic Synthetic Methods

CHAPTER 1
PREPARATION OF ALKYNES

SECTION 1: ALKYNES FROM ALKYNES

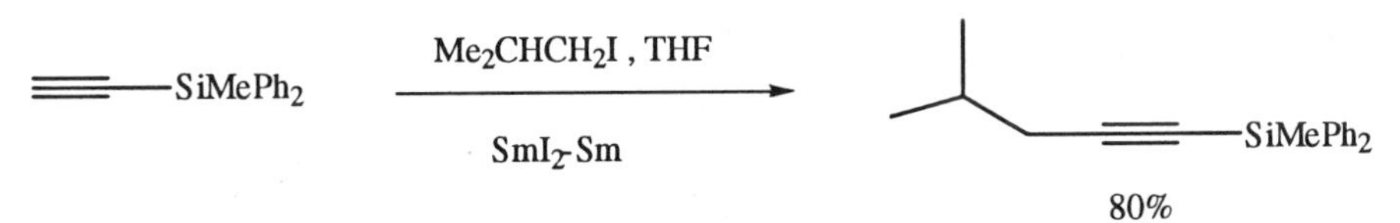

Murakami, M.; Hayashi, M.; Ito, Y. *Synlett*, ***1994***, 179

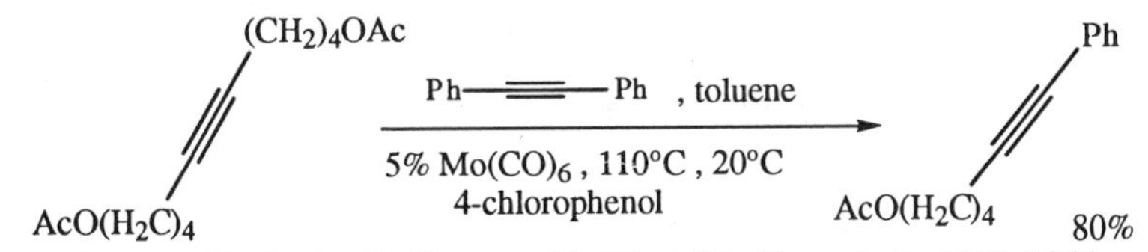

Kaneta, N.; Hikichi, K.; Asaka, S.; Uemura, M.; Mori, M. *Chem. Lett.*, ***1995***, 1055

REVIEW:

"Palladium And/Or Copper-Mediated Cross-Coupling Reactions Between 1-Alkynes And Vinyl, Aryl, 1-Alkynyl, 1,2-Propadienyl, Propargyl And Allylic Halides Or Related Compounds. A Review," Rossi, R.; Carpita, A.; Bellina, F. *Org. Prep. Proceed. Int.*, ***1995***, *27*, 129

SECTION 2: ALKYNES FROM ACID DERIVATIVES

NO ADDITIONAL EXAMPLES

SECTION 3: ALKYNES FROM ALCOHOLS AND THIOLS

NO ADDITIONAL EXAMPLES

SECTION 4: ALKYNES FROM ALDEHYDES

NO ADDITIONAL EXAMPLES

SECTION 5: ALKYNES FROM ALKYLS, METHYLENES AND ARYLS

NO ADDITIONAL EXAMPLES

SECTION 6: ALKYNES FROM AMIDES

NO ADDITIONAL EXAMPLES

SECTION 7: ALKYNES FROM AMINES

NO ADDITIONAL EXAMPLES

SECTION 8: ALKYNES FROM ESTERS

NO ADDITIONAL EXAMPLES

SECTION 9: ALKYNES FROM ETHERS, EPOXIDES AND THIOETHERS

NO ADDITIONAL EXAMPLES

SECTION 10: ALKYNES FROM HALIDES AND SULFONATES

NO_2 HO

Br

Pd/C , PPh_3 , CuI , DME
K_2CO_3

HO NO_2

98%

Bleicher, L.; Cosford, N.D.P. *Synlett*, ***1995***, 1115

AcO I I

H Bu

CuI , $Pd(PPh_3)_4$, NEt_3 , PhMe

AcO I Bu

95%

Bates, R.W.; Gabel, C.J.; Ji, J. *Tetrahedron Lett.*, ***1994***, *35*, 6993

SECTION 11: ALKYNES FROM HYDRIDES

For examples of the reaction RC≡CH → RC≡C-C≡CR[1], see section 300 (Alkyne-Alkyne).

NO ADDITIONAL EXAMPLES

SECTION 12: ALKYNES FROM KETONES

2 PhC(O)SiMe$_3$ —(Yb, THF, HMPA; -10°C, 20 min)→ Ph—≡—Ph 67%

Taniguchi, Y.; Fujii, N.; Makioka, Y.; Takai, K.; Fujiwara, Y. *Chem. Lett.*, ***1993***, 1165

PhC(O)Me —(TMSC(Li)N$_2$, THF; -78°C → reflux)→ Ph—≡—Me 58%

Miwa, K.; Aoyama, T.; Shioiri, T. *Synlett*, ***1994***, 107

SECTION 13: ALKYNES FROM NITRILES

NO ADDITIONAL EXAMPLES

SECTION 14: ALKYNES FROM ALKENES

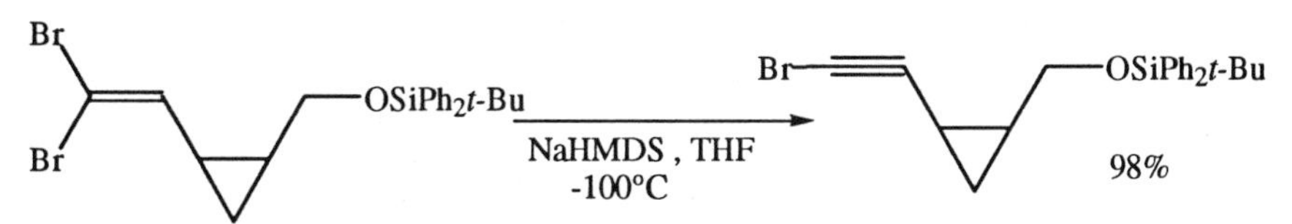

Grandjean, D.; Pale, P.; Chuche, J. *Tetrahedron Lett.*, ***1994***, *35*, 3529

SECTION 15: ALKYNES FROM MISCELLANEOUS COMPOUNDS

Ph$_3$P=CH–C(O)–*t*-Bu —(FVP (750°C))→ ≡—*t*-Bu 82%

Aitken, R.A.; Atherton, J.I. *J. Chem. Soc., Perkin Trans. 1.*, ***1994***, 1281

From arylCO derivatives:

Aitken, R.A.; Horsburgh, C.E.R.; McCreadie, J.G.; Seth, S. *J. Chem. Soc., Perkin Trans. 1.*, ***1994***, 1727

SECTION 15A: PROTECTION OF ALKYNES

NO ADDITIONAL EXAMPLES

CHAPTER 2
PREPARATION OF ACID DERIVATIVES

SECTION 16: ACID DERIVATIVES FROM ALKYNES

NO ADDITIONAL EXAMPLES

SECTION 17: ACID DERIVATIVES FROM ACID DERIVATIVES

CO_2H OMe — BuLi, THF-HMPA; OTs — HO_2C MeO + HO_2C MeO

(10 : 1) 82%

Kusumoto, T.; Ichikawa, S.; Asaka, K.; Sato, K.; Hiyama, T. *Tetrahedron Lett.*, ***1995***, *36*, 1071

CO_2H CO_2H — KF, microwave, 4 min; OAc — O O O

94%

Villemin, D.; Labiad, B.; Loupy, A. *Synth. Commun.*, ***1993***, *23*, 419

SECTION 18: ACID DERIVATIVES FROM ALCOHOLS AND THIOLS

$C_5H_{11}CH_2OH$ —[O_2 , $Pb(OAc)_2$•3 H_2O , Pd-C , 1h / NaOH , H_2O , 150°C]→ $C_5H_{11}CH_2OH$ 90%

Akada, M.; Nakano, S.; Sugiyama, T.; Tchitoh, K.; Nakao, H.; Akita, M.; Moro-Oka, Y. *Bull. Chem. Soc. Jpn.*, ***1993***, *66*, 1511

TMSO (bicyclic cyclopropane) —[1. $PhI(OAc)_2$, AcOH rt , 8h / 2. H_2O]→ HO_2C (alkenyl chain) 92%

Kirihara, M.; Yokoyama, S.; Kakuda, H.; Momose, T. *Tetrahedron Lett.*, ***1995***, *36*, 6907

SECTION 19: ACID DERIVATIVES FROM ALDEHYDES

PhCHO —[H_2O_2 , HCO_2H]→ $PhCO_2H$ 93%

Dodd, R.H.; Le Hyaric, M. *Synthesis*, ***1993***, 295

Ph–CH=CH–CHO —[$NaClO_2$, aq. MeCN / 10°C → rt]→ Ph–CH=CH–CO_2H 93%

Babu, B.R.; Balasubramanian, K.K. *Org. Prep. Proceed. Int.*, ***1994***, *26*, 123

PhCHO —[NaN_3-MnO_2/$SiCl_4$, CH_2Cl_2 , 0°C]→ Ph–C(=O)–N_3 89%

Elmorsy, S.S. *Tetrahedron Lett.*, ***1995***, *36*, 1341

SECTION 20: ACID DERIVATIVES FROM ALKYLS, METHYLENES AND ARYLS

NO ADDITIONAL EXAMPLES

SECTION 21: ACID DERIVATIVES FROM AMIDES

NO ADDITIONAL EXAMPLES

SECTION 22: ACID DERIVATIVES FROM AMINES

NO ADDITIONAL EXAMPLES

SECTION 23: ACID DERIVATIVES FROM ESTERS

Other reactions useful for the hydrolysis of esters may be found in Section 30A (Protection of Carboxylic Acids).

Ph—CO_2Me —(Et_2AlN_3)→ PhC(O)N_3 68%

Rawal, V.H.; Zhong, H.M. *Tetrahedron Lett.*, ***1994***, *35*, 4947

$PhCO_2Me$ —(1. 2 eq. KOH/Aliquat, microwave; 2. HCl)→ $PhCO_2H$

Loupy, A.; Pigeon, P.; Ramdani, M.; Jacquault, P. *Synth. Commun.*, ***1994***, *24*, 159

Ph–CH=CH–CH_2OCHO —(CO (3 atm), 8% PPh_3, 96h, rt; 2% $Pd_2(dba)_3CHCl_3$, toluene)→ Ph–CH=CH–CH_2CO_2H

92% (*E* only)

Yamamoto, A. *Bull. Chem. Soc. Jpn.*, ***1995***, *68*, 433

SECTION 24: ACID DERIVATIVES FROM ETHERS, EPOXIDES AND THIOETHERS

Ph-epoxide —(1. 10% $[Bi(mandelate)_2]_2O$, DMSO, 80°C, 30 min, air; 2. H_3O^+)→ Ph-COOH 60%

Zevaco, T.; Duñach, E.; Postel, M. *Tetrahedron Lett.*, ***1993***, *34*, 2601

4-O_2N-C_6H_4-CH_2OSiMe_3 —($AgBrO_3$, $AlCl_3$, MeCN; reflux, 0.35h)→ 4-O_2N-C_6H_4-CO_2H 95%

Firouzabadi, H.; Mohammadpoor-Baltork, I. *Synth. Commun.*, ***1994***, *24*, 1065

SECTION 25: ACID DERIVATIVES FROM HALIDES AND SULFONATES

MeI —(CO_2, H_2, 150°C, 1d; $Ni(cod)_2$ + $Co_2(CO)_8$)→ $MeCO_2H$

Fukuoka, A.; Gotoh, N.; Kobayashi, N.; Hirano, M.; Komiya, S. *Chem. Lett.*, ***1995***, 567

REVIEW:

"Preparation Of Thiol Acids, Thiol Esters And Amides By Reactions Of Carbonyl Sulfides With Grignard Reagents," Katritzky, A.R.; Moutou, J.-L.; Yang, Z. *Org. Prep. Proceed. Int.*, ***1995**, 27*, 361

SECTION 26: ACID DERIVATIVES FROM HYDRIDES

NO ADDITIONAL EXAMPLES

SECTION 27: ACID DERIVATIVES FROM KETONES

Cl ; $Na_2CO_3 \cdot 3/2\ H_2O_2$, 60°C ; 3h ; CO_2H ; CO_2H ; 86%

Yang, D.T.C.; Cao, Y.H.; Kabalka, G.W. *Synth. Commun.*, ***1995**, 25*, 3695

SECTION 28: ACID DERIVATIVES FROM NITRILES

NO ADDITIONAL EXAMPLES

SECTION 29: ACID DERIVATIVES FROM ALKENES

Ph ; $Fe(CO)_5$, KOH , CO (1 atm) ; H_2O/iPrOH , 55°C , 4d ; CO_2H ; Ph ; Me ; + ; Ph ; CO_2H

(96 : 4) 45%

Brunet, J.-J.; Neibecker, D.; Srivastava, R.S. *Tetrahedron Lett.*, ***1993**, 34*, 2759

HCO_2H , CO (6.8 atm) , DME ; $Pd(OAc)_2$ - dppb , 150°C , 16h ; CO_2H ; 86%

El Ali, B.; Alper, H. *J. Org. Chem.*, ***1993**, 58*, 3595

Ph ; HCOOH , 6.8 atm CO ; 10% Pd/C , dppb , DME ; 150°C , 1d ; Ph ; CO_2H ; + ; CO_2H ; Ph

(76 : 24) 65%

El Ali, B.; Vasapollo, G.; Alper, H. *J. Org. Chem.*, ***1993**, 58*, 4739

$C_6H_{13}CH=CH_2$ $\xrightarrow[\text{acetone, rt}]{\text{cat. } OsO_4 \text{, Jones reagent}}$ $C_6H_{13}-CO_2H$ 85%

Henry, J.R.; Weinreb, S.M. *J. Org. Chem.*, ***1993***, *58*, 4745

$\xrightarrow[\text{HCOOH, DME}]{\text{Pd/C, dppb, } PPh_3 \text{, CO}}$ CO_2H

60%

Vasapollo, G.; Somasunderam, A.; El Ali, B.; Alper, H. *Tetrahedron Lett.*, ***1994***, *35*, 6203

$PhCH{=}CH_2$ $\xrightarrow[70°C\text{, 1d}]{O_2\text{, 2\% Co(acac)}_3\text{, THF}}$ $PhCO_2H$ + PhCHO

(86 14) 99%

Reetz, M.T.; Töllner, K. *Tetrahedron Lett.*, ***1995***, *36*, 9461

SECTION 30: ACID DERIVATIVES FROM MISCELLANEOUS COMPOUNDS

$PhIO_2$ $\xrightarrow[2.\ H_3O^+]{1.\ 0.1\%\ Na_2[PdCl_4]\text{, CO, } H_2O\text{, } Na_2CO_3\text{, 40°C, 6.5h}}$ PhCOOH 71%

Grushin, V.V.; Alper, H. *J. Org. Chem.*, ***1993***, *58*, 4794

SECTION 30A: PROTECTION OF CARBOXYLIC ACID DERIVATIVES

PhC(=O)O–CH(Ph)–C(=O)Ph $\xrightarrow{5\ TBAF{\bullet}xH_2O,\ 5BnSH}$ PhCOOH + BrC(=O)Ph

82% quant

Ueki, M.; Aoki, H.; Katoh, T. *Tetrahedron Lett.*, ***1993***, *34*, 2783

PhC(=O)O–CH₂Ph $\xrightarrow[\textbf{microwave}\text{, } CH_2Cl_2\text{, rt}]{\text{acidic } Al_2O_3\text{, 7 min}}$ $PhCO_2H$ 92%

Varma, R.S.; Chatterjee, A.K.; Varma, M. *Tetrahedron Lett.*, ***1993***, *34*, 4603

HO (prenyl alcohol); cat. DMAP, rt, 6h; I_2, cyclohexane, rt; 75%

Cossy, J.; Albouy, A.; Scheloske, M.; Gomez Pardo, D. *Tetrahedron Lett.*, ***1994***, *35*, 1539

DCC, 0.1 DMAP, THF; HO–CH(CH$_3$)–C$_6$H$_4$–OMe; 4°C → rt, 17h; 93%; Mpeoc group; quant; 1% TFA in CH_2Cl_2, 5 min, rt

Bernatowicz, M.S.; Chao, H.-G.; Matsueda, G.R. *Tetrahedron Lett.*, ***1994***, *35*, 1651

PhO_2C–CH$_2$–C(O)–Ph; $CuCl_2$, O_2, aq. DMF; Ph—CO_2H; 45%

Ram, R.N.; Singh, L. *Tetrahedron Lett.*, ***1995***, *36*, 5401

CO_2Bn; 1. NBS, $(BzO)_2$, CCl_4, heat; 2. H_2O; CO_2H; 83%

Anson, M.S.; Montana, J.G. *Synlett*, ***1994***, 219

REVIEW:

"Recent Developments in Chemical Deprotection of Ester Functional Groups," Salomon, C.J.; Mata, E.G.; Mascaretti, O.A. *Tetrahedron*, ***1993***, *49*, 3691

Other reactions useful for the protection of carboxylic acids are included in Section 107 (Esters from Carboxylic Acids and Acid Halides) and Section 23 (Carboxylic Acids from Esters).

CHAPTER 3
PREPARATION OF ALCOHOLS

SECTION 31: ALCOHOLS AND THIOLS FROM ALKYNES

C_8H_{17}—≡ → (1. BH_3•NEt_2Ph; 2. BuLi; 3. BnBr; 4. H_2O_2, OH^-) → C_8H_{17}CH$_2$CH(OH)CH$_2$Ph 62%

Reddy, Ch.K.; Periasamy, M. *Tetrahedron*, ***1993***, *49*, 8897

SECTION 32: ALCOHOLS AND THIOLS FROM ACID DERIVATIVES

C_4H_9CH$_2$COOH → (BuLi, $CeCl_3$, THF; 16h, -78°C) → C_4H_9CH$_2$CH(OH)Bu 83%

Ahn, Y.; Cohen, T. *Tetrahedron Lett.*, ***1994***, *35*, 203

(valine: CO_2H, H, NH_2) → ($NaBH_4$ - I_2, THF; reflux) → (valinol: OH, H, NH_2) 94%

McKennon, M.J.; Meyers, A.I.; Drauz, K.; Schwarm, M. *J. Org. Chem.*, ***1993***, *58*, 3568

Ph—COOH → ($Zn(BH_4)_2$, THF, reflux) → Ph—CH_2OH 90%

Narasimhan, S.; Madhavan, S.; Prasad, K.G. *J. Org. Chem.*, ***1995***, *60*, 5314

PhCH_2C(O)O—N (pyridine-2-thione) → (1. S_8, hν, 0°C; 2. $NaBH_4$, MeOH) → PhCH_2SH + 2-mercaptopyridine (HS)

90%

Barton, D.H.R.; Castagnino, E.; Jaszberenyi, J.Cs. *Tetrahedron Lett.*, ***1994***, *35*, 6057

$PhCO_2H$ —(SmI_2-H_2O, 1 min)→ $PhCH_2OH$ 89%

Kamochi, Y.; Kudo, T. *Chem. Lett.*, ***1993***, 1495

$C_{14}H_{29}COOH$ —(H_2 (100 atm), DME, 160°C; 5% Ru/Al_2O_3—$Mo(CO)_6$)→ $C_{14}H_{29}CH_2OH$ 95%

He, D.-H.; Wakasa, N.; Fuchikami, T. *Tetrahedron Lett.*, ***1995***, *36*, 1059

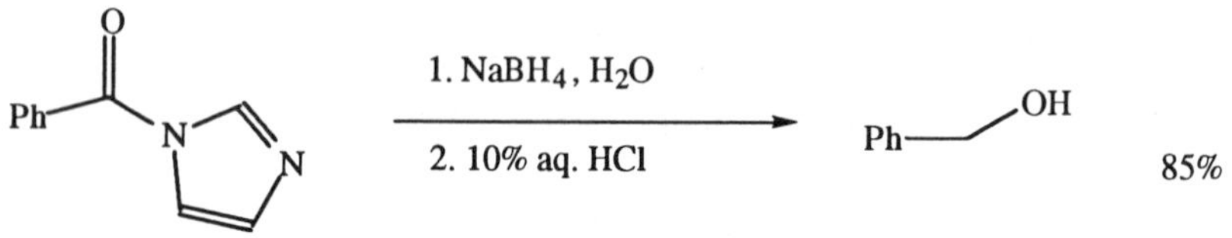

1. $NaBH_4$, H_2O
2. 10% aq. HCl

Ph OH 85%

Sharma, R.; Voymov, G.H.; Ovaska, T.V.; Marquez, V.E. *Synlett*, ***1995***, 839

SECTION 33: ALCOHOLS AND THIOLS FROM ALCOHOLS AND THIOLS

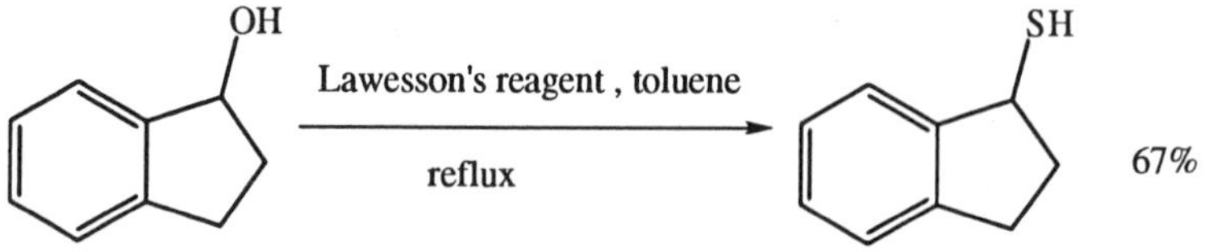

OH —(Lawesson's reagent, toluene; reflux)→ SH 67%

Nishio, T. *J. Chem. Soc., Perkin Trans. 1.*, ***1993***, 1113

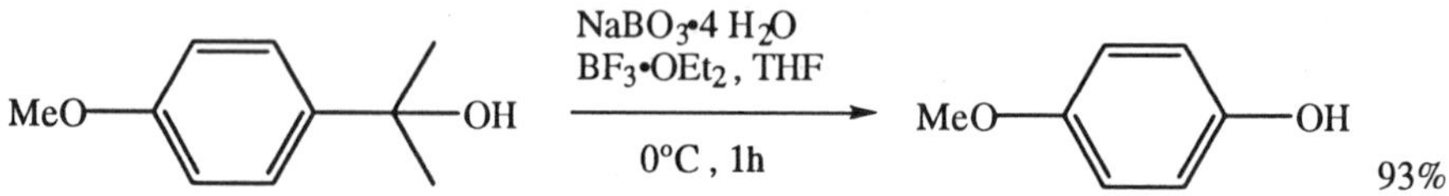

MeO OH —($NaBO_3$•4 H_2O, BF_3•OEt_2, THF; 0°C, 1h)→ MeO OH 93%

Kabalka, G.W.; Reddy, N.K.; Narayana, C. *Tetrahedron Lett.*, ***1993***, *34*, 7667

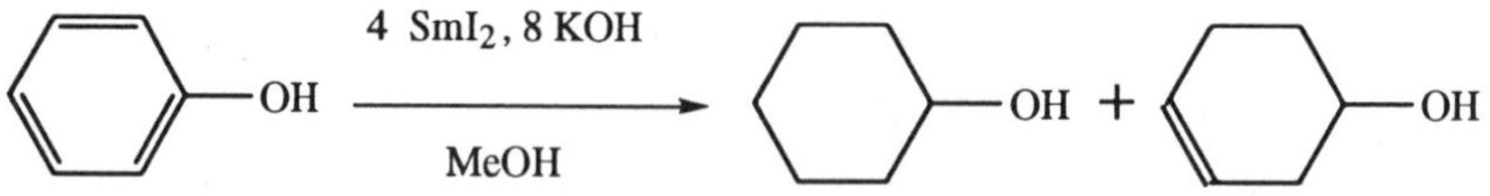

OH —(4 SmI_2, 8 KOH; MeOH)→ OH + OH

10% 73%

Kamochi, Y.; Kudo, T. *Tetrahedron Lett.*, ***1994***, *35*, 4169

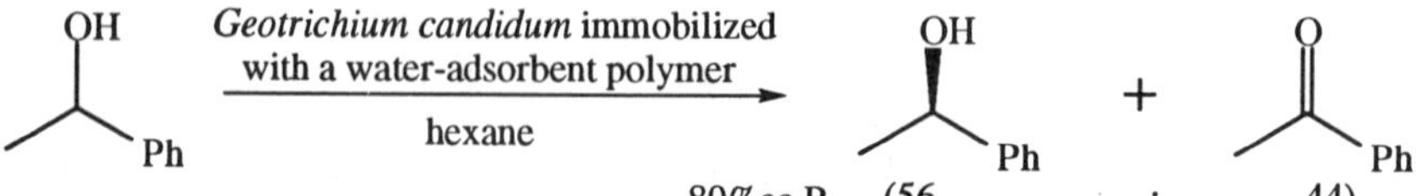

OH Ph —(*Geotrichium candidum* immobilized with a water-adsorbent polymer; hexane)→ OH Ph + O Ph

89%ee,R (56 : 44)

Nakamura, K.; Inoue, Y.; Ohno, A. *Tetrahedron Lett.*, ***1994***, *35*, 4375

SECTION 34: ALCOHOLS AND THIOLS FROM ALDEHYDES

The following reaction types are included in this section:
A. Reductions of Aldehydes to Alcohols
B. Alkylation of Aldehydes, forming Alcohols.

Coupling of Aldehydes to form Diols is found in Section 323 (Alcohol-Alcohol).

MeO—C6H4—CHO → (H_2O_2, EtOH, 50°C; $MgSO_4$, $MeReO_3$, 1d) → MeO—C6H4—OH 74%

Yamazaki, S. *Chem. Lett.*, ***1995***, 127

SECTION 34A: REDUCTIONS OF ALDEHYDES TO ALCOHOLS

Ph—CHO → (Na_2Te, NMP, PhH; 80°C, 0.8h) → PhCH2OH 59%

Suzuki, H.; Nakamura, T. *J. Org. Chem.*, ***1993***, *58*, 241

(piperidino)N—AlHEt2⁻ Na⁺ → (1. PhCHO, THF, 3h; 2. H_3O^+) → $PhCH_2OH$ 50% + PhCHO 50%

Yoon, N.M.; Ahn, J.H.; An, D.K.; Shon, Y.S. *J. Org. Chem.*, ***1993***, *58*, 1941

PhCH=CHCHO → (1. $(iPrO)_2TiCl_2/BnEt_3NBH_4$, CH_2Cl_2, -20°C; 2. aq. K_2CO_3) → PhCH=CHCH2OH 90%

Ravikumar, K.S.; Baskaran, S.; Chandrasekaran, S. *J. Org. Chem.*, ***1993***, *58*, 5981

O2N—C6H4—CHO → ($BH_2CN \cdot NH_2C_3H_7$; EtOH) → O2N—C6H4—CH2OH 87%

Das, M.K.; Maiti, P.K.; Bhaumik, A. *Bull. Chem. Soc. Jpn.*, ***1993***, *66*, 810

PhCH=CHCHO → ($[LiAlH_4/InCl_3/LiH] = LiInH_4$) → PhCH=CHCH2OH 92%

Yamada, M.; Tanaka, K.; Araki, S.; Butsugan, Y. *Tetrahedron Lett.*, ***1995***, *36*, 3169

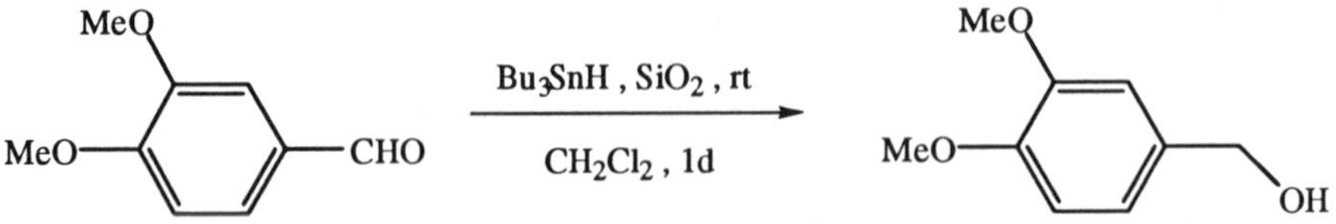

MeO MeO CHO

Bu_3SnH, SiO_2, rt

CH_2Cl_2, 1d

MeO MeO OH

81%

Figadère, B.; Chaboche, C.; Franck, X.; Peyrat, J.-F. *J. Org. Chem.*, ***1995***, *60*, 7138

REVIEW:

"β-Hydroxy Diisopinocampheylborane As A Mild, Chemoselective Reducing Agent For Aldehydes," Cha, J.S.; Kim, E.J.; Kwon, O.O.; Kwon, S.Y.; Seo, W.W.; Chang, S.W. *Org. Prep. Proceed. Int.*, ***1995***, *27*, 541

SECTION 34B: ALKYLATION OF ALDEHYDES, FORMING ALCOHOLS

ASYMMETRIC ALKYLATIONS

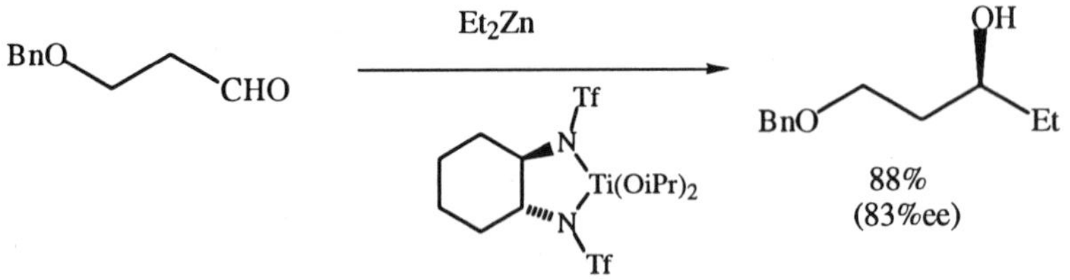

BnO CHO

Et_2Zn

Tf N $Ti(OiPr)_2$ N Tf

OH BnO Et

88%
(83%ee)

Knochel, P.; Brieden, W.; Rozema, M.; Eisenberg, C. *Tetrahedron Lett.*, ***1993***, *34*, 5881

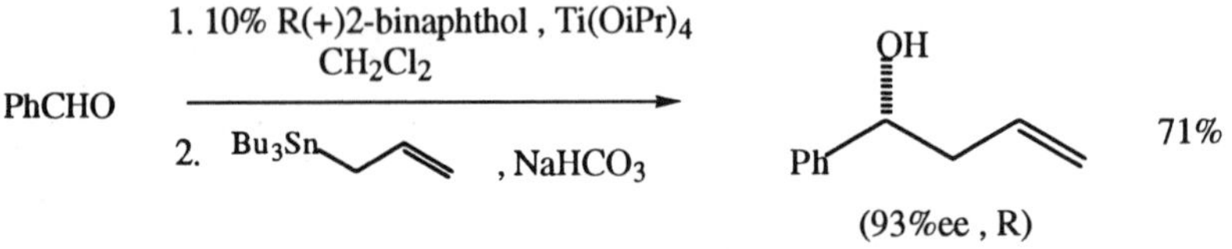

PhCHO

1. 10% R(+)2-binaphthol, $Ti(OiPr)_4$
CH_2Cl_2

2. Bu_3Sn , $NaHCO_3$

OH Ph

71%

(93%ee, R)

Keck, G.E.; Geraci, L.S. *Tetrahedron Lett.*, ***1993***, *34*, 7827

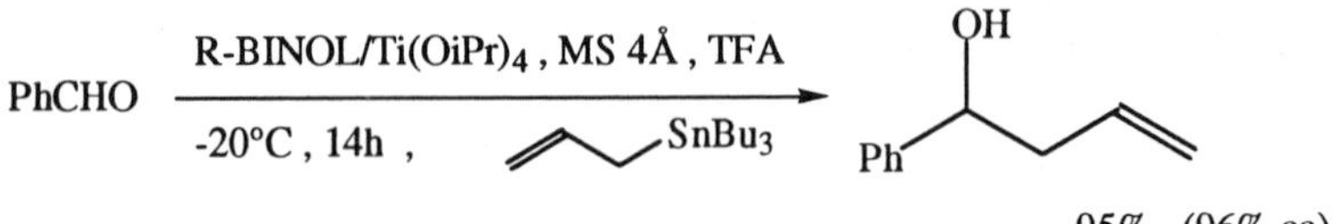

PhCHO

R-BINOL/$Ti(OiPr)_4$, MS 4Å, TFA

-20°C, 14h, $SnBu_3$

OH Ph

95% (96% ee)

Keck, G.E.; Krishnamurthy, D.; Grier, M.C. *J. Org. Chem.*, ***1993***, *58*, 6543

CCl_3CHO , R-BINOL-$TiCl_2$
MS 4Å , 0°C , CH_2Cl_2

OH + OH

(62 CCl_3 : 38) 78% CCl_3

[>95%ee] [>95%ee]

Mikami, K.; Yajima, T.; Terada, M.; Uchimaru, T. *Tetrahedron Lett.*, ***1993***, *34*, 7591

PhCHO

$SnBu_3$, MS4Å , CH_2Cl_2

20% (binol)$TiCl_2$, rt , 2d

OH

Ph

96% (82% ee , S)

Costa, A.L.; Piazza, M.G.; Tagliavini, E.; Trombini, C.; Umani-Ronchi, A. *J. Am. Chem. Soc.*, ***1993***, *115*, 7001

1. [$Ti(OiPr)_4$/1,1'-bi-2-naphthol , MS4Å CH_2Cl_2, reflux]

PhCHO

-78°C , 10 min

2. $SnBu_3$, -20°C , 70h

OH

Ph

88% (95% ee, R)

Keck, G.E.; Tarbet, K.H.; Geraci, L.S. *J. Am. Chem. Soc.*, ***1993***, *115*, 8467

1. Et_2Zn , 25°C , 10% LiO NH

PhCHO

2. 2N HCl

OH

Ph

80% ee

Mehler, T.; Martens, J.; Wallbaum, S. *Synth. Commun.*, ***1993***, *23*, 2691

1. $ZnEt_2$, 10% CPh_2OH N H

PhCHO

2. 2N HCl , -20°C

OH

100% ee

Wallbaum, S.; Martens, J. *Tetrahedron Asymmetry*, ***1993***, *4*, 637

With a chiral oxazoline additive, 60% ee

Allen, J.V.; Frost, C.G.; Williams, J.M.J. *Tetrahedron Asymmetry*, ***1993***, *4*, 649

PhCHO → (1. BuLi; 2. N-Me azetidine-CPh$_2$OH; 3. $ZnEt_2$; 4. 2N HCl) → OH product

98% ee , S

Behnen, W.; Mehler, T.; Martens, T. *Tetrahedron Asymmetry*, ***1993***, *4*, 1413

PhCHO → (Et_2Zn , toluene , hexane; OH Me OTBS, Ph, N, Ph, Et Et) → OH product

95% (85% ee , R)

de Vries, E.F.J.; Brussee, J.; Kruse, C.G.; van der Gen, A. *Tetrahedron Asymmetry*, ***1993***, *4*, 1987

PhCHO → ($ZnEt_2$, 5% iPr_2NH; 5% OH Ph) → OH Ph product

quant. (46% ee , R)

Sheng Jian, L.; Yaozhong, J.; Aiqiao, M.; Guishu, Y. *J. Chem. Soc., Perkin Trans. 1.*, ***1993***, 885

PhCHO → (Et_2Zn . PhMe , rt; 5% morpholine-N / O / HO / O catalyst) → HO H Ph product

90% (96% ee , R)

Cho, B.T.; Kim, N. *Tetrahedron Lett.*, ***1994***, *35*, 4115

CHO / TIPSO → (Me_2Zn , PhMe , 0°C , $Ti(Ot\text{-}Bu)_4$, 8% NHTf NHTf) → HO / TIPSO product

76% (93% ee)

with $Ti(OiPr)_4$ - 80% yield, 0% ee

Nowotng, S.; Vettel, S.; Knochel, P. *Tetrahedron Lett.*, ***1994***, *35*, 4539

Bu_3Sn (allenyl) —[PhCHO, (R)-Binol / 0.5 $Ti(OiPr)_4$, -20°C]→ HO (homopropargylic alcohol), Ph + HO (allenyl alcohol), Ph

(14 : 1) 48%

>99% ee

Keck, G.E.; Krishnamurthy, D.; Chen, X. *Tetrahedron Lett.*, ***1994***, *35*, 8323

PhCHO —[$ZnEt_2$, hexane/toluene, 10% chiral ligand / 20°C, 28h]→ OH

72% (80% ee)

Andrés, J.M.; Martínez, M.A.; Pedrosa, R.; Pérez-Encabo, A. *Tetrahedron Asymmetry*, ***1994***, *5*, 67

PhCHO —[$ZnEt_2$, 5% chiral ligand / 16h]→ OH

70% (68% ee, R)

Ishizaki, M.; Fujita, K.; Shimamoto, M.; Hoshino, O. *Tetrahedron Asymmetry*, ***1994***, *5*, 411

PhCHO —[Et_2Zn / Me, N (piperidine), Ph, SH]→ Ph, OH

94% (100% ee)

Kang, J.; Lee, J.W.; Kim, J.I. *J. Chem. Soc. Chem. Commun.*, ***1994***, 2009

CHO, O, Ph —[Et_2Zn, 18h / CPh_2OH, N, Me]→ OH, O, Ph

quant. (93% ee)

Watanabe, M.; Soai, K. *J. Chem. Soc., Perkin Trans. 1.*, ***1994***, 3125

PhCHO → (Et_2Zn, 22°C; 10% bicyclic pyrrolidine catalyst N-Bn, CH(Mes)OLi) → Ph–CH(OH)–Et

70-90% (86% ee, S)

Stingl, K.; Martens, J. *Liebigs Ann. Chem.*, ***1994***, 491

PhCHO → (Et_2Zn, chiral ferrocene catalyst; toluene) → 1-phenylpropan-1-ol (OH)

85% (55% ee, R)

Fukuzawa, S.; Tsudzuki, K. *Tetrahedron Asymmetry,* ***1995****, 6,* 1039

PhCHO → (N-Bu pyrrolidinylmethylpyrrolidine, (allyl)$_2SnBr_2$; CH_2Cl_2, -78°C) → Ph–CH(OH)–$CH_2CH=CH_2$

93% (79% ee)

Kobayashi, S.; Nishio, K. *Tetrahedron Lett.*, ***1995****, 36*, 6729

$C_7H_{15}CHO$ → (Bu_3Sn-allyl, CH_2Cl_2, MS; 20% Zr-BINOL derivative) → C_7H_{15}–CH(OH)–$CH_2CH=CH_2$

58% (87% ee)

Bedeschi, P.; Casolari, S.; Costa, A.L.; Tagliavini, E.; Umani-Ronchi, A. *Tetrahedron Lett.*, ***1995****, 36*, 7897

Dioxinone silyl dienol ether (OTMS) → (1. chiral Ti complex, ether, Ph–CH=CH–CHO; 2. TFA, THF) → hydroxy dioxinone product (Ph, OH, O, O, O)

97% (80% ee)

Singer, R.A.; Carreira, E.M. *J. Am. Chem. Soc.*, ***1995****, 117*, 12360

NON-ASYMMETRIC ALKYLATIONS

Ph—CHO → ($Sn(CH_2CH=CH_2)_4$, D-arabinose, rt; 5% $Sc(OTf)_3$, aq. MeCN) → Ph–CH(OH)–$CH_2CH=CH_2$

98%

Hachiya, I.; Kobayashi, S. *J. Org. Chem.*, ***1993****, 58*, 6958

PhCHO , Et_3Zn , THF
cat. $Pd(PPh_3)_4$, rt , 3h

Ph OH

88%

Yasui, K.; Goto, Y.; Yajima, T.; Taniseki, Y.; Fugami, K.; Tanaka, A.; Tamaru, Y. *Tetrahedron Lett.*, ***1993***, *34*, 7619

PhCHO

diethyl ether , -50°C

t-Bu—N N—*t*-Bu

OH

Ph Et

98%

(<1:>99 *syn;anti*)

Wissing, E.; Havennith, R.W.A.; Boersma, J.; Smeets, W.J.J.; Spek, A.L.; van Koten, G. *J. Org. Chem.*, ***1993***, *58*, 4228

BuCHO
+
t-$BuCH_2CHO$

Bu Ph OH + *t*-BuH_2C Ph OH

$PhTi(OiPr)_3$, -40°C (3.3 : 1) 60%
2 MAPH , PhLi , -78°C (1 : 10.8) 60%

MAPH = ArOAr, Ar = 2,5-diphenylphenyl

Maruoka, K.; Saito, S.; Concepcion, A.B.; Yamamoto, H. *J. Am. Chem. Soc.*, ***1993***, *115*, 1183

C_7H_{15}—CHO

(allyl)$_4$Sn , THF/aq. HCl

20°C , 1h

OH

C_7H_{15}

>99%

Yanagisawa, A.; Inoue, H.; Morodome, M.; Yamamoto, H. *J. Am. Chem. Soc.*, ***1993***, *115*, 10356

PhCHO

allyl-$SnBu_3$, CH_2Cl_2 , 25°C

5% $Ir(CO)(PPh_3)_2ClO_4$, 30h

OH

Ph

83%

Nuss, J.M.; Rennels, R.A. *Chem. Lett.*, ***1993***, 197

PhCHO

1. allyl-I , $SnCl_2$•2 H_2O

NaI , DMF

2. 30% NH_4F , H_2O

OH

Ph

89%

Imai, T.; Nishida, S. *Synthesis*, ***1993***, 395

Ph⌒⌒CHO → 1. $Bu_3SnSiMe_3$, THF , rt, 3h , Bu_4N CN; 2. H_3O^+ → Ph⌒⌒CH(OH)$SnBu_3$ 82%

Bhatt, R.K.; Ye, J.; Falck, J.R. *Tetrahedron Lett.*, ***1994***, *35*, 4081

PhCHO → allyl-$SnBu_3$, 5% $Yb(OTf)_3$ → PhCH(OH)CH$_2$CH=CH$_2$ 85%

Aspinall, H.C.; Browning, A.F.; Greeves, N.; Ravenscroft, P. *Tetrahedron Lett.*, ***1994***, *35*, 4639

Ph⌒⌒CHO → allyl-Br , DMF , rrt; GeI_2 , 1h → homoallylic alcohol (OH) 85%

Hashimoto, Y.; Kagoshima, H.; Saigo, K. *Tetrahedron Lett.*, ***1994***, *35*, 4805

TIPSO, OTIPS, SMe → $C_8H_{17}CHO$, Me_2AlCl , hexane, -78°C , 17h → OTIPS, OTIPS, SMe, C_8H_{17}, OH 90% (>98% ee)

chirality transfer ene reaction

Masaya, K.; Tanino, K.; Kuwajima, I. *Tetrahedron Lett.*, ***1994***, *35*, 7965

Bu_3Sn-dioxolane (OMe, OMe) → 1. BuLi , THF , -78°C; 2. PhCHO → HO, Ph, dioxolane (OMe, OMe) 90% (70:30 de)

Colombo, L.; DiGiacomo, M.; Brusotti, G.; Delougu, G. *Tetrahedron Lett.*, ***1994***, *35*, 2063

$PhSO_2Et$ → 1. Li (powder) , naphthalene; 2. Li (powder) , naphthalene; 3. PhCHO → Et–CH(OH)–Ph 61%

Guijarro, D.; Yus, M. *Tetrahedron Lett.*, ***1994***, *35*, 2965

$SiMe_3$ / Br → 1. Ga, KI, LiCl, PhCHO, THF, reflux; 2. H_3O^+ → HO, Ph, $SiMe_3$ + HO, Ph, $SiMe_3$

(91 : 9) 95%

Han, Y.; Huang, Y.-Z. *Tetrahedron Lett.*, ***1994***, *35*, 9433

$C_6H_{13}CHO$ → C_3H_7 / $SnBu_3$, $BuSnCl_3$, $CHCl_3$, 0°C → HO, C_6H_{13}, C_3H_7 + HO, C_6H_{13}, C_3H_7

(98 : 2) 86%

Miyaki, H.; Yamamura, K. *Chem. Lett.*, ***1994***, 897

PhCHO → 1. Br, 0.5 CuCl, THF, rt; 1.5 $SnCl_2 \cdot 2H_2O$, 1d; 2. 30% aq. NH_4F → OH, Ph

98%

Imai, T.X; Nishida, S. *J. Chem. Soc. Chem. Commun.*, ***1994***, 273

Ph / Cl → EtCHO, Zn, aq. NH_4Cl, THF → Ph, Et, OH

98% (73:27)

Sjöholm, R.; Rairama, R.; Ahonen, M. *J. Chem. Soc. Chem. Commun.*, ***1994***, 1217

SO_2Ph → cat. $Pd(PPh_3)_4$, PhCHO; Et_2Zn, THF, reflux → OH, Ph

86%

Clayden, J.; Julia, M. *J. Chem. Soc. Chem. Commun.*, ***1994***, 1905

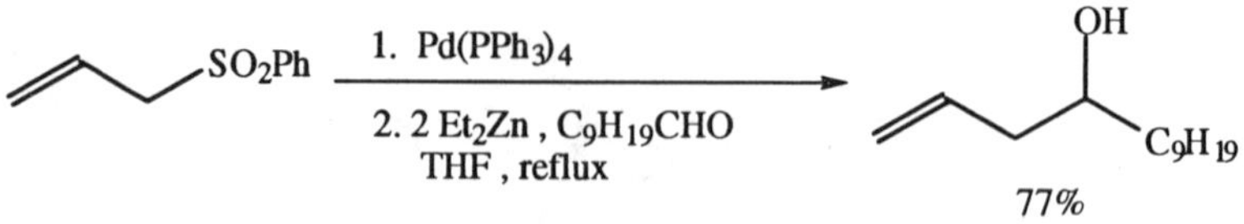

1. $Pd(PPh_3)_4$
2. 2 Et_2Zn, $C_9H_{19}CHO$ THF, reflux

SO_2Ph → OH, C_9H_{19} 77%

Clayden, J.; Julia, M. *J. Chem. Soc. Chem. Commun.*, ***1994***, 2261

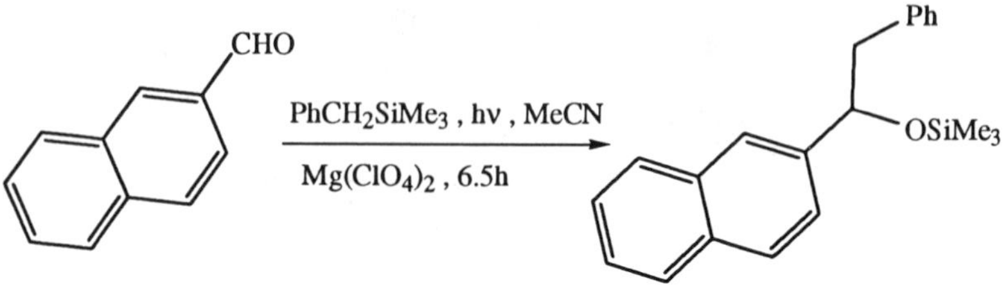

CHO → $PhCH_2SiMe_3$, hν, MeCN; $Mg(ClO_4)_2$, 6.5h → Ph, $OSiMe_3$

Fukuzumi, S.; Okamoto, T.; Otera, J. *J. Am. Chem. Soc.*, ***1994***, *116*, 5503

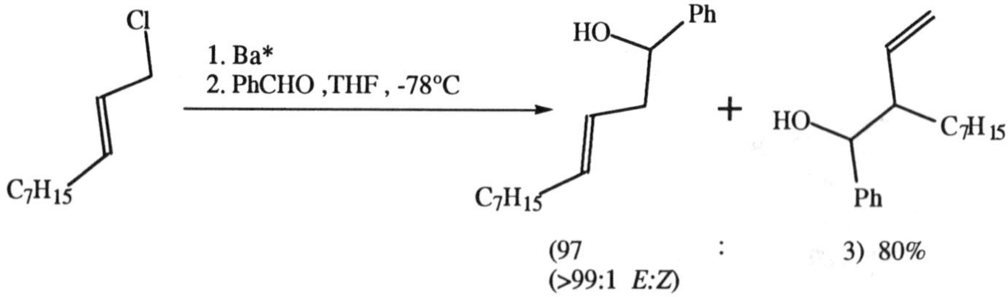

Cl, C_7H_{15} → 1. Ba* 2. PhCHO, THF, -78°C → HO, Ph, C_7H_{15} + HO, C_7H_{15}, Ph

(97 : 3) 80%
(>99:1 *E:Z*)

Yanagisawa, A.; Habaue, S.; Yasue, K.; Yamamoto, H. *J. Am. Chem. Soc.*, ***1994***, *116*, 6130

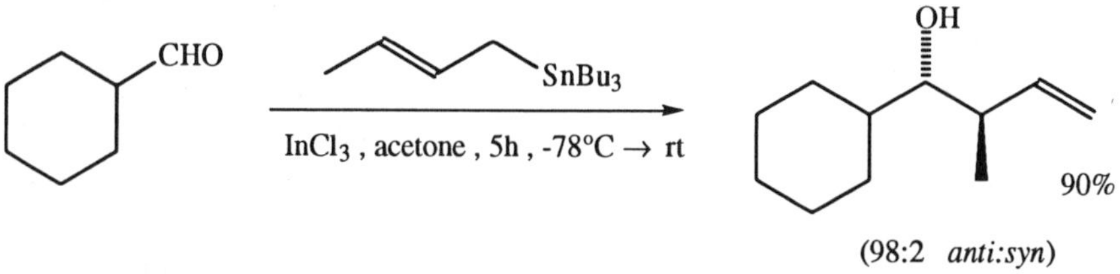

CHO + $SnBu_3$; $InCl_3$, acetone, 5h, -78°C → rt → OH 90%

(98:2 *anti:syn*)

Marshall, J.A.; Hinkle, K.W. *J. Org. Chem.*, ***1995***, *60*, 1920

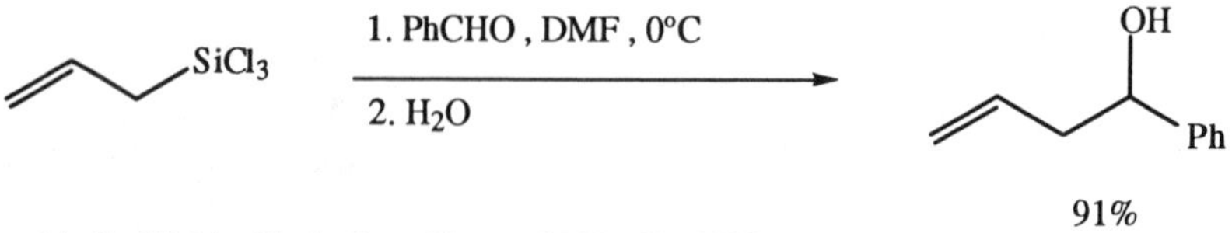

$SiCl_3$ → 1. PhCHO, DMF, 0°C 2. H_2O → OH, Ph

91%

Kobayashi, S.; Nishio, K. *J. Org. Chem.*, ***1995***, *60*, 6620

PhCHO → ($C_8H_{17}ZrCpCl$, 20% $ZnBr_2$; THF, 3h, 25°C) → H_2O → Ph–CH(OH)–C_8H_{17} 88%

Zheng, B.; Srebnik, M. *J. Org. Chem.*, ***1995***, *60*, 3278

PhCHO → (In, H_2O, 30 min; Br–CH_2–C(=CH_2)–CH_2–Br) → HO(H)(Ph)C–CH_2–C(=CH_2)–CH_2–C(H)(OH)Ph 75%

Li, C.-J. *Tetrahedron Lett.*, ***1995***, *36*, 517

Me–C≡C–Ph → 1. Ti(OiPr)$_4$, iPrMgCl; 2. cyclohexyl–CHO; 3. H_2O → (16 : 84) 81%

Harada, K.; Urabe, H.; Sato, F. *Tetrahedron Lett.*, ***1995***, *36*, 3203
Nakagawa, T.; Kasatkin, A.; Sato, F. *Tetrahedron Lett.*, ***1995***, *36*, 3207

cyclohexyl–CHO + cycloheptanone → 1. TMSOTf, CH_2Cl_2, Me_2S, -78°C; 2. $Bu_3SnCH_2CH=CH_2$, -40°C; 3. TBAF → (>99 : <1) 98%

Kim, S.; Kim, S.H. *Tetrahedron Lett.*, ***1995***, *36*, 3723

PhCHO → (CH_2=CH–CH_2Br, aq. NH_4Cl; sonoelectro-produced Zn powder) → Ph–CH(OH)–CH_2–CH=CH_2 82%

Durant, A.; Delplancke, J.-L.; Winand, R.; Reisse, J. *Tetrahedron Lett.*, ***1995***, *36*, 4257

PhCHO → (Br–CH_2–CH=CH_2, Mg°, $CuCl_2$•2 H_2O; THF, 6h) → Ph–CH(OH)–CH_2–CH=CH_2 99%

Sarangi, C.; Nayak, A.; Nanda, B.; Das, N.B.; Sharma, R.P. *Tetrahedron Lett.*, ***1995***, *36*, 7119

Me_3Si Br

CHO

In , H_2O , rt

OH

$SiMe_3$

62% (75:25 *E:Z*)

Isaac, M.B.; Chan, T.-H. *Tetrahedron Lett.*, ***1995***, *36*, 8957

CHO O O , THF

BiCl$_3$-Zn , reflux

Ph

O HO Me OH Ph + O HO Me OH Ph

(74 : 26) 85%

Miyoshi, N.; Kukuma, T.; Wada, M. *Chem. Lett.*, ***1995***, 999

Br

1. Ti(OiPr)$_2$, 2 iPrMgBr

2. PhCHO

Ph OH 94%

Kasatkin, A.; Nakagawa, T.; Okamoto, S.; Sato, F. *J. Am. Chem. Soc.*, ***1995***, *117*, 3881

SPh

1. LDBB , THF , -78°C

2. $MnCl_2$, THF , -78°C

3. MeO—C$_6$H$_4$—CHO

-78°C → 0°C

HO OMe

84% (44:56)

LDBB = 4,4'-di-*t*-butylbiphenylide

Ahn, Y.; Doubleday, W.W.; Cohen, T. *Synth. Commun.*, ***1995***, *25*, 33

MPMO O O O

PhCHO , 10% Pd(PhCN)$_2$Cl$_2$

3 SnCl$_2$•2 H_2O , DMI , 25°C

HO Ph HO MPMO

91%

Kang, S.-K.; Park, D.-C.; Park, C.-H.; Jang, S.-B. *Synth. Commun.*, ***1995***, *25*, 1359

PhCHO → (Br-allyl, Sn, Me_3SiCl; Bu_4NBr, MeOH, rt, 4h) → Ph-CH(OH)-CH$_2$CH=CH$_2$ 89%

Zhou, J.-Y.; Yao, X.-B.; Chen, Z.-G.; Wu, S.-H. *Synth. Commun.*, ***1995***, *25*, 3081

PhCHO → (allyl-$SnBu_3$, THF, rt, 4d; $PtCl_2(PPh_3)_2$) → Ph-CH(OH)-CH$_2$CH=CH$_2$ 90%

Nakamura, H.; Asao, N.; Yamamoto, Y. *J. Chem. Soc. Chem. Commun.*, ***1995***, 1273

$C_6H_{13}CHO$ → (allyl-Br; $SnBr_2$) → C_6H_{13}-CH(OH)-CH$_2$CH=CH~ + C_6H_{13}-CH(OH)-CH(CH$_3$)CH=CH$_2$

CH_2Cl_2-H_2O (9.1 : 9) 48%
H_2O, Bu_4NBr (1 : 99) 48%

Masuyama, Y.; Kishida, M.; Kurusu, Y. *J. Chem. Soc. Chem. Commun.*, ***1995***, 1405

REVIEW:

"Synthetic Organoindium Chemistry: What Makes Indium So Appealing," Cintas, P. *Synlett*, ***1995***, 1087

SECTION 35: ALCOHOLS AND THIOLS FROM ALKYLS, METHYLENES AND ARYLS

No examples of the reaction $RR^1 \rightarrow ROH$ (R^1 = alkyl, aryl, etc.) occur in the literature. For reactions of the type $RH \rightarrow ROH$ (R = alkyl or aryl) see Section 41 (Alcohols and Phenols from Hydrides).

NO ADDITIONAL EXAMPLES

SECTION 36: ALCOHOLS AND THIOLS FROM AMIDES

C_6H_{13}-CH$_2$-C(=O)-N(pyrrolidine) → ($LiBH_3P$-Pyr, THF, 25°C; Pyr = pyrrolidino) → C_6H_{13}-CH$_2$CH$_2$-OH 71%

Fisher, G.B.; Fuller, J.C.; Harrison, J.; Goralski, C.T.; Singaram, B. *Tetrahedron Lett.*, ***1993***, *34*, 1091

SECTION 37: ALCOHOLS AND THIOLS FROM AMINES

NO ADDITIONAL EXAMPLES

SECTION 38: ALCOHOLS AND THIOLS FROM ESTERS

H–C(=O)–OBu —[$P(Cy)_3$, 180°C, 15h]→ HO—Bu 68%

Vega, F.R.; Clément, J.-C.; des Abbayes, H. *Tetrahedron Lett.*, ***1993***, *34*, 8117

OMe, MeO, OAc —[Mg, MeOH, rt, 2.5h]→ OMe, MeO, OH

98%

Xu, Y.-C.; Lebeau, E.; Walker, C. *Tetrahedron Lett.*, ***1994***, *35*, 6207

Ph—CO_2Me —[1. PMHS, $Ti(OiPr)_4$, THF; 2. NaOH]→ $PhCH_2OH$ 86$

PMHS = polymethylhydrosiloxane

Breedon, S.W.; Lawrence, N.J. *Synlett*, ***1994***, 833

C_9H_{19}—CO_2Et —[1. 2.5 eq. PHMS, 25% $Ti(OiPr)_4$ 65°C, 1d; 2. aq. NaOH, THF]→ C_9H_{19}–CH_2OH 93%

Reding, M.T.; Buchwald, S.L. *J. Org. Chem.*, ***1995***, *60*, 7884

$C_7H_{15}CH_2SAc$ —[BER, $Pd(OAc)_2$, MeOH, reflux]→ $C_7H_{15}CH_2SH$

BER = borohydride exchange resin 98%

Choi, J.; Yoon, N.M. *Synth. Commun.*, ***1995***, *25*, 2655

SECTION 39: ALCOHOLS AND THIOLS FROM ETHERS, EPOXIDES AND THIOETHERS

t-$BuMe_2Si$–epoxide —[$Bu_2CuCNLi_2$, THF]→ t-$BuMe_2Si$–CH(Bu)–CH_2OH + t-$BuMe_2Si$–CH(OH)–CH_2–Bu

products are converted to alkenes via Swern oxidation and elimination (95 : 5) 80-90%

Chauret, D.C.; Chong, J.M. *Tetrahedron Lett.*, ***1993***, *34*, 3695

Al(Hg), 10% aq. $NaHCO_3$

95% EtOH, rt
ultrasound

80%

also used Zn rather than Al(Hg)

yield is 23% without ultrasound

Mirando Moreno, M.J.S.; Sáe Melo, M.L.; Campos Neves, A.S. *Tetrahedron Lett.*, ***1993***, *34*, 353
Salvador, J.A.R.; Sáe Melo, M.L.; Campos Neves, A.S. *Tetrahedron Lett.*, ***1993***, *34*, 361

PhI, $Pd(PPh_3)_2Cl_2$, Zn, $ZnCl_2$

THF, 60°C, 4h, NEt_3

86%

Duan, J.-P.; Cheng, C.-H. *Tetrahedron Lett.*, ***1993***, *34*, 4019

1. LiI, MeCN, Amberlyst 15
rt, 30 min

2. Bu_3SnH, AIBN, 80°C
toluene, 2h

99x88%

Federici, C.; Righi, G.; Rossi, L.; Bonini, C.; Chiummiento, L.; Funicello, M. *Tetrahedron Lett.*, ***1994***, *35*, 797

H_2, Pd-C, 16h

86%

Bach, T. *Tetrahedron Lett.*, ***1994***, *35*, 1855

, 10% $Pd(OAc)_2$

5 eq. NaO_2CH, 2 eq. Bu_4NCl
3 eq. iPr_2NEt, 80°C, 1d
N,N-dimethylacetamide

(69:31 *E:Z*) 70%

Larock, R.C.; Ding, S. *J. Org. Chem.*, ***1993***, *58*, 804

BnO, H, O, H, CO_2Me → Mg , MeOH → BnO, HO, H, CO_2Me

98%

Pak, C.S.; Lee, E.; Lee, G.H. *J. Org. Chem.*, ***1993***, *58*, 1523

C_8H_{17} (epoxide) → Pd/C , HCO_2NH_4 , EtOH; 23°C , 2d → C_8H_{17}, OH

81%

Dragovich, P.S.; Prins, T.J.; Zhou, R. *J. Org. Chem.*, ***1995***, *60*, 4922

t-Bu–C6H4–O–allyl → 1. 10% $Pd(PPh_3)_4$, $NaBH_4$ THF; 2. 1 N HCl → *t*-Bu–C6H4–OH

98%

Beugelmans, R.; Bourdet, S.; Bigot, A.; Zhu, J. *Tetrahedron Lett.*, ***1994***, *35*, 4349

TsO, O → Te , $LiBHEt_3$, rt; 3h → OH

84%

Kumar, A.; Dittmer, D.C. *Tetrahedron Lett.*, ***1994***, *35*, 5583

PhOMe → L-Selectride , THF , 67°C , 12h → PhOH 94%

Majetich, G.; Zhang, Y.; Wheless, K. *Tetrahedron Lett.*, ***1994***, *35*, 8727

Ph (epoxide) → 1. $TMSN_3$, W(N*t*-Bu)$_2$(NHI*t*-Bu)$_2$, 5d; 2. H+ → Ph, N_3, OH 95%

[with Cr(N*t*-Bu)$_2$Cl$_2$; 12h — 95%]

Leung, W.-H.; Chow, E.K.F.; Wu, M.-C.; Kum, P.W.Y.; Yeung, L.-L. *Tetrahedron Lett.*, ***1995***, *36*, 107

$5\%\ SmI_2(THF)_2$, CH_2Cl_2; Et_2NH, rt, 18h

(100 : 0) 63%

regioselectivity of addition reversed with Me_3SiN_3 to form azide-OTMS

Van de Weghe, P.; Collin, J. *Tetrahedron Lett.*, ***1995***, *36*, 1649

$TMSN_3$, Ph_4SbOH, PhH; reflux, 2h

67% 19%

Fujiwara, M.; Tanaka, M.; Baba, A.; Ando, H.; Souma, Y. *Tetrahedron Lett.*, ***1995***, *36*, 4849

OPh → 1. e^-, Ni(II), Mg anode, DMF, rt; 2. H_2O → PhOH 99%

Olivero, S.; Duñach, E. *J. Chem. Soc. Chem. Commun.*, ***1995***, 2497

Additional examples of ether cleavages may be found in Section 45A (Protection of Alcohols and Thiols).

SECTION 40: ALCOHOLS AND THIOLS FROM HALIDES AND SULFONATES

O_2N–C₆H₄–Br → Na_3SPO_3, MeOH; reflux, overnight → O_2N–C₆H₄–SH 80%

Bieniarz, C.; Cornwall, M.J. *Tetrahedron Lett.*, ***1993***, *34*, 939

1. Li dispersion, ether reflux; 2. PhCHO, -10°C 1h

also works with ketones

Takahashi, M.; Hatano, K.; Kimura, M.; Watanabe, T.; Oriyama, T.; Koga, G. *Tetrahedron Lett.*, ***1994***, *35*, 579

Ph—I → 1. Me_3ZnLi, THF, -78°C; 2. Me_3Al, CH_2Cl_2, 0°C, 1h → Ph–CH(OH)–Ph

55%

Kondo, Y.; Takazawa, N.; Yamazaki, C.; Sakamoto, T. *J. Org. Chem.*, ***1994***, *59*, 4717

PhCHO → PCl_5 → $PhCHCl_2$ → 1. Bu_3B, *t*-BuLi, -78°C; 2. $NaBO_3$•4 H_2O, 2h → Ph–CH(OH)–Bu

64%

Kabalka, G.W.; Lin, N.-S.; Yu, S. *Tetrahedron Lett.*, ***1995***, *36*, 8545

$Me_2C(Br)C(=O)NHBn$ → Ag_2O, MeCN, H_2O; 12h → $Me_2C(OH)C(=O)NHBn$ 90%

Yang, R.-Y.; Dai, L.-X. *Synth. Commun.*, ***1994***, *24*, 2229

Br, CO_2H → 1. HCO_2NHEt_3; 2. H_2O → OH, CO_2H

80%

Alexander, J.; Renyer, M.L.; Veerapanane, H. *Synth. Commun.*, ***1995***, *25*, 3875

SECTION 41: ALCOHOLS AND THIOLS FROM HYDRIDES

Benzene → 1. e^-, CF_3CO_2H, CH_2Cl_2, NEt_3; 2. H_2O → PhOH

73%

Fujimoto, K.; Maekawa, H.; Tokuda, Y.; Matsubara, Y.; Mizuno, T.; Nishiguchi, I. *Synlett*, ***1995***, 661

SECTION 42: ALCOHOLS AND THIOLS FROM KETONES

The following reaction types are included in this section:

A. Reductions of Ketones to Alcohols
B. Alkylations of Ketones, forming Alcohols

Coupling of ketones to give diols is found in Section 323 (Alcohol → Alcohol).

SECTION 42A: REDUCTION OF KETONES TO ALCOHOLS

ASYMMETRIC REDUCTION

BH_3, THF, 80°C

B-Me

>90% (92% ee, R)

Quallich, G.J.; Woodall, T.M. *Tetrahedron Lett.*, ***1993***, *34*, 4145

1. 10% ..., BH_3•SMe-2

2. 1N HCl

85% (76%ee,R)

Bolm, C.; Elder, M. *Tetrahedron Lett.*, ***1993***, *34*, 6041

G. candidium

cyclopentanol H_2O, 70h

immobilized *G. candidium*

hexane, 3h cyclopentanol

(7:93 *syn:anti*) *G. candidium* = *Geotrichum candidum* (95:5 *syn:anti*)

Nakamura, K.; Takano, S.; Ohno, A. *Tetrahedron Lett.*, ***1993***, *34*, 6087

0.6 BH_3•SMe_2, THF rt, <1h

10%

82% (27%ee,S)

Burns, B.; Studley, J.R.; Wills, M. *Tetrahedron Lett.*, ***1993***, *34*, 7105

1. 10%

Me_3SiCl

2. 2N HCl

92% (84% ee, R)

Bolm, C.; Seger, A.; Felder, M. *Tetrahedron Lett.*, ***1993***, *34*, 8079

1. Ph(CH(OH)CH$_2$)N(Bn)(CH$_2$CH(OH))Ph , BuLi , THF
2. SmI_3 , THF , ICH_2CH_2I
3. quench

quant. (97% ee , R)

Evans, D.A.; Nelson, S.G.; Gagné, M.R.; Muci, A.R. *J. Am. Chem. Soc.*, ***1993***, *115*, 9800

1. P-TADDOL complex , Ph_2SiH_2 , PhH
$[Rh(cod)Cl]_2$
2. TsOH , MeOH

50% (29.5:70.5 *R:S*)

TADDOL = α,α,α',α'-tetraaryl-4,5-dimethoxy-1,3-dioxolane

Jakaki, J.-i.; Schweizer, W.B.; Seebach, D. *Helv. Chim. Acta*, ***1993***, *76*, 2654

Bakers yeast (cell free extract)
NaDPH , 8h

62% (>99% ee, S)

Ishihara, K.; Sakai, T.; Tsuboi, S.; Utaka, M. *Tetrahedron Lett.*, ***1994***, *35*, 4569

$[(COD)Rh(DiPFc)]^+$ OTf^-
60 psi H_2 , 25°C , 4h

DiPFc = 1,1'-bis-(diisopropylphosphino)ferrocene

quant.

Burk, M.J.; Harper, T.G.P.; Lee, J.R.; Kalberg, C. *Tetrahedron Lett.*, ***1994***, *35*, 4963

10%
BH_3•THF , THF , rt

>95% (91.7% ee , S)

Hong, Y.; Gao, Y.; Nie, X.; Zepp, C.M. *Tetrahedron Lett.*, ***1994***, *35*, 6631

1. Ph_2SiH_2 , 5% Rh-diferrocenyl-dichalcoginide
THF , 0°C
2. 0.5M HCl/MeOH

67% (50% ee , R)

Nishibayashi, Y.; Singh, J.D.; Segawa, K.; Kukuzawa, S.i.; Uemura, S. *J. Chem. Soc. Chem. Commun.*, ***1994***, 1375

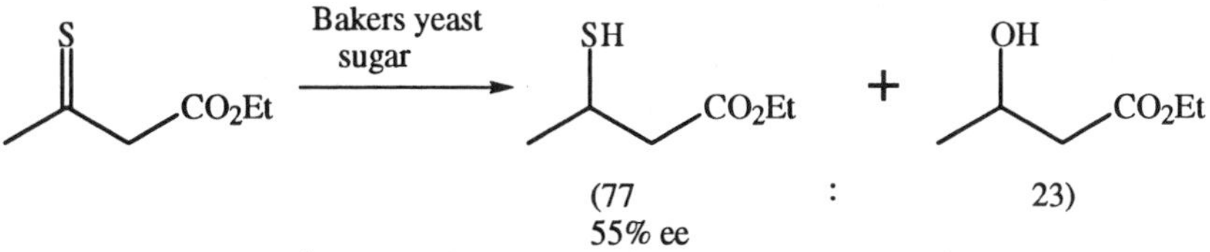
Bakers yeast
sugar
S
CO_2Et
SH
CO_2Et
+
OH
CO_2Et
(77 : 23)
55% ee

Nielsen, J.K.; Madsen, J.Ø. *Tetrahedron Asymmetry*, ***1994***, *5*, 403

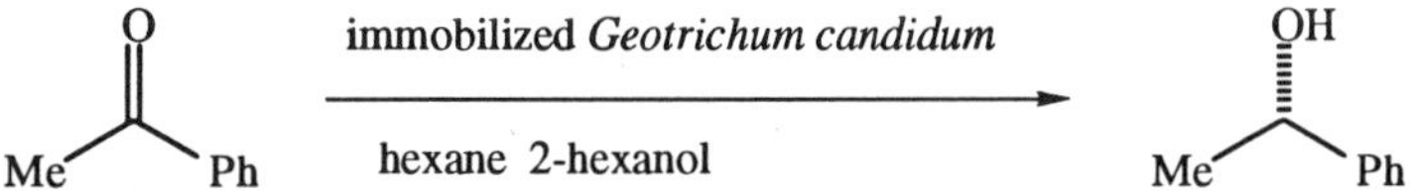
immobilized *Geotrichum candidum*
hexane 2-hexanol
O
Me
Ph
OH
Me
Ph

73% (>99% ee, S)

Nakamura, K.; Inoue, Y.; Ohno, A. *Tetrahedron Lett.*, ***1995***, *36*, 265

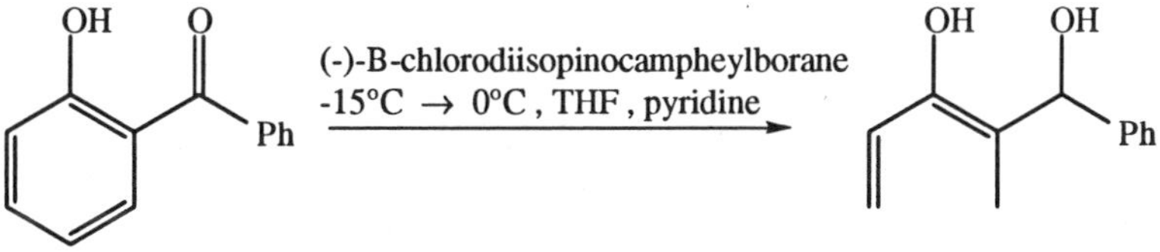
(-)-B-chlorodiisopinocampheylborane
-15°C → 0°C, THF, pyridine
OH
O
Ph
OH
OH
Ph

99% (96.4% ee)

Shieh, W.-C.; Cantrell Jr., W.R.; Carlson, J.A. *Tetrahedron Lett.*, ***1995***, *36*, 3797

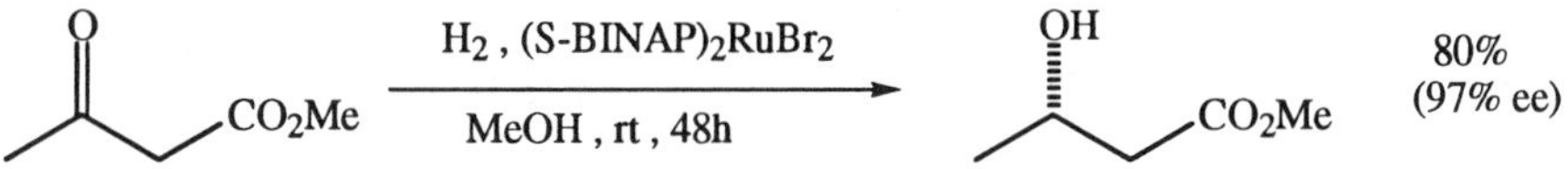
H_2, (S-BINAP)$_2$RuBr$_2$
MeOH, rt, 48h
O
CO_2Me
OH
CO_2Me
80%
(97% ee)

Genêt, J.P.; Ratovelomanana-Vidal, V.; Caño de Andrade, M.C.; Pfister, X.; Guerreiro, P.; Lenoir, J.Y. *Tetrahedron Lett.*, ***1995***, *36*, 4801

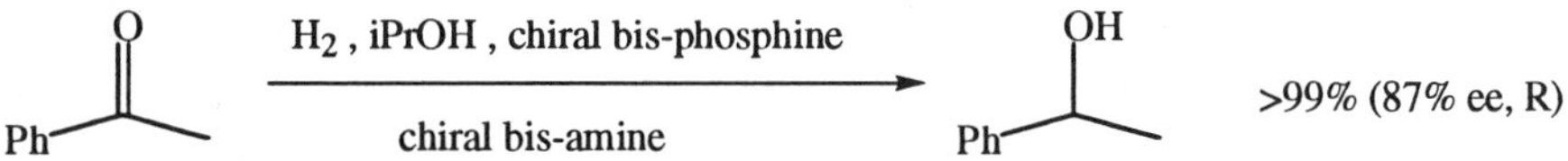
H_2, iPrOH, chiral bis-phosphine
chiral bis-amine
O
Ph
OH
Ph
>99% (87% ee, R)

Ohkuma, T.; Ooka, H.; Hashiguchi, S.; Ikariya, T.; Noyori, R. *J. Am. Chem. Soc.*, ***1995***, *117*, 2675

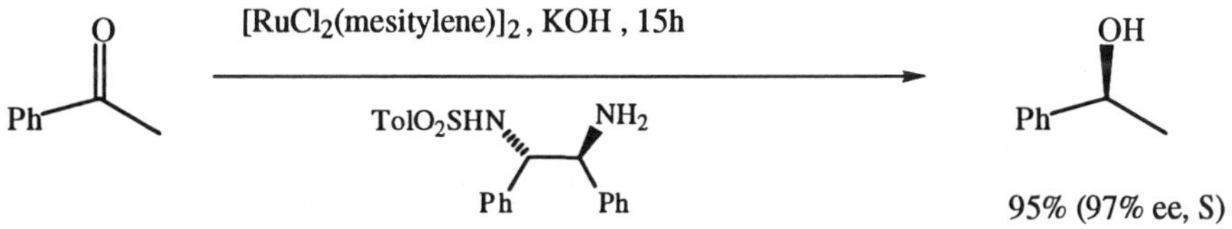
[RuCl$_2$(mesitylene)]$_2$, KOH, 15h
O
Ph
TolO$_2$SHN
NH$_2$
Ph
Ph
OH
Ph
95% (97% ee, S)

Hashiguchi, S.; Fujii, A.; Takehara, J.; Ikariya, T.; Noyori, R. *J. Am. Chem. Soc.*, ***1995***, *117*, 7562

1. Ph_2SiH_2, THF, 1% $RhCl(NBD)_2L$ chiral ferrocenyl phosphine ligand
2. H_3O^+

90% (87% ee, S)

Hayashi, T.; Hayashi, C.; Uozumi, Y. *Tetrahedron Asymmetry,* ***1995****, 6,* 2503

1. 0.1 chiral diol, hexane, -3°C, catecholborane
2. H_3O^+

quant. (82% ee)

Giffels, G.; Dreisbach, C.; Kragl, U.; Weigerding, M.; Waldmann, H.; Wandrey, C. *Angew. Chem. Int. Ed. Engl.,* ***1995****, 34,* 2005

$NaBH_4$, EtOH
Co (β-oxoaldiminato) complex

99% (73% ee, S)

Nagata, T.; Yorozu, K.; Yamada, T.; Mukaiyama, T. *Angew. Chem. Int. Ed. Engl.,* ***1995****, 34,* 2145

NON-ASYMMETRIC REDUCTION

1. $CdCl_2$/Mg (powder) THF, 15 min
2. H_2O

92%

Bordoloi, M. *Tetrahedron Lett.,* ***1993****, 34,* 1681

t-Bu—cyclohexanone → t-Bu—cyclohexanol

$LiAlH(OCEt_2CMe_3)_3$
THF

with $LiAlH(OCMe_3)_3$ - 90:10 *trans:cis*

5:95 *trans:cis*

Boireau, G.; Deberly, A.; Toneva, R. *Synlett,* ***1993****,* 585

Cu(II) exchanged cationic resin
$NaBH_4$, EtOH, 0°C

98%

Sarkar, A.; Rao, B.R.; Ram, B. *Synth. Commun.,* ***1993****, 23,* 291

O

THF, 25°C, 3h

N-BH_3 Li

OH

92%

Fuller, J.C.; Staangeland, E.L.; Goralski, C.T.; Singaram, B. *Tetrahedron Lett.*, ***1993***, *34*, 257

O

iPrOH, SiO_2-Zr catalyst

reflux

OH

87%

Inada, K.; Shibagaki, M.; Nakanishi, Y.; Matsushita, H. *Chem. Lett.*, ***1993***, 1795

O OMe

Ph

$TiCl_4$, Me_4NBH_4

CH_2Cl_2, -78°C

OH OMe

Ph

\+

OH OMe

Ph

other reducing agents also used (>99 : 1) 93%

Sarko, C.R.; Guch, I.C.; DiMare, M. *J. Org. Chem.*, ***1994***, *59*, 705

O

t-Bu

$LiBH_3N(n\text{-}C_3H_7)_2$

THF, 0°C

H

OH

t-Bu

99%

Harrison, J.; Fuller, J.C.; Goralski, C.T.; Singaram, B. *Tetrahedron Lett.*, ***1994***, *35*, 5201

O

1. $Cp_2TiCl_2/NaBH_4$, 5 min

2. 1 N NaOH

OH

90%

Barden, M.C.; Schwartz, J. *J. Org. Chem.*, ***1995***, *60*, 5963

O

O Ph

$(MeO)_3SiH$, LiOMe, ether

-20°C, 9h

O

HO Ph

\+

O

HO Ph

(8 : 92) quant.

wth $(MeO)_3SiH$/HMPA/LiOMe/0°C/2h → (90:10) 98%

Hojo, M.; Fujii, A.; Murakami, C.; Aihara, H.; Hosomi, A. *Tetrahedron Lett.*, ***1995***, *36*, 571

H_2 (8 atm), cat. $RuCl_2(Binap)(dmf)_n$

iPrOH, diamine, KOH, 28°C

Ph Ph HO

Ohkuma, T.; Ooka, H.; Ikariya, T.; Noyori, R. *J. Am. Chem. Soc.*, ***1995***, *117*, 10417

O cat. $[NiCl_2(PPh_3)_2$, cat. NaOH OH

82%

Ph iPrOH, Ar, 30h, heat Ph

Iyer, S.; Varghese, J.P. *J. Chem. Soc. Chem. Commun.*, ***1995***, 465

SECTION 42B: ALKYLATION OF KETONES, FORMING ALCOHOLS

Aldol reactions are listed in Section 330 (Ketone-Alcohol)

Cl Me Me O

e^-, $SmCl_3$, DMF, 20°C OH 74%

Hebri, H.; Duñach, E.; Périchon, J. *Tetrahedron Lett.*, ***1993***, *34*, 1475

3 eq. $Me_2S{=}CH_2$, THF OH

O -10°C → rt

91%

Harnett, J.J.; Alcaraz, L.; Mioskowski, C.; Martel, J.P.; LeGall, T.; Shin, D.-S.; Falck, J.R. *Tetrahedron Lett.*, ***1994***, *35*, 2009

O e^-, OH CH_3

OH

C_6H_{13} CH_3 DMF, Bu_4NBF_4 C_6H_{13} OH

89%

Shono, T.; Morishima, Y.; Moriyoshi, N.; Ishifune, M. *J. Org. Chem.*, ***1994***, *59*, 273

1. 2.2 eq. SmI_2, THF
2 eq. *t*-BuOH
2. H_3O^+

OH

53% (1:1)

Molander, G.A.; McKie, J.A. *J. Org. Chem.*, ***1994***, *59*, 3186

0.05 $CeCl_3$, 1h, THF

O 1.1 MgBr

OH

93%

Dimitrov, V.; Bratavanov, S.; Simova, S.; Kostova, K. *Tetrahedron Lett.*, ***1994***, *35*, 6713

O

CO_2Et

Br, In, 10h

aq. MeOH

OH

CO_2Et

76%

Li, C.-J.; Lu, Y.-Q. *Tetrahedron Lett.*, ***1995***, *36*, 2721

Br, Zn_{dust}

O

THF, 1.5h

OH

uses commercial Zn dust

80%

Ranu, B.C.; Majee, A.; Das, A.R. *Tetrahedron Lett.*, ***1995***, *36*, 4885

PhCHO

1.8 Et_2Zn, $Ti(OiPr)_4$, hexane

-23°C, 0.1 chiral catalyst

OH

Ph Et

97%
99% ee, S

Zhang, X.; Guo, C. *Tetrahedron Lett.*, ***1995***, *36*, 4947

O

C_5H_{11}

$Me_3Ti{\bullet}MeLi$, ether

-50°C

HO Me

C_5H_{11}

78%

selective for conjugated ketones in the presence of non-conjugated ketones

Markó, I.E.; Leung, C.W. *J. Am. Chem. Soc.*, ***1994***, *116*, 371

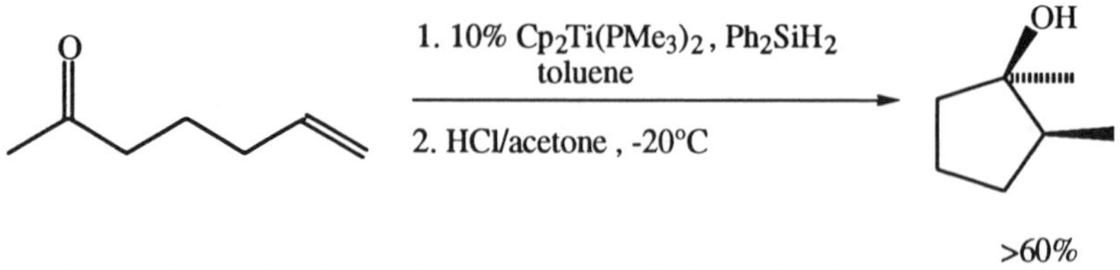

1. 10% $Cp_2Ti(PMe_3)_2$, Ph_2SiH_2 toluene

2. HCl/acetone, -20°C

OH

>60%

Kablaoui, N.M.; Buchwald, S.L. *J. Am. Chem. Soc.*, ***1995***, *117*, 6785

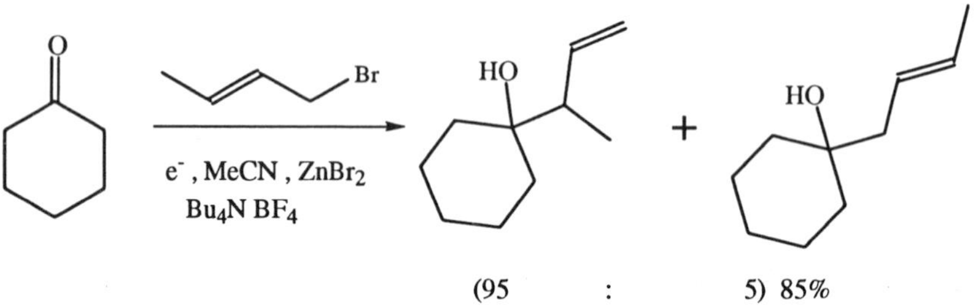

O

Br

e^-, MeCN, $ZnBr_2$
$Bu_4N\ BF_4$

HO HO

(95 : 5) 85%

Rollin, Y.; Derien, S.; Duñach, E.; Gebehenne, C.; Perichon, J. *Tetrahedron*, ***1993***, *49*, 7723

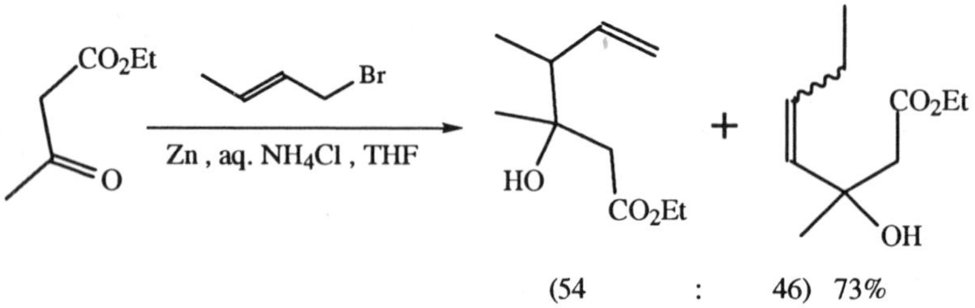

CO_2Et

Br

Zn, aq. NH_4Cl, THF

O

HO CO_2Et CO_2Et OH

(54 : 46) 73%

Ahonen, M.; Sjöholm, R. *Chem. Lett.*, ***1995***, 341

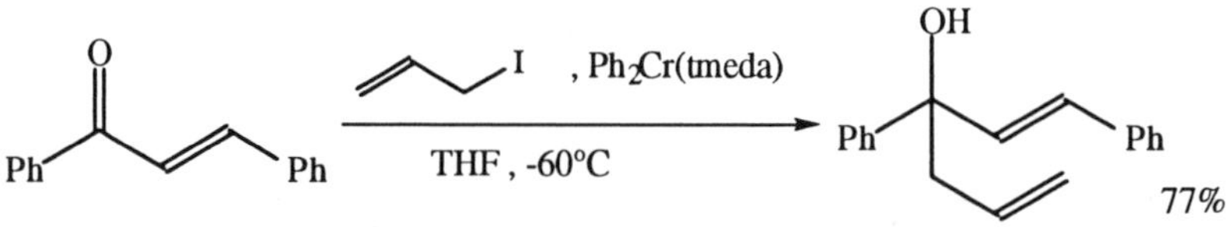

O

I, Ph_2Cr(tmeda)

THF, -60°C

Ph Ph

OH

Ph Ph

77%

Wipf, P.; Lim, S. *J. Chem. Soc. Chem. Commun.*, ***1993***, 1655

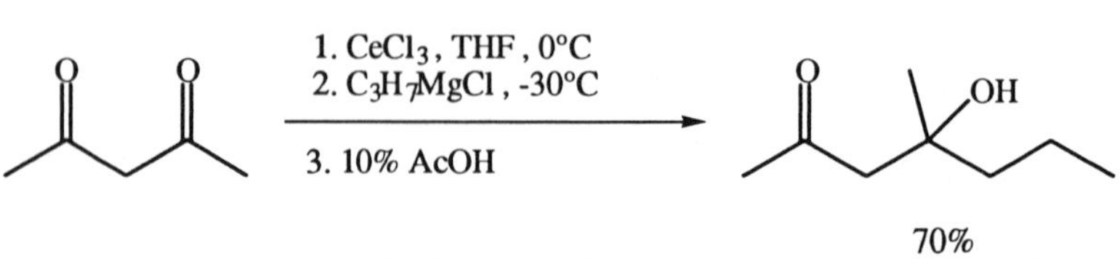

O O

1. $CeCl_3$, THF, 0°C
2. C_3H_7MgCl, -30°C

3. 10% AcOH

O OH

70%

Bartoli, G.; Marcantoni, E.; Petrini, M. *Angew. Chem. Int. Ed. Engl.*, ***1993***, *32*, 1061

SECTION 43: ALCOHOLS AND THIOLS FROM NITRILES

MeCN → 1. Li (excess), PhCHO, -30°C; 5% 4,4'-*bis-t-B*u biphenyl; 2. H_2O, HCl → Me–CH(OH)–Ph 39%

Guijarro, D.; Yus, M. *Tetrahedron*, ***1994***, *50*, 3441

SECTION 44: ALCOHOLS AND THIOLS FROM ALKENES

1. $Ti(BH_4)_3$, 30 min; CH_2Cl_2, -20°C; 2. 10% K_2CO_3

(1 : 1.2) 82%

Kumar, K.S.R.; Baskaran, S.; Chandrasekaran, S. *Tetrahedron Lett.*, ***1993***, *34*, 171

$Zn(BH_4)_2$, DMF; SiO_2, rt, 30 min

(70 : 30) 80%

Ranu, B.C.; Charkraborty, R.; Saha, M. *Tetrahedron Lett.*, ***1993***, *34*, 4659

1. Mg*, THF, rt; 2. ethylene oxide, THF, -78°C; 3. CO_2, 0·C → rt; 4. H_3O^+, 0°C → 40°C

69% or 86% with H_3O^+ at -78°C

Mg* = 8 Li, naphthalene, 4 $MgCl_2$, THF, 3 Li

Sell, M.S.; Xiong, H.; Rieke, R.D. *Tetrahedron Lett.*, ***1993***, *34*, 6007, 6011

1. 2 eq. PhO_2BH, THF; Nb catalyst, 25°C, 1d; 2. H_2O_2, NaOH

71%

Burgess, K.; Jaspars, M. *Tetrahedron Lett.*, ***1993***, *34*, 6813
also with Cp_2TiCl_4, see
Burgess, K.; van der Donk, W.A. *Tetrahedron Lett.*, ***1993***, *34*, 6817

1. Ph_3SiSH, AIBN
2. TFA

Br — 74% — SH

Haché, B.; Gareau, Y. *Tetrahedron Lett.*, ***1994***, *35*, 1837

buffered AD-mix-β, 0°C

t-BuOH-H_2O, $MeSO_2NH_2$

OH, I, OH 70%

(70%ee)

Vanhessche, K.P.M.; Wang, Z.-M.; Sharpless, K.B. *Tetrahedron Lett.*, ***1994***, *35*, 3473

1. catecholborane, 10% SmI_3, 18h
2. oxidation

C_8H_{17} → C_8H_{17}, OH 79% (50:1 1°:2°)

Evans, D.A.; Muci, A.R.; Stürmer, R. *J. Org. Chem.*, ***1993***, *58*, 5307

1. , -25°C
2. oxidation

BH_2, CMe_2 → OH, CMe_2 + OH

new asymmetric borane

74% (76% ee, S)

Dhokte, U.P.; Brown, H.C. *Tetrahedron Lett.*, ***1994***, *35*, 4715

1. Et_2BH
2. Et_2Zn
3. PhCHO, $Ti(OiPr)_4$

8% NHTf, NHTf

OH, Ph

67% (80% ee)

Schwink, L.; Knochel, P. *Tetrahedron Lett.*, ***1994***, *35*, 9007

1. 4 EtMgCl, ether, 25°C, 12m
5% Cp_2ZrCl_2
2. O_2, 0°C

OH, C_9H_{19} → OH, C_9H_{19}, OH

70% (95:5)

Houri, A.F.; Didiuk, M.T.; Xu, Z.; Horan, N.R.; Hoveyda, A.H. *J. Am. Chem. Soc.*, ***1993***, *115*, 6614

SiO_2, $Zn(BH_4)_2$, DME, rt

C_5H_{11} → C_5H_{11} OH

95%

Ranu, B.C.; Sarkar, A.; Saha, M.; Chakraborty, R. *Tetrahedron*, ***1994***, *50*, 6579

1. Co(tdcpp), O_2, Et_3SiH
iPrOH, CH_2Cl_2
2. $P(OMe)_3$

Ph → Ph, OH

tdcpp = 5,10,15,20-tetrakis(2,6-dichlorophenyl)porphinato 81%

Matsushita, Y.; Sugamoto, K.; Matsui, T. *Chem. Lett.*, ***1993***, 925

1. $HSiCl_3$, MeO^-/MOP, 40°C, 1d
$PdCl(\eta^3\text{-}C_3H_5)_2$
2. H_2O_2, KF, $KHCO_3$

C_6H_{13} → C_6H_{13}, OH

83x71% (93% ee)

Uozumi, Y.; Kitayama, K.; Hayashi, T.; Yanagi, K.; Fukuyo, E. *Bull. Chem. Soc. Jpn.*, ***1995***, *68*, 713

SECTION 45: ALCOHOLS AND THIOLS FROM MISCELLANEOUS COMPOUNDS

1. MOM-Cl, iPr_2NEt, CH_2Cl_2
2. 3.5 eq. Dibal

O, N → HO

93%

Meyers, A.I.; Shimano, M. *Tetrahedron Lett.*, ***1993***, *34*, 4893

$Cr(CO)_5$, OMe, O

hν, CO, THF

OMe, OH, O

65%

Merlic, C.A.; Roberts, W.M. *Tetrahedron Lett.*, ***1993***, *34*, 7379

$C_{12}H_{25}$—$SiMe_2SiMe_3$ → (1. TBAF, THF, rt; 2. H_2O_2, $KHCO_3$, MeOH, 40°C) → $C_{12}H_{25}$—OH 65%

Suginome, M.; Matusnaga, S.; Ito, Y. *Synlett*, ***1995***, 941

SECTION 45A: PROTECTION OF ALCOHOLS AND THIOLS

Me_2t-BuSiO$(CH_2)_5$OSiEt$_3$ → (PdO•H_2O, MeOH; cyclohexane, 30 min) → HO$(CH_2)_5$OSiEt$_3$ 90%

Cormier, J.F.; Issac, M.B.; Chen, L-F. *Tetrahedron Lett.*, ***1993***, *34*, 243

OH OH → (Py, CH_2Cl_2; Cl_3CO–C(O)–$OCCl_3$) → 87%

Burk, R.M.; Roof, M.B. *Tetrahedron Lett.*, ***1993***, *34*, 395

OH → (10% FSO_3H, rt, 30 min; xs allyl-SiMe$_3$) → OSiMe$_3$ 95%

Lipshutz, B.H.; Burgess-Henry, J.; Roth, G.P. *Tetrahedron Lett.*, ***1993***, *34*, 995

Ph–CH$_2$CH$_2$–OH → (Bu_4N^+ $^-OSO_2$-O-OSO_2O^- Bu_4N^+; 0.67 h, THF) → Ph–CH$_2$CH$_2$–O–(2-tetrahydrofuranyl) 95%

Jung, J.C.; Choi, H.C.; Kim, Y.H. *Tetrahedron Lett.*, ***1993***, *34*, 3581

OH → (t-BuPh$_2$SiCl, NH_4NO_3; DMF, 1d) → OSiPh$_2t$-Bu 87%

CAUTION: NH_4NO_3 & NH_4ClO_4 are potentially explosive

[w/ $AgNO_3$/BnOH/15 min - 83%]

Hardinger, S.A.; Wijaya, N. *Tetrahedron Lett.*, ***1993***, *34*, 3821

Bu–CH₂–OH → (dihydropyran, CH_2Cl_2, rt; $[Rh(MeCN)_3(triphos)][OTf]_2$) → Bu–CH₂–OTHP 99.5%

Ma, S.; Venanzi, L.M. *Tetrahedron Lett.*, ***1993***, *34*, 5269

Bn–(CH₂)₆–OH → (2-methyl-1-butene, $BF_3{\bullet}OEt_2$; CH_2Cl_2, 20°C) → Bn–(CH₂)₆–O–C(CH₃)₂CH₂CH₃ 98% ROTAM group

← see *Tetrahedron Lett.*, ***1988***, *29*, 2951

Figadère, B.; Franck, X.; Cavé, A. *Tetrahedron Lett.*, ***1993***, *34*, 5893

1. $BF_3{\bullet}2$ AcOH, CH_2Cl_2 rt, 20 sec
2. H_2O_2, KF, $NaHCO_3$ MeOH, THF, rt, 3d

Ph_2Si (CH₂C(CH₃)=CHCH₃), Ph, CO_2Me → Ph–CH(OH)–CH₂–CO_2Me 77%

Fleming, I.; Winter, S.B.D. *Tetrahedron Lett.*, ***1993***, *34*, 7287

1. "Cp_2Zr"
2. H_3O^+

Ph–CH(CH₃)–O–CH₂CH=CH₂ → Ph–CH(CH₃)–OH 97%

Cp_2Zr = Cp_2ZrCl_2/BuLi - see *Tetrahedron Lett.*, ***1986***, *27*, 2829

Ito, H.; Taguchi, T.; Hanzawa, Y. *J. Org. Chem.*, ***1993***, *58*, 774

$PhCH_2OH$ → ($MoO_2(acac)_2$, $CH_2(OMe)_2$, reflux) → $PhCH_2OCH_2OMe$ 85%

Kantam, M.L.; Santhi, P.L. *Synlett*, ***1993***, 429

BnO–$(CH_2)_{10}$–OTBDPS → (7 $BCl_3{\bullet}SMe_2$, ether, 1d) → HO–$(CH_2)_{10}$–OTBDPS 93%

Congreve, M.S.; Davison, E.C.; Fuhry, M.A.M.; Holmes, A.B.; Payne, A.N.; Robinson, R.A.; Ward, S.E. *Synlett*, ***1993***, 663

O O CMe3 → OH

Al_2O_3 microwave, 9 min

89%

Ley, S.V.; Mynett, D.M. *Synlett*, ***1993***, 793

HO OH → BnO OH

1. $Bu_2Sn(OMe)_2$, PhH, reflux Dean-Stark
2. $PhCH_2Br$

80%

Boons, G.-J.; Castle, G.H.; Clase, J.A.; Grice, P.; Ley, S.V.; Pinel, C. *Synlett*, ***1993***, 913

Bu OH → Bu O O

dihydropyran, hexane

HY zeolite, reflux, 1.5h

85%

Kumar, P.; Dinesh, C.U.; Reddy, R.R.; Pandey, B. *Synthesis*, ***1993***, 1069

OH → OTHP

dihydropyran, CAN, MeCN

rt, 30 min

91%

Maity, G.; Roy, S.C. *Synth. Commun.*, ***1993***, *23*, 1667

OH → OTHP

dihydropyran, $CHCl_3$

$MoO_2(acac)_2$, reflux, 4h

68%

Kantam, M.L.; Santhi, P.L. *Synth. Commun.*, ***1993***, *23*, 2225

AcO $(CH_2)_3OAc$ → HO $(CH_2)_3OAc$

alumina, 30 sec

microwave

with 2.5 min of microwave, obtain 92% bis deacetylation

93%

Varma, R.S.; Varma, M.; Chatterjee, A.K. *J. Chem. Soc., Perkin Trans. 1.*, ***1993***, 999

Bu OH → Bu OMOM

MOM-Cl, Na-X-zeolite

CH_2Cl_2, reflux

90%

Kumar, P.; Raju, S.V.N.; Reddy, R.S.; Pandey, B. *Tetrahedron Lett.*, ***1994***, *35*, 1289

OTHP → OH (quant.)

1. 5% EtSH, 10% $BF_3\bullet OEt_2$, CH_2Cl_2, -20°C → 0°C
2. aq. $NaHCO_3$

$OSiMe_2t$-Bu

cleavage of THP ethers in the presence of other acid labile functional groups

Nambiar, K.P.; Mitra, A. *Tetrahedron Lett.*, ***1994***, *35*, 3033

Bu–CH2–OH → (DHP, Al_2O_3, $ZnCl_2$; rt, 15 min) 90%

Ranu, B.C.; Saha, M. *J. Org. Chem.*, ***1994***, *59*, 8269

OTBDMS, OH ← (MeCN, K_2CO_3, Kriptofix 222; 55°C, 2h) OTBDMS, OTBDMS → (PPTS; EtOH, 50°C, 1h) OH, OTBDMS

94% 85%

Prakash, C.; Saleh, S.; Blair, I.A. *Tetrahedron Lett.*, ***1994***, *35*, 7565

$HOOC(CH_2)_{15}OH$ → (TMSHN–C(O)–NHTMS; cat. TBAF, CH_2Cl_2, 0°C) $HOOC(CH_2)_{15}OSiMe_3$ quant

other silazanes can be used as well

Tanabe, Y.; Murakami, M.; Kitaichi, K.; Yoshida, Y. *Tetrahedron Lett.*, ***1994***, *35*, 8409

Ph–C6H4–CH2OH → (HIP-OH, DEAD, PPh_3) Ph–C6H4–CH2OHIP; ← (LiNaphth, THF, -78°C)

HIP-OH = $(CF_3)_2CPhOH$ 86%

Cho, H.-S.; Yu, J.; Falck, J.R. *J. Am. Chem. Soc.*, ***1994***, *116*, 8354

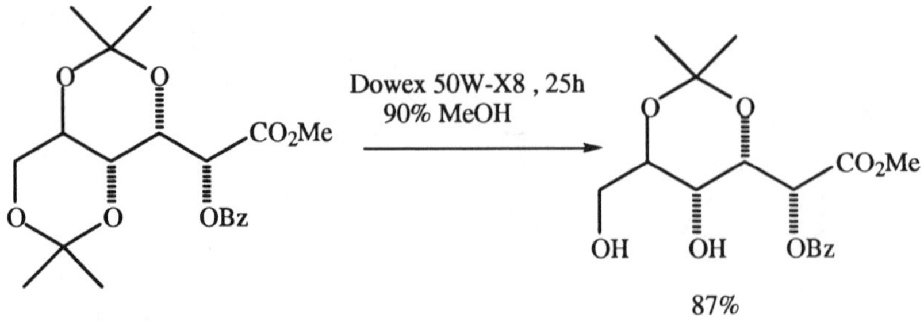

Dowex 50W-X8 , 25h
90% MeOH

CO_2Me OBz → OH OH OBz CO_2Me

87%

Park, K.H.; Yoon, Y.J.; Lee, S.G. *Tetrahedron Lett.*, ***1994***, *35*, 9737

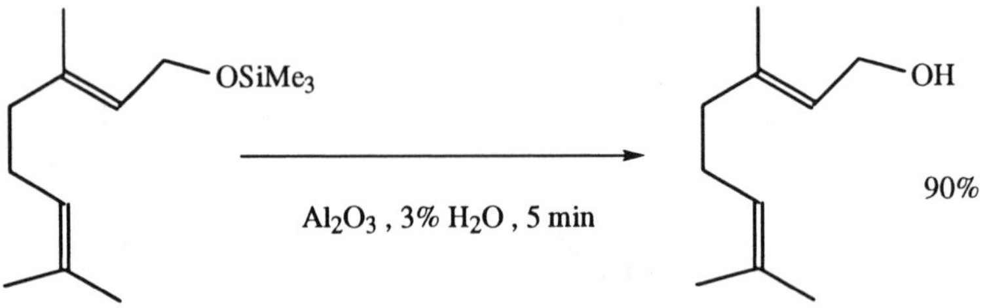

$OSiMe_3$ → OH

Al_2O_3 , 3% H_2O , 5 min

90%

Feixas, J.; Capdevila, A.; Guerrero, A. *Tetrahedron*, ***1994***, *50*, 8539

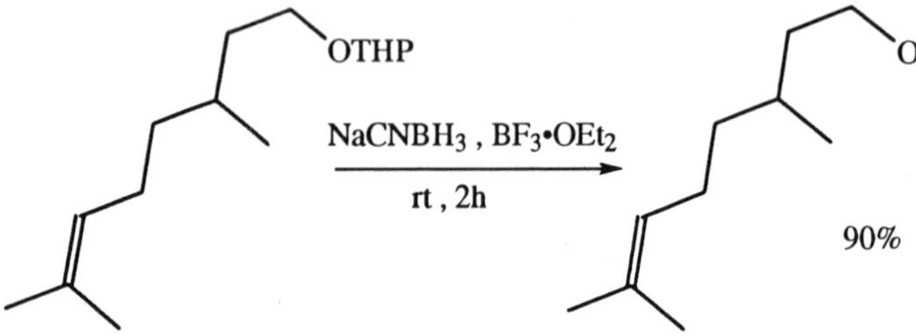

OTHP → OTHP

$NaCNBH_3$, $BF_3{\bullet}OEt_2$
rt , 2h

90%

Srikrishna, A.; Sattigeri, J.A.; Viswajanani, R.; Yelamaggad, C.V. *J. Org. Chem.*, ***1995***, *60*, 2260

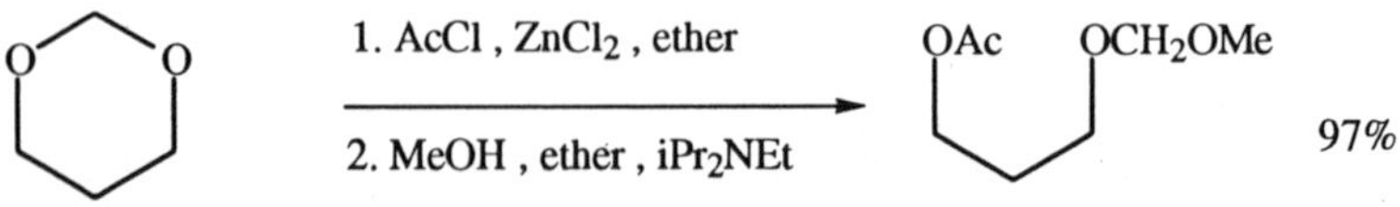

1. AcCl , $ZnCl_2$, ether
2. MeOH , ether , iPr_2NEt

OAc OCH_2OMe

97%

Bailey, W.F.; Zarcone, L.M.J.; Rivera, A.D. *J. Org. Chem.*, ***1995***, *60*, 2532

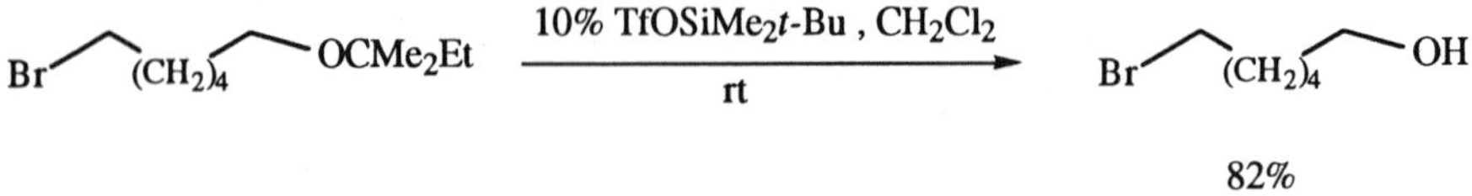

Br $(CH_2)_4$ $OCMe_2Et$ → Br $(CH_2)_4$ OH

10% $TfOSiMe_2t$-Bu , CH_2Cl_2
rt

82%

Franck, X.; Figadère, B.; Cavé, A. *Tetrahedron Lett.*, ***1995***, *36*, 711

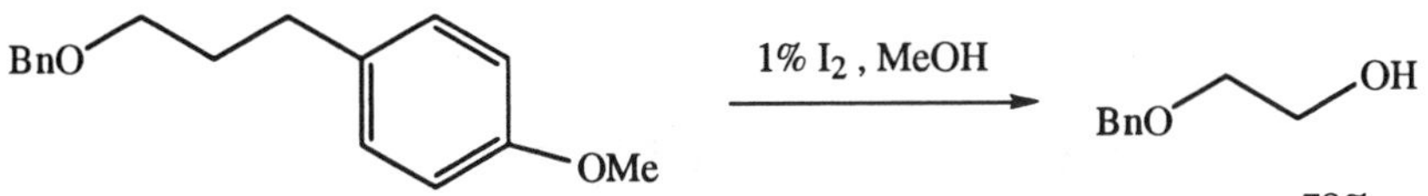

BnO(CH2)3-C6H4-OMe → ($1\%\ I_2$, MeOH) → BnO(CH2)2OH 78%

Vaino, A.R.; Szarek, W.A. *Synlett*, ***1995***, 1157

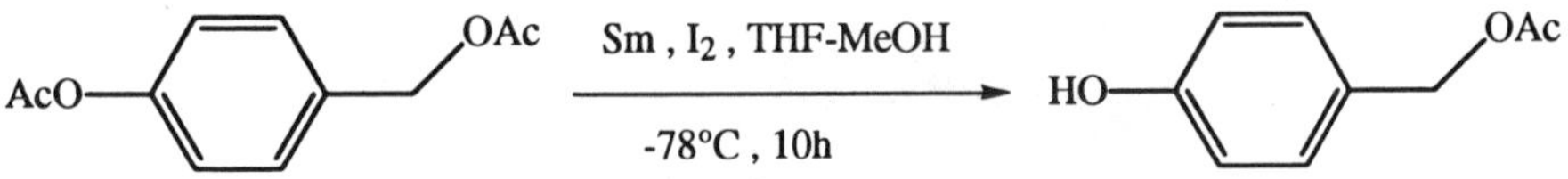

AcO–C6H4–CH2OAc → (Sm, I_2, THF-MeOH; -78°C, 10h) → HO–C6H4–CH2OAc 98%

Yanada, R.; Negoro, N.; Bessho, K.; Yanada, K. *Synlett*, ***1995***, 1261

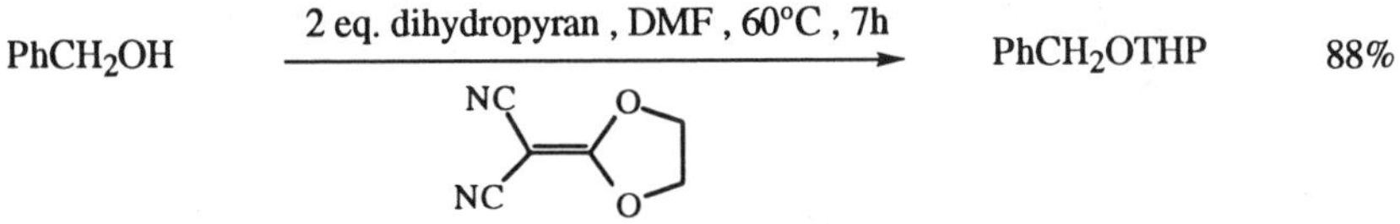

$PhCH_2OH$ → (2 eq. dihydropyran, DMF, 60°C, 7h; (NC)2C=C(OCH2CH2O)) → $PhCH_2OTHP$ 88%

Miura, T.; Masaki, Y. *Synth. Commun.*, ***1995***, *25*, 1981

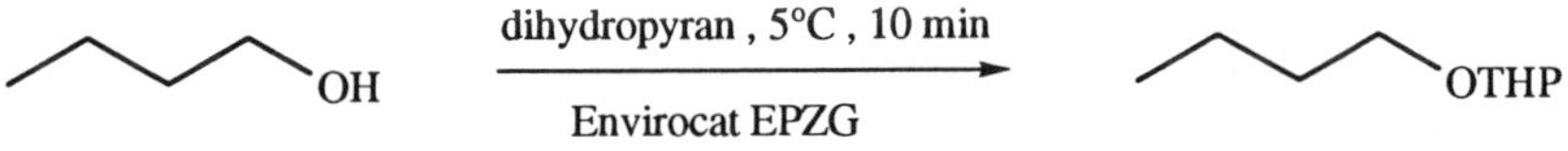

n-BuOH → (dihydropyran, 5°C, 10 min; Envirocat EPZG) → n-BuOTHP 88%

Bandgar, B.P.; Jagtap, S.R.; Aghade, B.B.; Wadgaonkar, P.P. *Synth. Commun.*, ***1995***, *25*, 2211

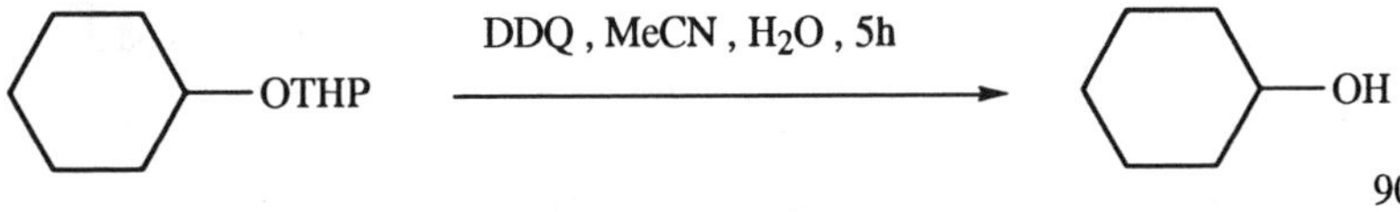

Cyclohexyl-OTHP → (DDQ, MeCN, H_2O, 5h) → Cyclohexyl-OH 90%

Raina, S.; Singh, V.K. *Synth. Commun.*, ***1995***, *25*, 2395

REVIEW:

"Silyl Ethers as Protective Groups for Alcohols. Oxidative Deprotection and Stability Under Alcohol Oxidation Conditions," Muzart, J. *Synthesis*, ***1993***, 11

CHAPTER 4
PREPARATION OF ALDEHYDES

SECTION 46: ALDEHYDES FROM ALKYNES

NO ADDITIONAL EXAMPLES

SECTION 47: ALDEHYDES FROM ACID DERIVATIVES

PhC(O)Cl → $AlH(Ot\text{-}Bu)_3$, diglyme-THF, -78°C → PhCHO 81%

Cha, J.S.; Brown, H.C. *J. Org. Chem.*, ***1993***, *58*, 4732

$PhCH_2CH_2C(O)Cl$ → 1. PBu_3 2. Zn-Cu, $MeSO_3H$ 3. 10 K_2CO_3 → $PhCH_2CH_2CHO$ quant.

Maeda, H.; Maki, T.; Ohmori, H. *Tetrahedron Lett.*, ***1995***, *36*, 2247

$C_5H_{11}C(O)Cl$ → 1. 2 eq. piperidine, THF, rt 2. 2 eq. $NaEt_2AlH_2$, rt, 1h → $C_5H_{11}CHO$ 74%

Yoon, N.M.; Choi, K.I.; Gyoung, Y.S.; Jun, W.S. *Synth. Commun.*, ***1993***, *23*, 1775

SECTION 48: ALDEHYDES FROM ALCOHOLS AND THIOLS

2-MeO-$C_6H_4CH_2OH$ → MnO_2, Bentonite, microwave → 2-MeO-C_6H_4CHO 60%

Martinez, L.A.; García, O.; Delgado, F.; Alvarez, C.; Patiño, R. *Tetrahedron Lett.*, ***1993***, *34*, 5293

TCM (trichloromelamine)
CH_2Cl_2, rt, 20h

$PhCH_2OH \longrightarrow PhCHO$ 83%

also oxidizes 2° alcohols to ketones

Kondo, S.; Ohira, M.; Kawasoe, S.; Kunisada, H.; Yuki, Y. *J. Org. Chem.*, ***1993***, *58*, 5003

5% $CpRu(PPh_3)_6Cl$, 10% $(C_2H_5)_3NHPF_6$, 8h

Ph⁀⁀OH → Ph⁀⁀OH 90%

Trost, B.M.; Kulawiec, R.J. *J. Am. Chem. Soc.*, ***1993***, *115*, 2027

hydrous zirconium (IV) oxide
Me_3SiCl, PhCHO, xylene

C_5H_{11}⁀⁀OH → C_5H_{11}⁀CHO 67%

Kuno, H.; Shibagaki, M.; Yakahashi, K.; Matsushita, H. *Bull. Chem. Soc. Jpn.*, ***1993***, *66*, 1699

HO O I O O

DMSO, rt

OH O → CHO O 88%

also for the preparation of ketones and 1,2-diketones

Frigerio, M.; Santagostino, M. *Tetrahedron Lett.*, ***1994***, *35*, 8019

Montmorillonite K-10, pentane
K_2FeO_4, 3h

Ph⁀OH → PhCHO 63%

Delaude, L.; Laszlo, P.; Lehance, P. *Tetrahedron Lett.*, ***1995***, *36*, 8505

PDC/Adogen 464

$PhCH_2OH \longrightarrow PhCHO$ 98%

Mohand, S.A.; Muzart, J. *Synth. Commun.*, ***1995***, *25*, 2373

SECTION 49: ALDEHYDES FROM ALKYNES

Conjugate reductions and Michael Alkylations of conjugated aldehydes are listed in Section 74 (Alkyls from Alkenes).

CHO; $Me-C(=S)-NH_2$; DMF; CHO; 70%

Phillips, O.A.; Eby, P.; Maiti, S.N. *Synth. Commun.*, ***1995***, *25*, 87

PhCHO → 1. Me_3SiI , KOH , H_2O 2. SiO_2 → $PhCH_2CHO$

Lemini, C.; Ordoñez, M.; Pérez-Flores, J.; Cruz-Almanza, R. *Synth. Commun.*, ***1995***, *25*, 2695

Ph CHO → 1. (dioxaborolane)$_2$CHLi 2. H_2O 3. $NaBO_3$ → Ph CHO 85%

Schummer, D.; Höfle, G. *Tetrahedron*, ***1995***, *51*, 11219

Related Methods: Aldehydes from Ketones (Section 57)
Ketones from Ketones (Section 177)
Also via: Alkenyl aldehydes (Section 341)

SECTION 50: ALDEHYDES FROM ALKYLS, METHYLENES AND ARYLS

O_2N–C$_6$H$_4$–CH_3 → 1. $(MeO)_2CHNMe_2$ DMF , 140°C 2. 3 eq. $NaIO_4$, THF → O_2N–C$_6$H$_4$–CHO 72x72%

Vetelino, M.G.; Coe, J.W. *Tetrahedron Lett.*, ***1994***, *35*, 219

H_3C–C$_6$H$_4$–CH_3 → $KBrO_3$, CAN , 95°C → H_3C–C$_6$H$_4$–CHO 70%

Ganin, E.; Amer, I. *Synth. Commun.*, ***1995***, *25*, 3149

SECTION 51: ALDEHYDES FROM AMIDES

NO ADDITIONAL EXAMPLES

SECTION 52: ALDEHYDES FROM AMINES

1. BuLi
2. H_3O^+

98 x 70%

Flippin, L.A.; Carter, D.S.; Dubree, H.J.P. *Tetrahedron Lett.*, ***1993***, *34*, 3255

Related Methods: Ketones from Amines (Section 172)

SECTION 53: ALDEHYDES FROM ESTERS

Bu_3SnH , PhH
0.4% $Pd(PPh_3)_4$

93%
1:99 (*E*:*Z*)

Kuniyasu, H.; Ogawa, A.; Sonoda, N. *Tetrahedron Lett.*, ***1993***, *34*, 2491

1. 5% Cp_2TiCl_2/BuLi
polymethyl-hydrosiloxane
THF
2. workup

65%

Barr, K.J.; Berk, S.C.; Buchwald, S.L. *J. Org. Chem.*, ***1994***, *59*, 4323

$NaAlH(NEt_2)_3$, THF , 6h

65%

Cha, J.S.; Kim, J.M.; Jeoung, M.K.; Kwon, O.O.; Kim, E.J. *Org. Prep. Proceed. Int.*, ***1995***, *27*, 95

$C_{11}H_{23}CO_2Me$ —[$Ca(BH_4)_2$, 1,5-cod , THF]→ $C_{11}H_{23}CH_2OH$ 95%

Narasimhan, S.; Prasad, K.G.; Madhavan, S. *Synth. Commun.*, ***1995***, *25*, 1689

$MeCO_2Me$ —[1. $PhMe_2SiLi$, THF , -110°C; 2. CrO_3 , DMSO , rt]→ MeCHO 70x74%

Fleming, I.; Ghosh, U. *J. Chem. Soc., Perkin Trans. 1.*, ***1994***, 257

REVIEW:

"Asymmetric Reductions of C-N Double Bonds. A Review," Zhu, Q.-C.; Hutchins, R.O. *Org. Prep. Proceed. Int.*, ***1994***, *26*, 193

SECTION 54: ALDEHYDES FROM ETHERS, EPOXIDES AND THIOETHERS

$LiClO_4$, ether; rt, 6h

CHO

60%

Palani, N.; Balasubramanian, K.K. *Tetrahedron Lett.*, ***1995***, *36*, 9527

Bu — Bu + Bu, CHO

MABR, CH_2Cl_2, -20°C	(0	:	100) 73%
SbF_3, PhH, 25°C	(85	:	15) 62%

Maruoka, K.; Murase, N.; Bureau, R.; Ooi, T.; Yamamoto, H. *Tetrahedron*, ***1994***, *50*, 3663

$C_{10}H_{21}$ — LiTMP, THF, 20°C → $C_{10}H_{21}$ CHO

83%

Yanagisawa, A.; Yasue, K.; Yamamoto, H. *J. Chem. Soc. Chem. Commun.*, ***1994***, 2103

Related Methods: Ketones from Ethers and Epoxides (Section 174)

SECTION 55: ALDEHYDES FROM HALIDES AND SULFONATES

Ph–Br → 1. iPr_3Si-SH/KH 2. TBAF, THF 3. H_2O → Ph–SH

88x73%

Miranda, E.I.; Diaz, M.J.; Rosado, I.; Soderquist, J.A. *Tetrahedron Lett.*, ***1994***, *35*, 3221

Rane, A.M.; Miranda, E.I.; Soderquist, J.A. *Tetrahedron Lett.*, ***1994***, *35*, 3225

C_8H_{17}—Br → 30 atm CO, $SiH(SiMe_3)_3$, PhH; AIBN, reflux → C_8H_{17}—CHO 80% + octane (16%)

Ryu, I.; Hasegawa, M.; Kurihara, A.; Ogawa, A.; Tsunoi, S.; Sonoda, N. *Synlett*, ***1993***, 143

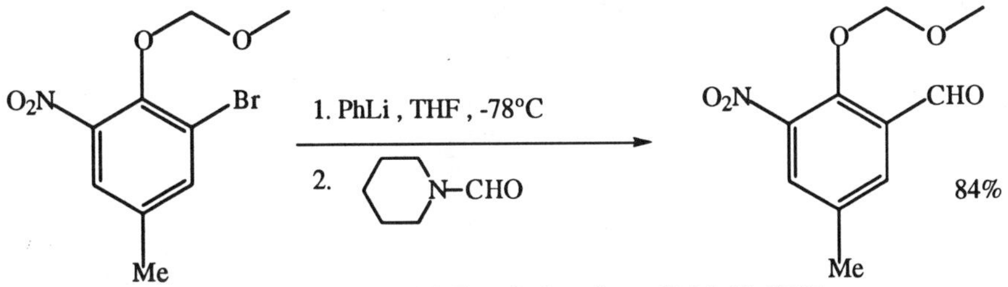

O_2N ... Br → 1. PhLi, THF, -78°C; 2. N-CHO (piperidine) → O_2N ... CHO, Me, 84%

Hardcastle, I.R.; Quayle, P.; Ward, E.L.M. *Tetrahedron Lett.*, ***1994***, *35*, 1747

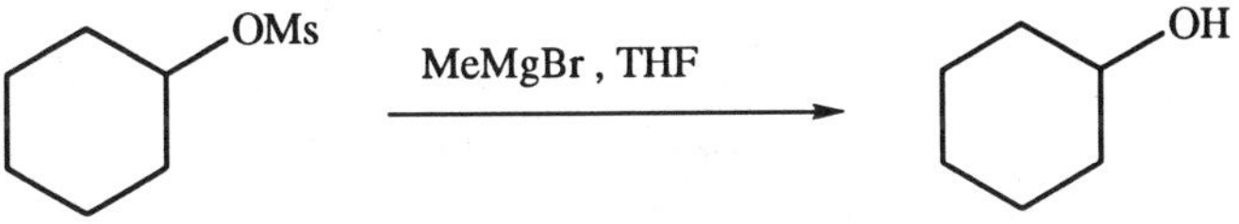

OMs → MeMgBr, THF → OH

Cossy, J.; Ranaivosata, J.-L.; Bellosta, V.; Wietzke, R. *Synth. Commun.*, ***1995***, *25*, 3109

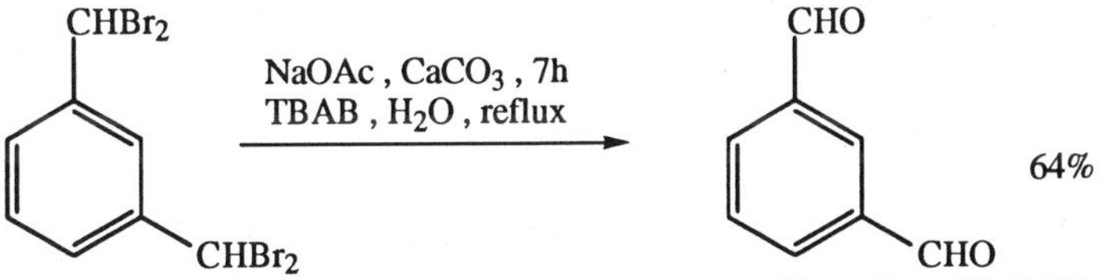

$CHBr_2$, $CHBr_2$ → NaOAc, $CaCO_3$, 7h; TBAB, H_2O, reflux → CHO, CHO 64%

Mataka, S.; Liu, G.-B.; Sawada, T.; Tori-i, A.; Tashiro, M. *J. Chem. Res. (S)*, ***1995***, 410

SECTION 56: ALDEHYDES FROM HYDRIDES

NO ADDITIONAL EXAMPLES

SECTION 57: ALDEHYDES FROM KETONES

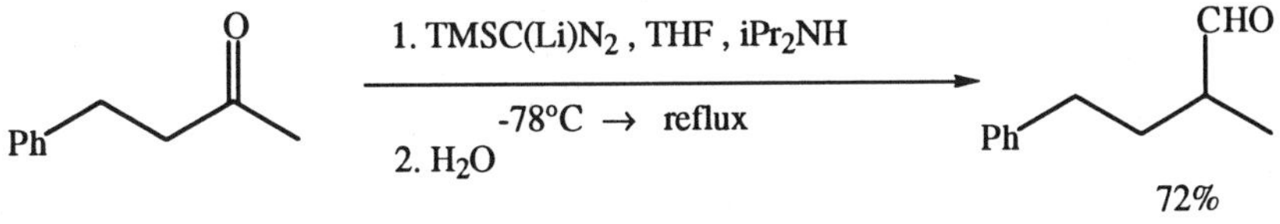

Ph, O → 1. TMSC(Li)N_2, THF, iPr_2NH, -78°C → reflux; 2. H_2O → Ph, CHO 72%

Miwa, K.; Aoyama, T.; Shioiri, T. *Synlett*, ***1994***, 109

SECTION 58: ALDEHYDES FROM NITRILES

PhCN → NaAlH$(NEt_2)_3$, THF, 25°C → PhCHO 97%

Cha, J.S.; Jeoung, M.K.; Kim, J.M.; Kwon, O.O.; Lee, J.C. *Org. Prep. Proceed. Int.*, ***1994***, *26*, 583

SECTION 59: ALDEHYDES FROM ALKENES

Ph Ph Ph — $Rh(acac)(CO)_2$, H_2/CO 100 atm, 60°C, PhH, 42h → Ph Ph Ph CHO + OHC Ph Ph Ph

(90 : 10)

> 99% conversion

Higashizima, T.; Sakai, N.; Nozaki, K.; Takaya, H. *Tetrahedron Lett.*, ***1994***, *35*, 2023

PhS — 4 eq. PhSH, Et_4NOTs, MeCN; e^- (0.05 F mol^{-1}) → PhS CHO 77%

Yoshida, J.; Nakatani, S.; Isoe, S. *J. Org. Chem.*, ***1993***, *58*, 4855

MPMO O O — 10% $PdCl_2/CuCl$ aq. DMF, O_2; 60°C, 1d → MPMO O O CHO 93%

Kang, S.-K.; Jung, K.-Y.; Chung, J.-U.; Namkoong, E.-Y.; Kim, T.-H. *J. Org. Chem.*, ***1995***, *60*, 4678

— H_2, CO (1600 psi); carbohydrate-$OPAr_2$ catalyst → CHO H Me + OHC

53% conversion 51%ee (96 : 4)

RajanBabu, T.V.; Ayers, T.A. *Tetrahedron Lett.*, ***1994***, *35*, 4295

Ph — CO, H_2; $[Rh(cod)IOAc]_2$ → Ph CHO + Ph CHO

(96 : 4) 94%

Doyle, M.P.; Shanklin, M.S.; Zlokazov, M.V. *Synlett*, ***1994***, 615

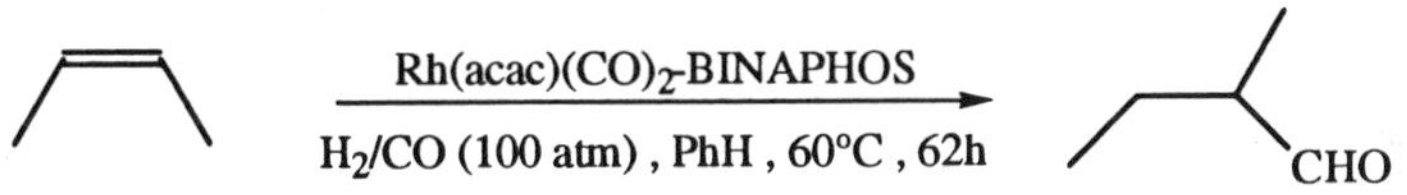

Rh(acac)(CO)$_2$-BINAPHOS; H_2/CO (100 atm), PhH, 60°C, 62h → CHO; 82% ee

Sakai, N.; Nozaki, K.; Takaya, H. *J. Chem. Soc. Chem. Commun.*, ***1994***, 395

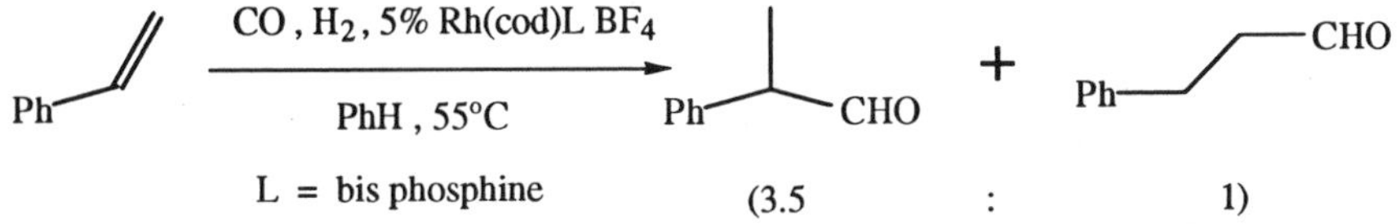

CO, H_2, 5% Rh(cod)L BF_4; PhH, 55°C; Ph → Ph–CH(CH$_3$)–CHO + Ph–CH$_2$CH$_2$–CHO

L = bis phosphine (3.5 : 1)

Chan, A.S.C.; Pai, C.-C.; Yang, T.-K.; Chen, S.-M. *J. Chem. Soc. Chem. Commun.*, ***1995***, 2031

Related Methods: Ketones from Alkenes (Section 179)

SECTION 60: ALDEHYDES FROM MISCELLANEOUS COMPOUNDS

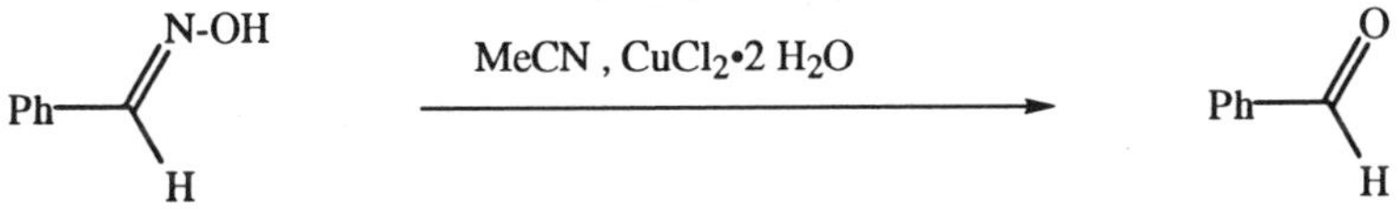

Ph–CH=N-OH → MeCN, $CuCl_2$•2 H_2O → Ph–CHO

Singh, L.; Ram, R.N. *Synth. Commun.*, ***1993***, *23*, 3139

Ph–CH=N–NH–C(O)–NH_2 → TMSCl, $NaNO_2$, CCl_4; Aliquat 336 → PhCHO 95%

Khan, R.H.; Mathur, R.K.; Ghosh, A.C. *J. Chem. Res. (S)*, ***1995***, 506

SECTION 60A: PROTECTION OF ALDEHYDES

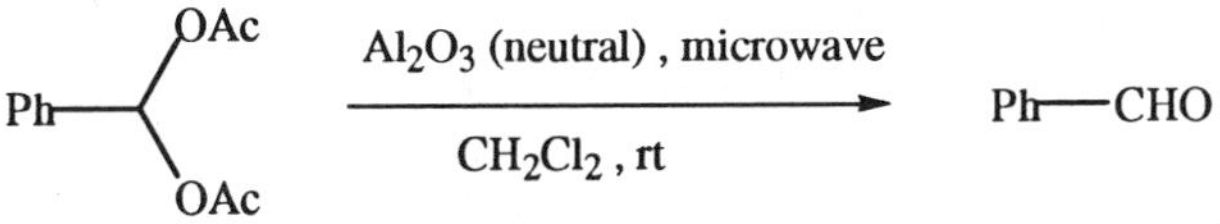

Ph–CH(OAc)$_2$ → Al_2O_3 (neutral), microwave; CH_2Cl_2, rt → Ph—CHO 98%

Varma, R.S.; Chatterjee, A.K.; Varma, M. *Tetrahedron Lett.*, ***1993***, *34*, 3207

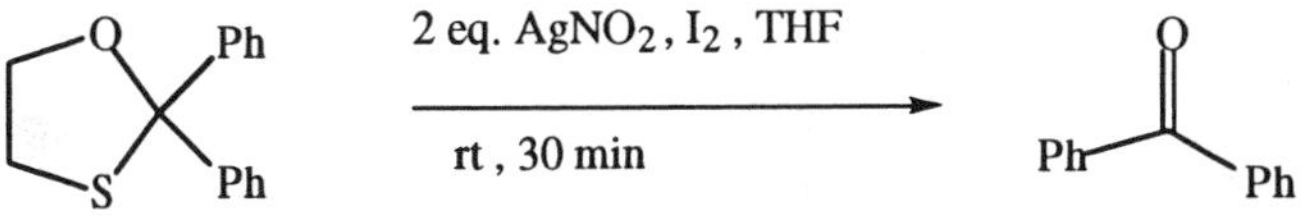

2,2-diphenyl-1,3-oxathiolane (O, S, Ph, Ph) → 2 eq. $AgNO_2$, I_2, THF; rt, 30 min → Ph–C(O)–Ph 96%

also works with dithianes — also for protected ketones

Nishide, K.; Yokota, K.; Nakamura, D.; Sumiya, T.; Node, M.; Ueda, M.; Fuji, K. *Tetrahedron Lett.*, ***1993***, *34*, 3425

Ph–CH(OAc)(OAc) → Ph—CHO 95%

t-BuOK , THF , rt , 20 min

Me_2N–C6H4–$O^- K^+$

Ku, Y.-Y.; Patel, R.; Sawick, D. *Tetrahedron Lett., **1993**, 34*, 8037

Ph–CH2–CH(SPh)(SPh) → Ph–CH2–CHO 95%

DDQ , H_2O , MeOH

hν (λ > 350 nm)

Mathew, L.; Sankararaman, S. *J. Org. Chem., **1993**, 58*, 7576

Ph–CH=CH–CHO → Ph–CH=CH–(1,3-dioxolan-2-yl) 97%

p-TsOH , $MgSO_4$, PhH

ethylene glycol , reflux , 4.5h

Lu, T.-J.; Yang, J.-F.; Sheu, L.-J. *J. Org. Chem., **1995**, 60*, 2931

Cyclohexyl–CHO → 2-cyclohexyl-1,3-dithiolane 99%

HS–CH2CH2–SH , CH_2Cl_2

$CoBr_2°SiO_2$, 3 min

also with ketones

Patney, H.K. *Tetrahedron Lett., **1994**, 35*, 5717

CH3CH2CH2–CHO → CH3CH2CH2–CH(SEt)(SEt) 92%

2 EtSH , HY zeolite , CH_2Cl_2

35°C

ketones can also be used

Kumar, P.; Reddy, R.S.; Singh, A.P.; Pandley, B. *Synthesis, **1993***, 67

Ph–CH2CH2–CHO → Ph–CH2CH2–(1,3-dioxolan-2-yl) 96%

ethylene glycol , TsOH

microwave , 2 min

ketones can also be used

Moghaddam, F.M.; Sharifi, A. *Synth. Commun., **1995**, 25*, 2457

Ph–CH(OMe)(OMe) $\xrightarrow{MoO_2(acac)_2\text{ , MeCN , 4h}}$ PhCHO

97%

Kantam, M.L.; Swapna, V.; Santhi, P.L. *Synth. Commun.*, ***1995***, *25*, 2529

PhCHO $\xrightarrow[CHCl_3\text{ , rt , overnight}]{\text{1,2-ethanediol , cat. CAN}}$ Ph–(1,3-dithiolan-2-yl)

85%

ketones can also be used

Mandal, P.K.; Roy, S.C. *Tetrahedron*, ***1995***, *51*, 7823

CHAPTER 5

PREPARATION OF ALKYLS, METHYLENES AND ARYLS

This chapter lists the conversion of functional groups into methyl, ethyl, propyl, etc. as well as methylene (CH_2), phenyl, etc.

SECTION 61: ALKYLS, METHYLENES AND ARYLS FROM ALKYNES

OH CO2Me I

1. $C_6H_{13}C{\equiv}CH$, cat. CuCl
cat. $Pd(PPh)_3)_4$, t-$BuNH_2$
PhMe , rt

2. PCC , CH_2Cl_2 , rt
3. $CHCl_3$, CSA

OH CO2Me C_6H_{13}

Ciufolini, M.A.; Weiss, T.J. *Tetrahedron Lett.*, ***1994***, *35*, 1127

Ph

2. 7 eq. TfOH , 22°C , 28h

2 eq. $Cp(CO)_3WH$

Ph 98%

Luan, L.; Song, J.-S.; Bullock, R.M. *J. Org. Chem.*, ***1995***, *60*, 7170

O O SCPh3

33% HBr-AcOH , CH_2Cl_2
0°C → rt

Br O S 72%

Masquelin, T.; Obrecht, D. *Tetrahedron Lett.*, ***1994***, *35*, 9387

PhC≡CH , 1,2-dichlorobenzene
110°C , 14h

zeolite HSZ-360

Ph

40% yield (49% conversion)

Sartori, G.; Bigi, F.; Pastorio, A.; Porto, C.; Arienti, A.; Maggi, R.; Moretti, N.; Gnappi, G. *Tetrahedron Lett.*, ***1995***, *36*, 9177

$Rh_2(C_7H_{15}CO_2)_4$

(3 : 1) 71%

Taber, D.F.; Rahimizadeh, M. *Tetrahedron Lett.*, ***1994***, *35*, 9139

1. $Rh(OAc)_2$, PhH
2. air

75%

Mueller, P.H.; Kassir, J.M.; Semones, M.A.; Weingarten, M.D.; Padwa, A. *Tetrahedron Lett.*, ***1993***, *34*, 4285

, NEt_3

PhH , 37°C , 18h

76%

Grissom, J.W.; Klingberg, D. *Tetrahedron Lett.*, ***1995***, *36*, 6607

1. $TaCl_5$, Zn , DME/PhH , 50°C , 2h
2. Py , THF
3. $HC{\equiv}C{-}(CH_2)_4{-}C{\equiv}CH$
4. aq. NaOH

82%

Takai, K.; Yamada, M.; Utimoto, K. *Chem. Lett.*, ***1995***, 851

SECTION 62: ALKYLS, METHYLENES AND ARYLS FROM ACID DERIVATIVES

NO ADDITIONAL EXAMPLES

SECTION 63: ALKYLS, METHYLENES AND ARYLS FROM ALCOHOLS AND THIOLS

OH MeO OH $AlCl_3$, PhOH MeO OMe 70%

Bijoy, P.; Subba Rao, G.S.R. *Tetrahedron Lett.*, ***1994***, *35*, 3341

CO_2Et CO_2Et C_3H_7OH, TMAD, Bu_3P PhH, DHTD-TBP C_3H_7 CO_2Et CO_2Et 64%

TMAD = $Me_2NCON{=}NCONMe_2$

DHTD = 4,7-dimethyl-3,5,7-hexahydro-1,2,4,7-tetrazocin-3,8-dione

Tsudoda, T.; Nagaku, M.; Nagino, C.; Kawamura, Y.; Ozaki, F.; Hioki, H.; Itô, S. *Tetrahedron Lett.*, ***1995***, *36*, 2531

$C_9H_{19}CH_2OH$ 1. PhNCS, NaH 2. Ph_3SiH, BEt_3 rt, 0.2h $C_9H_{19}CH_3$ 93%

Oba, M.; Nishiyama, K. *Tetrahedron*, ***1994***, *50*, 10193

SECTION 64: ALKYLS, METHYLENES AND ARYLS FROM ALDEHYDES

PhCHO BER-$Ni(OAc)_2$, MeOH $PhCH_3$ 91%

BER = borohydride exchange resin

Bandgar, B.P.; Kshirsagar, S.N.; Wadgaonkar, P.P. *Synth. Commun.*, ***1995***, *25*, 941

Related Methods: Alkyls, Methylenes and Aryls from Ketones (Section 72)

SECTION 65: ALKYLS, METHYLENES AND ARYLS FROM ALKYLS, METHYLENES AND ARYLS

CO_2H / CO_2H → CO_2Et / Ph

1. e^-, 70°C, dioxane/5% H_2SO_4
2. Ph-C≡C-CO_2Et, 220°C, 4.5h

86 x 77%

Ohno, T.; Ozaki, M.; Inagaki, A.; Hirashima, T.; Nishiguchi, I. *Tetrahedron Lett.*, ***1993**, 34*, 2601

I, N + Br, NO_2 → N, NO_2

5% $PdCl_2(PPh_3)_2$
130°C, DMSO
4 eq. Cu, 1h

94%

Shimizu, N.; Kitmura, T.; Watanabe, K.; Yamaguchi, T.; Shigyo, H.; Ohta, T. *Tetrahedron Lett.*, ***1993**, 34*, 3421

OTf

$Pd(PPh_3)_2Cl_2$

(2 : 1) 90%

Rice, J.E.; Cai, Z.-W. *J. Org. Chem.*, ***1993**, 58*, 1415

N-Cy → N-Cy

1. 2 eq. LTMP, THF
2. MeI

89%

Flippin, L.A.; Muchowski, J.M.; Carter, D.S. *J. Org. Chem.*, ***1993**, 58*, 2463

OH OH → OH OH

2 eq. $AlCl_3$, 2 eq. $FeCl_3$

50°C , 1h

OMe 48% OMe

Sartori, G.; Maggi, R.; Bigi, F.; Grandi, M. *J. Org. Chem.*, ***1993***, *58*, 7271

OMe → OMe + OMe

PhCHO , Ga_2Cl_4 , CS_2

0°C , 72h

Bn Bn

(44 : 56) 77%

Hashimoto, Y.; Hirata, K.; Kagoshima, H.; Kihara, N.; Hasegawa, M.; Saigo, K. *Tetrahedron*, ***1993***, *49*, 5969

O N H N H

5 eq. H_2SO_4 , reflux , 3.5 h

68%

yield based on moles of alkylating agent

Chung, K.H.; Kim, J.N.; Ryu, E.K. *Tetrahedron Lett.*, ***1994***, *35*, 2913

CO_2H → CO_2H Me

1. *s*-BuLi.TMEDA , -90°C , THF
2. MeI , -78°C
3. H_3O^+

65%

Mortier, J.; Moyroud, J.; Bennetau, B.; Cain, P.A. *J. Org. Chem.*, ***1994***, *59*, 4042

OAc BocHN CO_2Me → BocHN CO_2Me

Br

Zn dust

Alkylation also occurs with benzylic and allylic halides 97%

Abood, N.A.; Nosal, R. *Tetrahedron Lett.*, ***1994***, *35*, 3669

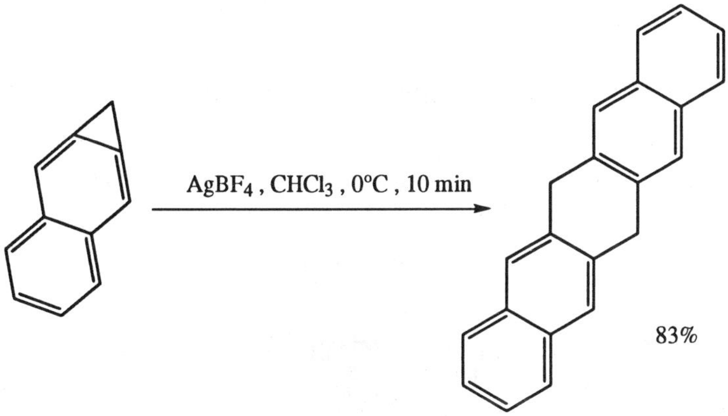

$AgBF_4$, $CHCl_3$, 0°C, 10 min

83%

Billups, W.E.; McCord, D.J.; Maughon, B.R. *Tetrahedron Lett.*, ***1994***, *35*, 4493

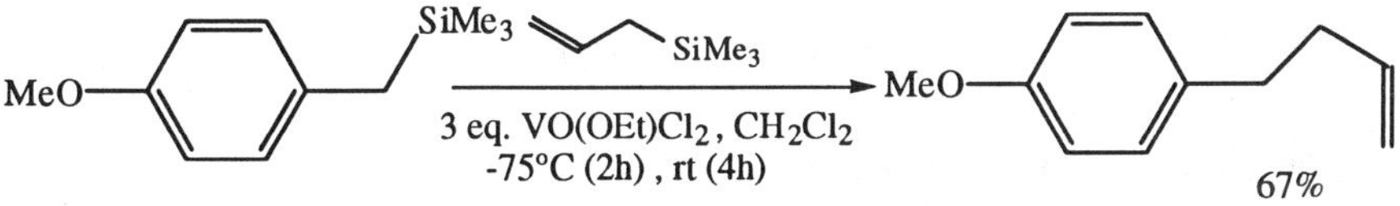

MeO SiMe3 SiMe3 MeO

3 eq. $VO(OEt)Cl_2$, CH_2Cl_2
-75°C (2h), rt (4h)

67%

Hirao, T.; Fujii, T.; Ohshiro, Y. *Tetrahedron Lett.*, ***1994***, *35*, 8005

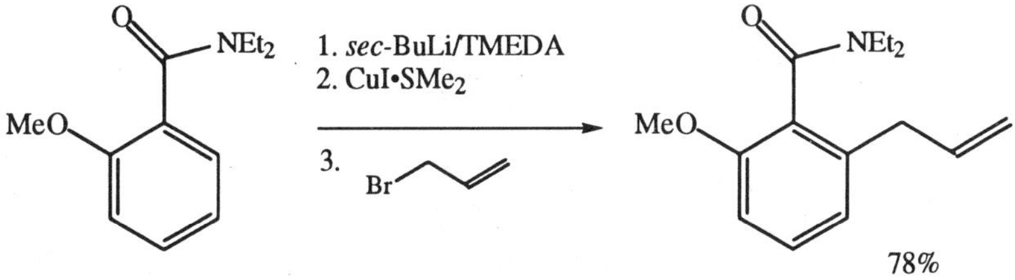

O NEt2 MeO

1. *sec*-BuLi/TMEDA
2. CuI•SMe_2
3. Br

O NEt2 MeO

78%

Casas, R.; Cavé, C.; d'Angelo, J. *Tetrahedron Lett.*, ***1995***, *36*, 1039

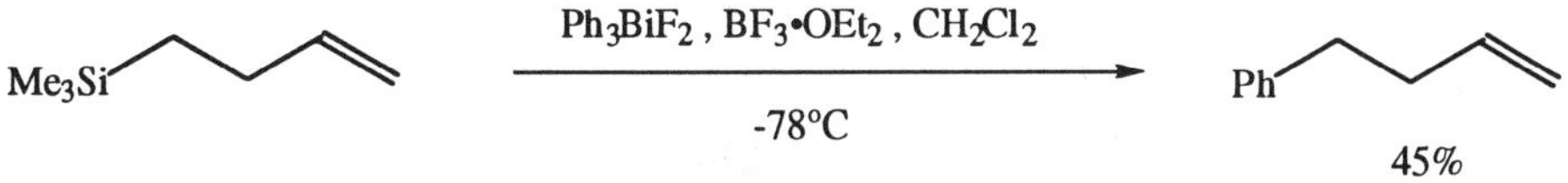

Me_3Si

Ph_3BiF_2, BF_3•OEt_2, CH_2Cl_2
-78°C

Ph

45%

Matano, Y.; Yoshimune, M.; Suzuki, H. *Tetrahedron Lett.*, ***1995***, *36*, 7475

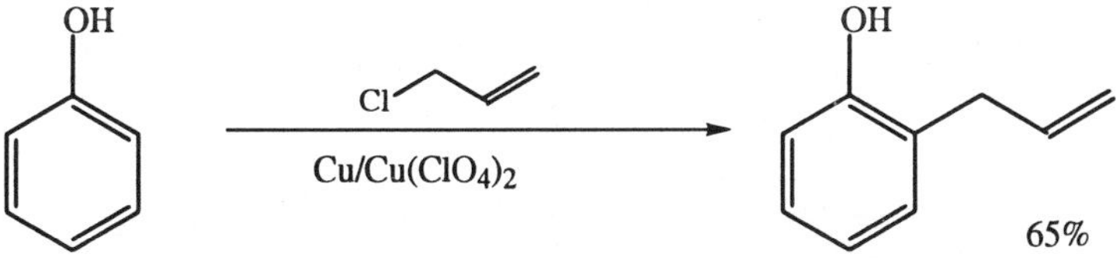

OH

Cl
$Cu/Cu(ClO_4)_2$

OH

65%

Baruah, J.B. *Tetrahedron Lett.*, ***1995***, *36*, 8509

SECTION 66: ALKYLS, METHYLENES AND ARYLS FROM AMIDES

NO ADDITIONAL EXAMPLES

SECTION 67: ALKYLS, METHYLENES AND ARYLS FROM AMINES

NO ADDITIONAL EXAMPLES

SECTION 68: ALKYLS, METHYLENES AND ARYLS FROM ESTERS

Ph, C_5H_{11}, OCO_2Et → $PhB(OMe)_3Li$, THF, 60°C, 15h; cat. $[NiCl_2(dppf)]$ → Ph, C_5H_{11}, Ph

83%

Kobayashi, Y.; Ikeda, E. *J. Chem. Soc. Chem. Commun.*, ***1994***, 1789

O, O, PhO → 13% (SCu, NMe_2), ether, 0•C; BuMgI (120 min addition) → Bu, PhO

quant (34% ee)

van Klaveren, M.; Persson, E.S.M.; del Villar, A.; Grove, D.M.; Bäckvall, J.-E. *Tetrahedron Lett.*, ***1995***, *36*, 3059

N → 10% $Ni(acac)_2$, 20% $AlEt_3$ toluene, reflux, PPh_3; $PhB(OH)_2$ → Ph + Ph

(2.1 : 1) 70%

Trost, B.M.; Spagnol, M.D. *J. Chem. Soc., Perkin Trans. 1.*, ***1995***, 2083

SECTION 69: ALKYLS, METHYLENES AND ARYLS FROM ETHERS, EPOXIDES AND THIOETHERS

The conversion ROR → RR' (R' = alkyl, aryl) is included in this section.

1. Li , THF , -18°C
2. -18°C → -70°C (air → Ar)

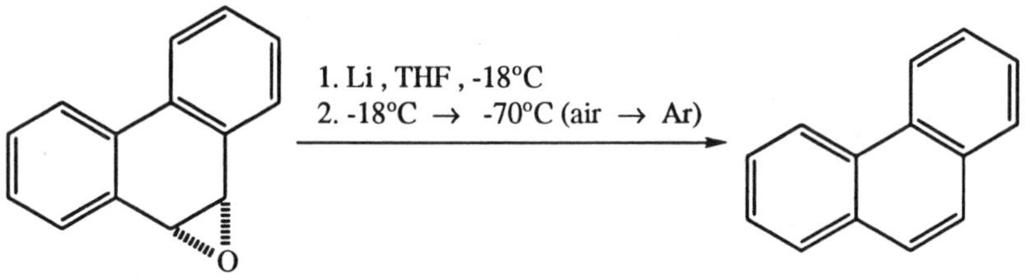

95%

Elmalak, O.; Rabinovitz, M.; Blum, J. *J. Heterocyclic Chem.*, ***1993***, *30*, 291

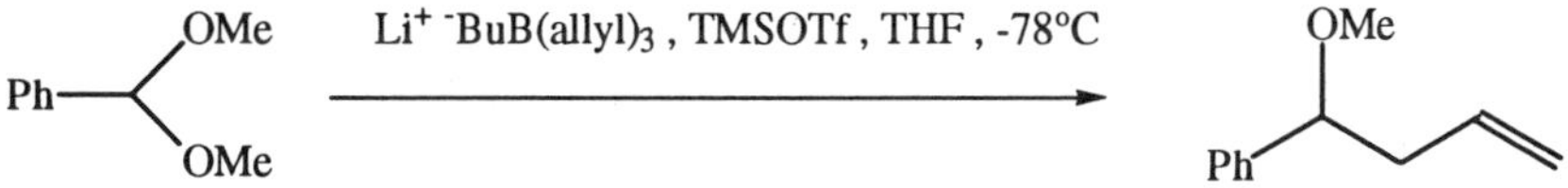

Ph–CH(OMe)(OMe)

Li^+ $^-BuB(allyl)_3$, TMSOTf , THF , -78°C

OMe Ph

94%

Hunter, R.; Michael, J.P.; Tomlinson, G.D. *Tetrahedron*, ***1994***, *50*, 871

SECTION 70: ALKYLS, METHYLENES AND ARYLS FROM HALIDES AND SULFONATES

The replacement of halogen by alkyl or aryl groups is included in this section. For the conversion of RX → RH (X = halogen) see Section 160 (Hydrides from Halides and Sulfonates).

OH Br

1. Cl
hν , C_6D_6 , $(Bu_3Sn)_2$
2. I_2 , DBU

OH

57%

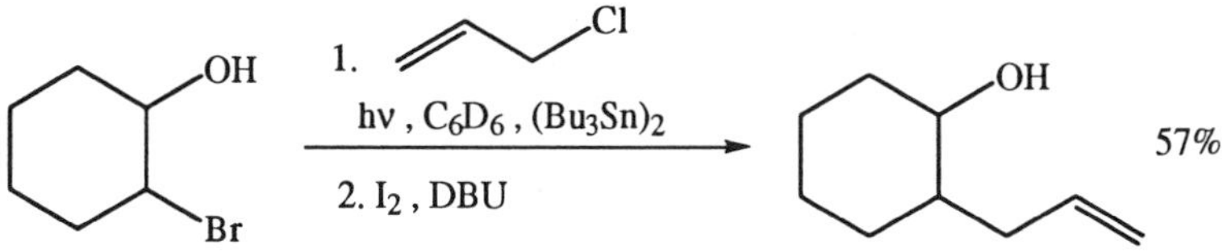

(80:20 , *trans:cis*)

Huval, C.C.; Singleton, D.A. *Tetrahedron Lett.*, ***1993***, *34*, 3041

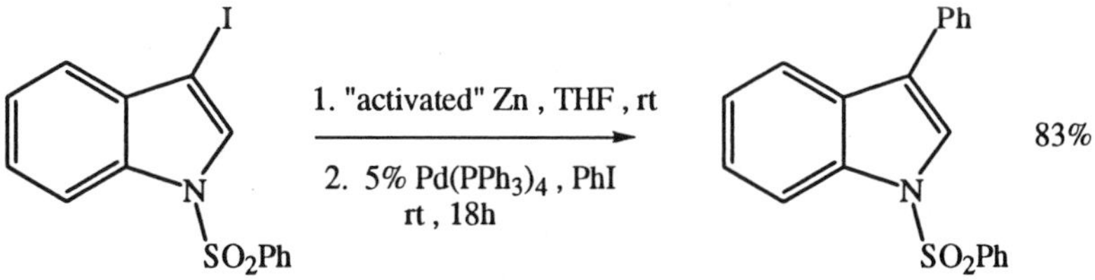

I N SO_2Ph

1. "activated" Zn , THF , rt
2. 5% $Pd(PPh_3)_4$, PhI
rt , 18h

Ph N SO_2Ph

83%

Sakamoto, T.; Kondo, Y.; Takazawa, N.; Yamanaka, H. *Tetrahedron Lett.*, ***1993***, *34*, 5955

1. Zn* , ether , 20°C
2. H_3O^+

I, Bu → Bu

60%
(73:27 *cis:trans*)

Meyer, C.; Marek, I.; Courtemanche, G.; Normant, J.-F. *Tetrahedron Lett.*, ***1993***, *34*, 6053

I → CO_2Me

CO_2Me / $SnBu_3$

$Pd(PPh_3)_4$, CuI-DMF
48h

87%

Levin, J.I. *Tetrahedron Lett.*, ***1993***, *34*, 6211

Me, Br, OMe + iPr, O, N, OMe, OMe

Mg , THF
heat

iPr, O, N, Me, OMe MeO

79%

(90:10 S:R)

Moorlag, H.; Meyers, A.I. *Tetrahedron Lett.*, ***1993***, *34*, 6989, 6993

1. Zn , DMSO , THF , 0°C
2. CuCN , THF , LiCl
3. PhC(O)Cl

O, O, I → O, O, Ph, O

93%

Knochel, P.; Chou, T.-S.; Jubert, C.; Rajagopal, D. *J. Org. Chem.*, ***1993***, *58*, 588

3 eq. PhMgCl , ether , 35°C
1.5 $ZnCl_2$•dioxane

10% $NiCl_2$(dppf)

I → Ph

72%

Park, K.; Yuan, K.; Scott, W.J. *J. Org. Chem.*, ***1993***, *58*, 4866

C_9H_{19}—Br → C_9H_{19}—Bu

BuMnCl , 3% $CuCl_4Li_2$

THF/NMP , rt , 1h

93%

Cahiez, G.; Marquais, S. *Synlett*, ***1993***, 45

PhCHO , -78° → rt

95%

Barluenga, J.; Montserrat, J.M.; Flórez, J. *J. Org. Chem.*, ***1993***, *58*, 5976

1. BuLi , -40°C , hexane:THF

2. allyl bromide , rt

40%

Wu, X.; Chen, T.-A.; Zhu, L.; Rieke, R.D. *Tetrahedron Lett.*, ***1993***, *34*, 3673

1. e^- , $ZnBr_2$, $Ni(BF_4)_2bpy_3$ DMF , Bu_4NBr , 2,2'-bipy

2. I–C6H4–OMe $Pd(PPh_3)_2Cl_2$, DMF

83%

Sibille, S.; Ratovelomanana, V.; Nédélec, J.Y.; Périchon, J. *Synlett*, ***1993***, 425

Bu_2Zn-2 LiCl , HMPA

93%

Arai, M.; Kawasuji, T.; Nakamura, E. *Chem. Lett.*, ***1993***, 357

1. CH_2=$CHCH_2SnBu_3$, AIBN
2. HBr , hν
3. Bu_3SnH , AIBN

51x97x51%

Zhang, W.; Hua, Y.; Hoge, G.; Dowd, P. *Tetrahedron Lett.*, ***1994***, *35*, 3865

1.5% KOH , $NaBH_4$ 90°C , ACVA

75%

Rai, R.; Collum, D.B. *Tetrahedron Lett.*, ***1994***, *35*, 6221

OSO_2CF_3 ; SiF_2Ph , TASF ; $Pd(PPh_3)_4$, THF , 60°C

64% (58%e , R)

Hatanaka, Y.; Goda, K.; Hiyama, T. *Tetrahedron Lett.*, ***1994***, *35*, 1279

OTf ; Zn , KI , 10% $NiCl_2(PPh_3)_2$; 67°C , THF , 4h

92%

Jutand, A.; Mosleh, A. *Synlett*, ***1993***, 568

ClZn ; O ; O ; 5% $Ni(acac)_2$, THF ; PPh_3 , 10% Dibal ; F—⟨⟩—OTf ; rt ; F ; O ; O

68%

Quesnelle, C.A.; Familoni, O.B.; Snieckus, V. *Synlett*, ***1994***, 349

O ; Cl ; PhI , $NiBr_2(bipy)$, DMF ; e^- , Bu_4NBF_4 ; O ; Ph

54%

Drandetti, M.; Sibille, S.; Nédélec, J.-Y.; Périchon, J. *Synth. Commun.*, ***1994***, *24*, 145

C_6H_{13} ; OTf ; OTf ; $Me_2(CN)CuLi_2$•THF ; -78°C → 25°C ; C_6H_{13}

60%

Martínez, A.G.; Barcina, J.O.; Díez, B.R.; Subramanian, L.R. *Tetrahedron*, ***1994***, *50*, 13231

Br; SiF_3; 5% $PdCl_2(PPh_3)_2$, 5% TBAF; THF , 120°C , 12h; O; O; 97%

w/ 5% $PdCl_2$(dppp)/TBAF/THF/20°C — 92% 4-(-CH_2CH=CHMe) — 7.2 E/Z

Hatanaka, Y.; Goda, K.; Hiyama, T. *Tetrahedron Lett.*, ***1994***, *35*, 6511

MeO_2C–C6H4–OSO_2CF_3 → (PhB(OH)$_2$, Pd(PPh$_3$)$_4$; 3 LiCl , aq. Na_2CO_3; Tol , EtoH) MeO_2C–C6H4–Ph

Percec, V.; Bae, J.-Y.; Hill, D.H. *J. Org. Chem.*, ***1995***, *60*, 1060

Me–C6H4–I → ($PhSnCl_3$, 0.01 $PdCl_2$; aq. KOH , 100°C, 3h) Me–C6H4–Ph 86%

Roshchin, A.I.; Bumagin, N.A.; Beletskaya, I.P. *Tetrahedron Lett.*, ***1995***, *36*, 125

I; Me; CO_2H → (1. MeZnBr , 5% $PdCl_2(PPh_3)_2$ DMF , 12h; 2. aq. NH_4Cl) Me; Me; CO_2H 95%

Abarbri, M.; Parrain, J.-L.; Duchêne, A. *Tetrahedron Lett.*, ***1995***, *36*, 2469

Br; O → ($NaBH_4$, Cp_2TiCl_2; DMF , 75°C) O 57%

Liu, Y.; Schwartz, J. *Tetrahedron*, ***1995***, *51*, 4471

Cl → ($ZnCl_2/SiO_2$, K_2CO_3; Al_2O_3, 30°C , 1h) 61%

Kodomari, M.; Nawa, S.; Miyoshi, T. *J. Chem. Soc. Chem. Commun.*, ***1995***, 1895

$B(OH)_2$; O_2N—C6H4—Br ; 10% $Pd(PPh_3)_4$, 10% aq. TlOH ; DMA , 20°C ; DMA = dimethylacetamide ; NO_2 ; 77%

Anderson, J.C.; Namli, H. *Synlett*, ***1995***, 765

Ph ; Br ; Et_2Zn , 20% LiI , THF , -35°C ; 7.5% $Ni(acac)_2$, 18h ; Ph ; Et ; 81%

Devasagayaraj, A.; Stüdemann, T.; Knochel, P. *Angew. Chem. Int. Ed. Engl.*, ***1995***, *34*, 2723

SECTION 71: ALKYLS, METHYLENES AND ARYLS FROM HYDRIDES

This section lists examples of the reaction of RH → RR' (R,R' = alkyl or aryl). For the reaction C=CH → C=C-R (R = alkyl or aryl), see Section 209 (Alkenes from Alkenes). For alkylations of ketones and esters, see Section 177 (Ketones from Ketones) and Section 113 (Esters from Esters).

I ; O ; $F_3Si(—CH_2)_6—H$, TBAF , THF ; 5% $Pd(PPh_3)_4$, 100°C ; $(CH_2)_6—H$; O ; 61%

Matsuhashi, H.; Juroboshi, M.; Hatanaka, Y.; Hiyama, T. *Tetrahedron Lett.*, ***1994***, *35*, 6507

CO_2H ; Cl ; 1. *s*-BuLi/TMEDA-THF ; -98°C → -78°C ; 2. MeI ; 3. H^+ ; CO_2H ; Me ; Cl ; 62%

Bennetau, B.; Mortier, J.; Moyroud, J.; Guesnet, J.-L. *J. Chem. Soc., Perkin Trans. 1.*, ***1995***, 1265

Me O Me ; $Si(OEt)_3$ (vinyl), toluene, 2h; 4% $Ru(H)_2(CO)(PPh_3)_3$, reflux → Me O Me, $Si(OEt)_3$ 93%

Kakiuchi, F.; Sekine, S.; Tanaka, Y.; Kamatani, A.; Sonoda, M.; Chatani, N.; Murai, S. *Bull. Chem. Soc. Jpn.*, ***1995***, *68*, 62

SECTION 72: ALKYLS, METHYLENES AND ARYLS FROM KETONES

The conversions $R_2C=O \rightarrow R\text{-}R$, R_2CH_2, R_2CHR', etc. are listed in this section.

CO_2Et, O ; 1. $TsNHNH_2$ 2. $NaBH_3CN \cdot ZnCl_2$ → CO_2Et + CO_2Et (3.5 : 1) 66%

Taber, D.F.; Wang, Y.; Stachel, S.J. *Tetrahedron Lett.*, ***1993***, *34*, 6209

CO_2Et, O ; $Me_3Al\text{-}TMSOSO_2CF_3$; CH_2Cl_2, 0°C → 25°C, 15h → CO_2Et 95%

Kim, C.U.; Misco, P.F.; Luh, B.Y.; Mansuri, M.M. *Tetrahedron Lett.*, ***1994***, *35*, 3017

MeO, MeO, CO_2H, O ; 1. MeI, KOH, DMSO rt, 5h; 2. H_2O → MeO, MeO 93%

Yli-Kauhaluoma, J.T.; Janda, K.D. *Tetrahedron Lett.*, ***1994***, *35*, 4509

EtO_2C, Me, O ; 6 NaOEt, 2 I_2; EtOH, -78°C → EtO_2C, I, Me, OH 66%

Hegde, S.G.; Kassim, A.M.; Ingrum, A.I. *Tetrahedron Lett.*, ***1995***, *36*, 8395

, t-BuOK

20% MeOH , rt , 1d

67%

Coe, J.W.; Vetelino, M.G.; Kemp, D.S. *Tetrahedron Lett.*, ***1994***, *35*, 6627

1. MeMgBr
2. $VCl_2(tmeda)_2$
3. Br

Ph Ph Ph Ph

50% 46%

Kataoka, Y.; Makihira, I.; Akiyama, H.; Tani, K. *Tetrahedron Lett.*, ***1995***, *36*, 6495

SECTION 73 ALKYLS, METHYLENES AND ARYLS FROM NITRILES

NO ADDITIONAL EXAMPLES

SECTION 74: ALKYLS, METHYLENES AND ARYLS FROM ALKENES

The following reaction types are included in this section:

A. Hydrogenation of Alkenes (and Aryls)
B. Formation of Aryls
C. Alkylations and Arylations of Alkenes
D. Conjugate Reduction of Conjugated Aldehydes, Ketones, Acids, Esters and Nitriles
E. Conjugate Alkylations
F. Cyclopropanations, including halocyclopropanations

SECTION 74A: Hydrogenation of Alkenes (and Aryls)

Reduction of aryls to dienes are listed in Section 377 (Alkene-Alkene).

SO_2Ph [(t-Bu_2Ph)PdPt-Bu]$_2$ / O_2 SO_2Ph

THF , H_2

83%

Cho, I.S.; Alper, H. *J. Org. Chem.*, ***1994***, *59*, 4027

"colloidal palladium"

H_2 , ether , 25°C

>99%

"colloidal palladium" = [TMS(OSiHMe)$_n$OTMS]Pd(hfacac)$_2$

Fowley, L.A.; Michos, D.; Luo, X.-L.; Crabtree, R.H. *Tetrahedron Lett.*, ***1993***, *34*, 3075

chiral catalyst , H_2 , 65°C

Ph Ph Ph Ph

91% (>99% ee)

Broene, R.D.; Buchwald, S.L. *J. Am. Chem. Soc.*, ***1993***, *115*, 12569

CO_2Me H_2 , rt , 1h , MeOH/H_2O CO_2Me

$RuCl_3$, TOA

TOA = trioctylamine

99%
(*cis/trans* = 1.5)

Fache, F.; Lehuede, S.; LeMaire, M. *Tetrahedron Lett.*, ***1995***, *36*, 885

[(*t*-Bu_2PH)Pd(P*t*-Bu)$_2$]$_2$, H_2 , O_2
THF/CH_2Cl_2 , 6h

Ph O Ph O

88%

Cho, I.S.; Lee, B.; Alper, H. *Tetrahedron Lett.*, ***1995***, *36*, 6009

OH OH

aq. $Ba(OH)_2$, NiAl(Raney alloy)

60°C , 1h

Br Br

54%

Tsukinoki, T.; Kakinami, T.; Iida, Y.; Ueno, M.; Ueno, Y.; Mashimo, T.; Tsuzuki, H.; Tashiro, M. *J. Chem. Soc. Chem. Commun.*, ***1995***, 209

H_2 , [$W_2(OCH_2t\text{-}Bu)_6(Py)_2$]

no yield

Barry, J.T.; Chisholm, M.H. *J. Chem. Soc. Chem. Commun.*, ***1995***, 1599

SECTION 74B: Formation of Aryls

OMe OMe OTf SnBu$_3$ LiCl , CuBr , inhibitor $Pd(PPh_3)_4$, dioxane , reflux OMe OMe OMe OMe 93%

Saá, J.M.; Martorell, G. *J. Org. Chem.*, ***1993***, *58*, 1963

170°C , $C_6H_4Cl_2$, 2h 1,4-CHP CO_2Me MeO_2C 11%

Grissom, J.W.; Calkins, T.L.; McMillen, H.A. *J. Org. Chem.*, ***1993***, *58*, 6556

Ph OAc 50 atm CO , Ac_2O $PdCl_2(PPh_3)_2$, NEt_3 PhH , 140°C , 3h OAc 69%

Ishii, Y.; Gao, C.; Xu, W.-X.; Iwasaki, M.; Hidai, M. *J. Org. Chem.*, ***1993***, *58*, 6818

$HSiMe_2t$-Bu , C) (50 atm) $Ru_3(CO)_{12}/PCy_3$, 140°C , 20h $OSiMe_2t$-Bu $OSiMe_2t$-Bu 45%

Chatani, N.; Fukumoto, Y.; Ida, T.; Murai, S. *J. Am. Chem. Soc.*, ***1993***, *115*, 11614

PhOH , PhMe , 180°C 7h O O 52%

Danheiser, R.L.; Gould, A.E.; de la Pradilla, R.F.; Helgason, A.L. *J. Org. Chem.*, ***1994***, *59*, 5514

20% Ni(0)-PPh_3 , HC≡CH
THF , 23°C , 12 h

N—Ts

91%

Sato, Y.; Nishimata, T.; Mori, M. *J. Org. Chem.*, ***1994***, *59*, 6133

PhCHO

1. $PhNH_2$
2. $Pd(OAc)_2$, MeCN
3. BuLi/PPh_3/ Br
4. 2% HCl

CHO

48%

Rama Rao, A.V.; Reddy, K.L.; Reddy, M.M. *Tetrahedron Lett.*, ***1994***, *35*, 5039

O

1.5 Si(iPr)$_3$
0.1 $SnCl_4$, 0°C
1.5 $(MeO)_3SiMe_3$

OH

78% Si(iPr)$_3$

Angle, S.R.; Boyce, J.P. *Tetrahedron Lett.*, ***1994***, *35*, 6461

H BocHN O N Me CO_2Me

t-BuOOCOPh , $Cu(OAc)_2$
CuBr , PhH , 80°C , 8.5h

H BocHN O N Me CO_2Me

56%

Tavares, F.; Meyers, A.I. *Tetrahedron Lett.*, ***1994***, *35*, 6803

N-OH Bn Ar

5% Pd/C , diethylene glycol
188°C , diethyl ether , 45 min

Ar = 4-methoxyphenyl

NH_2 Bn Ar

89%

Matsumoto, M.; Tomizuka, J.; Suzuki, M. *Synth. Commun.*, ***1994***, *24*, 1441

SECTION 74C: Alkylations and Arylations of Alkenes

MeO OMe

O

2 $BF_3 \cdot OEt_2$

25°C

MeO OMe

O

82%

Majetich, G.; Zhang, Y.; Feltman, T.L. *Tetrahedron Lett.*, ***1993***, *34*, 441
Majetich, G.; Zhang, Y.; Feltman, T.L.; Duncan Jr., S. *Tetrahedron Lett.*, ***1993***, *34*, 445

O—Ar

O

1. PhLi , Ph Ph MeO OMe

Toluene , -45°C , 18h

2. $LiEt_3BH$
3. MeOH
4. $NaBH_4$

OH

Ph

Ar = 2,6-di-*t*-Bu-4-OMephenyl

80% (84% ee)

Tomioka, K.; Shindo, M.; Koga, K. *Tetrahedron Lett.*, ***1993***, *34*, 681

H N Me O

Me—⟨⟩—I

DMF , Bu_4NCl , 1d
10% $Pd(OAc)_2$, 3 iPr_2NEt
5 HCO_2Na

Me

O

Me NH_2

66:34 *E:Z* , 99%

Larock, R.C.; Ding, S. *Tetrahedron Lett.*, ***1993***, *34*, 979

C_5H_{11}

OH

cat. $Pd(OAc)_2$, Ag_2CO_3
DMF , 45°C , overnight

CO_2Me

I

CO_2Me

C_5H_{11}

OH

75%

Jeffery, T. *Tetrahedron Lett.*, ***1993***, *34*, 1133

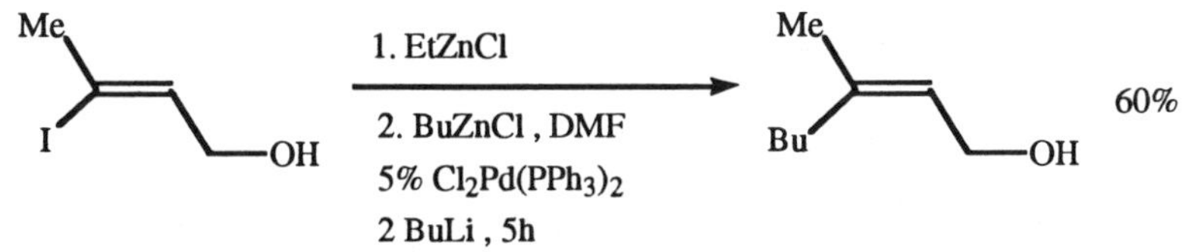

1. EtZnCl
2. BuZnCl, DMF
5% $Cl_2Pd(PPh_3)_2$
2 BuLi, 5h

60%

Negishi, E.; Ay, M.; Gulevich, Y.V.; Noda, Y. *Tetrahedron Lett.*, ***1993***, *34*, 1437

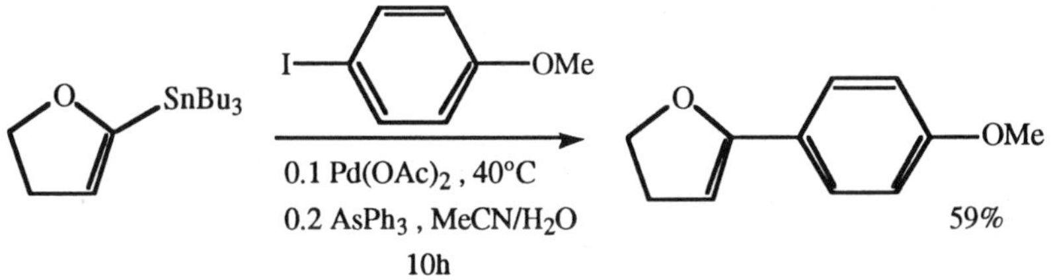

I—C_6H_4—OMe
$SnBu_3$
0.1 $Pd(OAc)_2$, 40°C
0.2 $AsPh_3$, $MeCN/H_2O$
10h

59%

Zhang, H-C.; Brakta, M.; Daves Jr., G.D. *Tetrahedron Lett.*, ***1993***, *34*, 1571

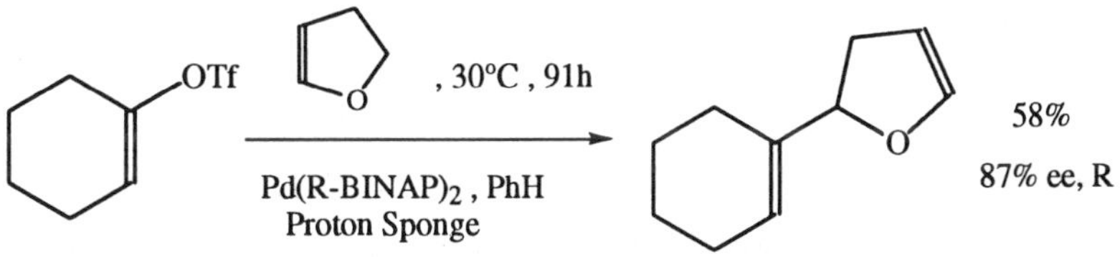

OTf
, 30°C, 91h
$Pd(R\text{-}BINAP)_2$, PhH
Proton Sponge

58%
87% ee, R

Ozawa, F.; Kobatake, T.; Hayashi, T. *Tetrahedron Lett.*, ***1993***, *34*, 2505

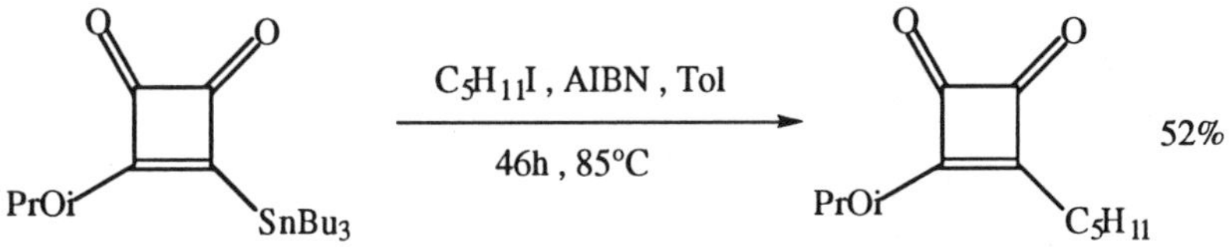

PrOi, $SnBu_3$
$C_5H_{11}I$, AIBN, Tol
46h, 85°C
PrOi, C_5H_{11}

52%

Kinney, W.A. *Tetrahedron Lett.*, ***1993***, *34*, 2715

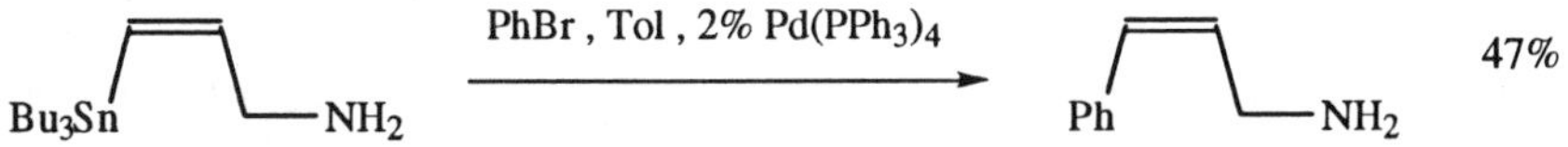

Bu_3Sn, NH_2
PhBr, Tol, 2% $Pd(PPh_3)_4$
Ph, NH_2

47%

Corriu, R.J.P.; Geng, B.; Moreau, J.J.E. *J. Org. Chem.*, ***1993***, *58*, 1443

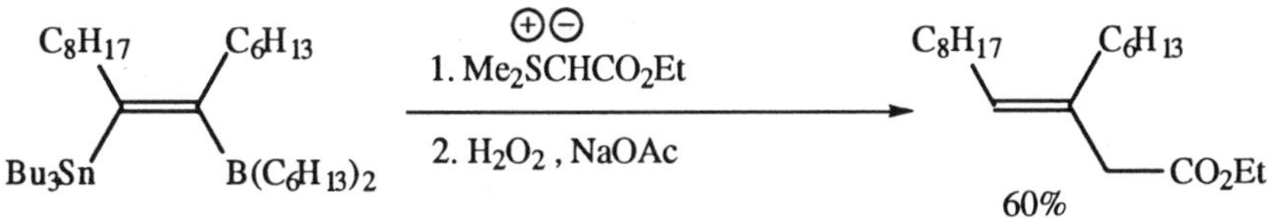

C_8H_{17}, C_6H_{13}, Bu_3Sn, $B(C_6H_{13})_2$
1. $Me_2\overset{\oplus}{S}\overset{\ominus}{C}HCO_2Et$
2. H_2O_2, NaOAc
C_8H_{17}, C_6H_{13}, CO_2Et

60%

Deng, M.-Z.; Li, N.-S.; Huang, Y.-Z. *J. Org. Chem.*, ***1993***, *58*, 1949

NHCy NHCy O O MeO MeO N N Bu_3SnH , AIBN Br MeO OMe

Rigby, J.H.; Qabar, M.N. *J. Org. Chem.*, ***1993***, *58*, 4473

H N O 1. [1. $PdCl_2$, LiCl , THF 0°C → 25°C 2. aq. NH_4Cl] 2. PhHgCl , O_2 Ph NH_2 O

60% (74:26 *E:Z*)

Larock, R.C.; Ding, S. *J. Org. Chem.*, ***1993***, *58*, 2081

$(CH_2)_8CO_2Me$ /9-BBN PhOTf , $Pd(PPh_3)_4$, 85°C K_3PO_4 , dioxane Ph $(CH_2)_8CO_2Me$

87%

Oh-e, T.; Miyaura, N.; Suzuki, A. *J. Org. Chem.*, ***1993***, *58*, 2201

Ph I 1. 2 Et_2Zn , 1.5% $PdCl_2(dppf)$, THF 2. CuCN•2 LiCl 3. CO_2Et Br Ph CO_2Et

73%

Stadmüller, H.; Lentz, R.; Tucker, C.E.; Stüdemann, T.; Dörner, W.; Knochel, P. *J. Am. Chem. Soc.*, ***1993***, *115*, 7027

PhO 3 eq. EtMgBr , Cp_2ZrCl_2 THF , rt , 6h 70%

Suzuki, N.; Kondakov, D.Y.; Takahashi, T. *J. Am. Chem. Soc.*, ***1993***, *115*, 8485

$N_2^+BF_4^-$; CO_2Et , EtOH ; 2% $Pd(OAc)_2$, 60°C , 1h ; MeO ; CO_2Et ; MeO ; 71%

Sengupta, S.; Bhattacharyya, S. *J. Chem. Soc., Perkin Trans. 1.*, ***1993***, 1943

OH ; Cl ; 5 eq. MeMgBr ; 0.05 $PdCl_2(PPh_3)_2$; OH ; 97%

Hoffmann, R.W.; Giesen, V.; Fuest, M. *Liebigs Ann. Chem.*, ***1993***, 629

Br ; MeO_2C ; CO_2Me ; e-, DMF, Et_4ClO_4 ; $Ni(teta)(ClO_4)_2$; MeO_2C ; CO_2Me ; 47%

Ozaki, S.; Horiguchi, I.; Matsushita, H.; Ohmori, H. *Tetrahedron Lett.*, ***1994***, *35*, 725

O ; C_3H_7 ; O ; Et ; LDA , TMS-Cl ; THF , -78°C ; O ; $SiMe_3$; C_3H_7 ; O ; Et ; 93%

Wright, S.W. *Tetrahedron Lett.*, ***1994***, *35*, 1841

CO_2Me ; PhI , Bu_4NHSO_4 , aq. MeOH ; 5% $[Pd(OAc)_2/2\ PPh_3]$; CO_2Me ; Ph ; 96%

Heck reaction in aqueous media

Jeffery, T. *Tetrahedron Lett.*, ***1994***, *35*, 3051

O ; Ph ; ⊖ ; PhI , SmI_2/THF , 15 min ; DMSO ; O ; Ph ; Ph ; 47%

Nazareno, M.A.; Rossi, R.A. *Tetrahedron Lett.*, ***1994***, *35*, 5185

MeHNCOO (allyl carbamate) → [$Pd(OAc)_2$, Bu_3P, 3h; K_2CO_3, DMF, 80°C; PhI] → Ph-substituted MeHNCOO product + Ph / OOCNHMe product

quant. (78 6)

Ono, K.; Fugami, K.; Tanaka, S.; Tamaru, Y. *Tetrahedron Lett.*, ***1994***, *35*, 4133

O, SPh → [2 $AlEt_3$, 0°C; 10h] → O, SPh, O 70%

Majetich, G.; Zhang, Y.; Liu, S. *Tetrahedron Lett.*, ***1994***, *35*, 4887

O, I → [$SnBu_3$, F; $Pd(PPh_3)_4$, THF, 65°C, 1h] → O, F

66%

Matthews, D.P.; Waid, P.P.; Sabol, J.S.; McCarthy, J.R. *Tetrahedron Lett.*, ***1994***, *35*, 5177

$SnBu_3$ → [TfO, OMe] → OMe 70%

Badone, D.; Cardamone, R.; Guzzi, U. *Tetrahedron Lett.*, ***1994***, *35*, 5477

I, O → [1. $SnBu_3$; 5% $Pd_2(dba)_3$, 20% $AsPh_3$; 2. 5% TFA, CH_2Cl_2] → O, NH_2

● = Rink amide resin 89%

Deshpande, M.S. *Tetrahedron Lett.*, ***1994***, *35*, 5613

2 eq. Me_3SiCH_2MgCl

5% $NiCl_2(PPh_3)_2$, DME
reflux, 1.5h

79%

Hevesi, L.; Hermans, B.; Allard, C. *Tetrahedron Lett.*, ***1994***, *35*, 6729

1. 0.3 η^3-$C_3H_5PdCl)_2$, $HSiPh_2F$
0.6 *o'*-diphenylphosphinobinaphthol derivative
20°C, 10h
2. MeLi

96% (66% ee, S)

Hatanaka, Y.; Goda, K.; Yamashita, F.; Hiyama, T. *Tetrahedron Lett.*, ***1994***, *35*, 7981

$Mn(OAc)_3$, CuOAc
ultrasound, 4h

47%

Bosman, C.; D'Annibale, A.; Resta, S.; Trogolo, C. *Tetrahedron Lett.*, ***1994***, *35*, 8049

$Bu_4N^+ (Ph_3SnF_2)^-$, THF
$Pd(PPh_3)_4$, reflux, 30 min

81%

Martinez, A.G.; Barcina, J.O.; Cerezo, A. de F.; Subramanian, L.R. *Synlett*, ***1994***, 1047

10% $ZrCl_2(\eta C_5H_2)$, 1.5 BuMgCl
rt, 6h

90%

Takahashi, T.; Kotora, M.; Kasai, K. *J. Chem. Soc. Chem. Commun.*, ***1994***, 2693

1. BuLi, THF, -70°C
2. $ZnBr_2$, -70°C → 20°C
3. H_3O^+

81%

$H_3C(H_2C)_4$ — $C_9H_{19}CMe_2SH$, Et_3SiH / tridecane , di-*t*-Bu hyponitrite → $AcH_2C(H_2C)_4$ — $SiEt_3$

65%

Dang, H.-S.; Roberts, B.P. *Tetrahedron Lett., 1995, 36*, 2875

Br — COOH — $PhSnCl_3$, aq. KOH , 3% $PdCl_2$ → Ph — COOH

83%

Rai, R.; Aubrecht, K.B.; Collum, D.B. *Tetrahedron Lett., 1995, 36*, 3169

MeO_2C, MeO_2C, I, O — Bu_3SnH , SiO_2 / CH_2Cl_2 , rt , 1d → O, CO_2Me, CO_2Me

81%

tandem radical cylization-Diels-Alder

Journet, M.; Malacria, M. *J. Org. Chem., 1995, 60*, 6885

CO_2Et, Et — Mg , MeOH , rt , 2h → CO_2Et, CO_2Et + CO_2Et, CO_2Et

(87 : 13) 90%

Chavan, S.P.; Ethiraj, K.S. *Tetrahedron Lett., 1995, 36*, 2281

OMe, O — $BF_3{\cdot}OEt_2$, C_6H_{12}, reflux , 2h → OMe, O

56%

Majetich, G.; Liu, S.; Siesel, D. *Tetrahedron Lett., 1995, 36*, 4749

HO, OH, MPMO — 10% $Pd(OAc)_2$, 20% Bu_3P, PhI , DMF , 100°C , 1.5h → Ph, HO, O, MPMO 85%

Kang, S.-K.; Jung, K.-Y.; Park, C.-H.; Namkoong, E.-Y.; Kim, T.-H. *Tetrahedron Lett.*, ***1995***, *36*, 6287

Br — Bu_3SnH, AIBN → + (95 : 5) 91%

Zhang, W.; Dowd, P. *Tetrahedron Lett.*, ***1995***, *36*, 8539

HO, PhO_2S, SO_2Ph — 2 PPh_3 , toluene/PhH, 2 DEAD → H, H, SO_2Ph, SO_2Ph 95%

Yu, J.; Cho, H.S.; Falck, J.R. *Tetrahedron Lett.*, ***1995***, *36*, 8577

O — $SiEt_3$, PhMe , reflux, 5% $RuH_2(CO)(PPh_3)_3$ → O, $SiEt_3$ 94%

Trost, B.M.; Imi, K.; Davies, I.W. *J. Am. Chem. Soc.*, ***1995***, *117*, 5371

1. Me_3Al , 8% Cl_2ZrCp_2 , DCE , 22°C, 2. O_2 → OH 92% (74% ee)

Kondakov, D.Y.; Negishi, E. *J. Am. Chem. Soc.*, ***1995***, *117*, 10771

$CH_2=CHCO_2Et$, 5% $PdCl_2(PPh_3)_2$; 3 LiCl, DMF, 110°C — 85%

Draper, T.L.; Bailey, T.R. *Synlett*, ***1995***, 157

$CH_2=CHSi(OEt)_3$, toluene, 135°C; $Ru(H)_2(CO)(PPh_3)_2$, 2d — 50%

Kakiuchi, F.; Tanaka, Y.; Sato, T.; Chatani, N.; Murai, S. *Chem. Lett.*, ***1995***, 679

Ph–CH=CH–$SiMe_3$ → Ph–CH=CH–Ph 86%

1. BCl_3, CH_2Cl_2, 0°C
2. catechol, PhH
3. PhBr, $Pd(PPh_3)_4$, NaOH

Farinola, G.M.; Fiandanese, V.; Mazzone, L.; Naso, F. *J. Chem. Soc. Chem. Commun.*, ***1995***, 2523

REVIEW:

"Recent Developments and New Perspectives In The Heck Reaction," Cabri, W.; Candiani, T. *Accts. Chem. Res.*, ***1995***, *28*, 2

SECTION 74D: Conjugate Reduction of α,β-Unsaturated Carbonyl Compounds and Nitriles

2% $[Pd(Pt\text{-}Bu)(PHt\text{-}Bu_2)]_2$, O_2, H_2 (10 psi); THF, 18 h, glass autoclave — 90%

Sommovigo, M.; Alper, H. *Tetrahedron Lett.*, ***1993***, *34*, 59

Ph–CH=CH–CO_2Ph → Ph–CH_2CH_2–CO_2Ph 80%

$EtOCH_2CH_2OH$, $Rh_6(CO)_{16}$; CO (1 atm), NEt_3, 30°C

Joh, T.; Fujiwara, K.; Takahashi, S. *Bull. Chem. Soc. Jpn.*, ***1993***, *66*, 978

$PhSiH_3$, 20°C , 16h
Mo complex

(62 : 38) 95% conversion

Schmidt, T. *Tetrahedron Lett.*, ***1994***, *35*, 3513

Ph–CH=CH–CHO → NaHTe , EtOH; rt , 5h → Ph–CH_2CH_2–CHO >98%

Yamashita, M.; Tanaka, Y.; Arita, A.; Nishida, M. *J. Org. Chem.*, ***1994***, *59*, 3500

CO_2Me ... $SnBu_3$ → TFA-Et_3SiH , 30 min; -78°C → 0°C → CO_2Me ... $SnBu_3$

various protonic and Lewis acids used

Zhao, Y.; Quayle, P.; Keo, E.A. *Tetrahedron Lett.*, ***1994***, *35*, 4179

CHO → NH_4^+ HCO_2^- , Pd-C; 25°C , 4h → CHO 92%

Ranu, B.C.; Sarkar, A. *Tetrahedron Lett.*, ***1994***, *35*, 8649

Ph–CH=CH–NO_2 → Bakers yeast , 31°C; 3d → Ph–CH_2CH_2–NO_2 43%

Takeshita, M.; Yoshida, S.; Kohno, Y. *Heterocycles*, ***1994***, *37*, 553

=O → limonene , 10% Pd/C , reflux → =O quant.

von Holleben, M.L.; Zucolotto, M.; Zini, C.A.; Oliveira, E.R. *Tetrahedron*, ***1994***, *50*, 973

$Na_2S_2O_4$, aq. dioxane , 50°C , 1h

Ph CHO

selective reduction of conjugated carbonyl C=C units

Dhillon, R.S.; Singh, R.P.; Kaur, D. *Tetrahedron Lett.*, ***1995***, *36*, 1107

Ph CO_2Et — Sm , I_2 , MeOH , 1 min → Ph CO_2Et 97%

Yanada, R.; Bessho, K.; Yanada, K. *Synlett*, ***1995***, 443

CO_2Me — BER-$CuSO_4$, MeOH , rt → CO_2Me 94%

BER = borohydride exchange resin

Sim, T.B.; Yoon, N.M. *Synlett*, ***1995***, 726

CO_2Et — $NaBH_4$-$BiCl_3$, 95% EtOH → CO_2Et 98%

Ren, P.-D.; Pan, S.-F.; Dong, T.-W.; Wu, S.-H. *Synth. Commun.*, ***1995***, *25*, 3395

Ph — Bakers yeast → Ph 42% (71% ee)

Kawai, Y.; Saitou, K.; Hida, K.; Ohno, A. *Tetrahedron Asymmetry*, ***1995***, *6*, 2143

SECTION 74E: Conjugate Alkylations

$OSiMe_3$ — alumina•$ZnCl_2$, 0°C , 5 min → 78%

Ranu, B.C.; Saha, M.; Bhar, S. *Tetrahedron Lett.*, ***1993***, *34*, 1989

PhS, $O^- Li^+$, CO_2Me, Et, CH_2Cl_2, 2d, -78°C, CO_2Me, SPh, +, CO_2Me, SPh

(15 : 1) 81%

Corey, E.J.; Houpis, I.N. *Tetrahedron Lett.*, ***1993***, *34*, 2421

Ph, N, O, O, O, 1. PhMgBr, CuBr•SMe_2, 2. NBS, THF, -78°C, Br, Ph, Ph, H, N, O, O, O, 89%

Li, G.; Jarosinski, M.A.; Hruby, V.J. *Tetrahedron Lett.*, ***1993***, *34*, 2565

O, Ph_2CuLi, THF, -78°C 20%, + TMSCl 81%, + TMSCN 83%, O, Ph

Lipshutz, B.H.; James, B. *Tetrahedron Lett.*, ***1993***, *34*, 6689

O, 3 eq. $SnBu_3$, 0.05 $(Bu_3Sn)_2$, 6 eq. $PhSCH(OMe)_2$, 0.3 M PhH, hν (sunlamp), O, OMe, MeO, 69%, (8:1 *trans:cis*)

Keck, G.E.; Kordik, C.P. *Tetrahedron Lett.*, ***1993***, *34*, 6875

O, [$Me(Boc)NCH_2SnBu_3$/*s*-BuLi TMEDA, CuCN, THF], TMSCl, O, NMeBoc, 98%

Dieter, R.K.; Alexander, C.W. *Synlett*, ***1993***, 407

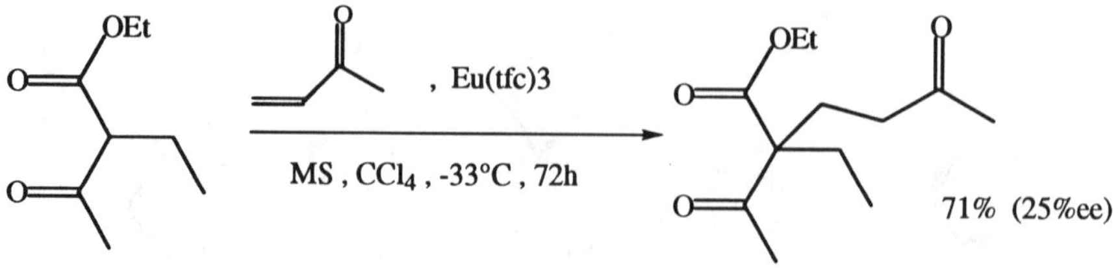

OEt, O, O, O, OEt, O, O, O

, Eu(tfc)3

MS , CCl_4 , -33°C , 72h

71% (25%ee)

$Eu(tfc)_3$ = [tris-(3-trifluoromethylhydroxymethylene)-d-camphorato] europium (III)

Bonadies, F.; Lattanzi, A.; Orelli, L.R.; Resci, S.; Screttri, A. *Tetrahedron Lett.*, ***1993***, *34*, 7649

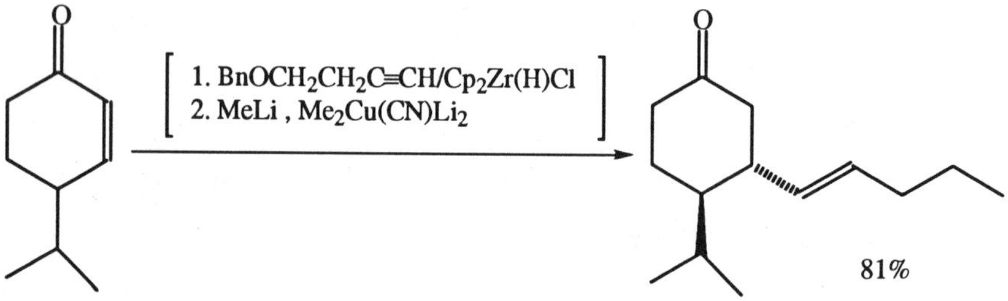

O, O

1. $BnOCH_2CH_2C{\equiv}CH/Cp_2Zr(H)Cl$
2. MeLi , $Me_2Cu(CN)Li_2$

81%

Lipshutz, B.H.; Wood, M.R. *J. Am. Chem. Soc.*, ***1993***, *115*, 12625

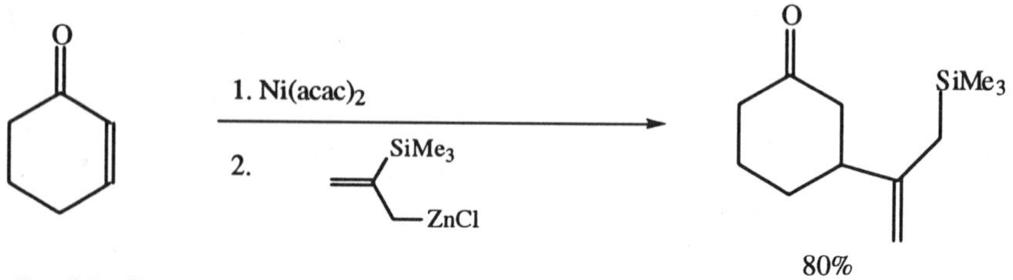

O, O

1. $Ni(acac)_2$
2. $SiMe_3$, ZnCl

$SiMe_3$

80%

Eshelby, J.J.; Crowley, P.J.; Parsons, P.J. *Synlett*, ***1993***, 279

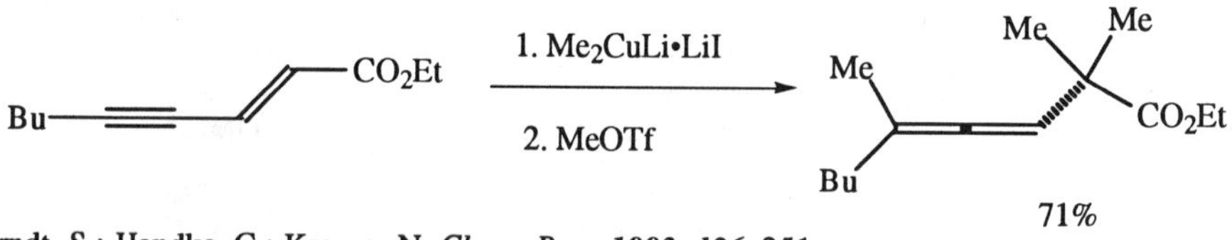

Bu, CO_2Et

1. $Me_2CuLi{\bullet}LiI$
2. MeOTf

Me, Me, Me, CO_2Et, Bu

71%

Arndt, S.; Handke, G.; Krause, N. *Chem. Ber.*, ***1993***, *126*, 251

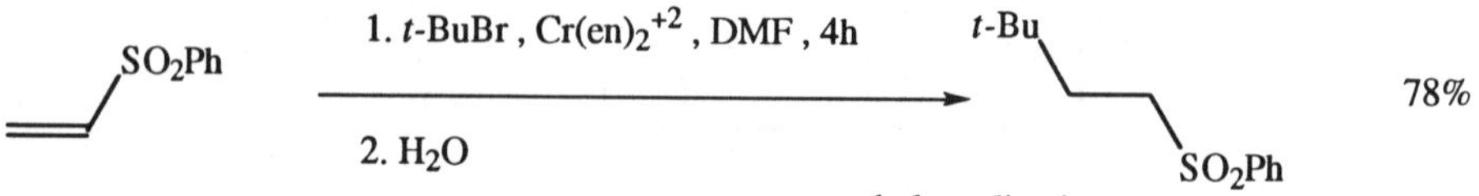

SO_2Ph

1. *t*-BuBr , $Cr(en)_2^{+2}$, DMF , 4h
2. H_2O

t-Bu, SO_2Ph

78%

en = ethylene diamine

Tashtoush, H.I.; Sustmann, R. *Chem. Ber.*, ***1993***, *126*, 1759

Ph$_3$Sb , cat. Pd(OAc)$_2$

AgOAc , AcOH , 25°C

Ph O

41%

Cho, C.S.; Tanabe, K.; Uemura, S. *Tetrahedron Lett.*, ***1994***, *35*, 1275

CHO

2 eq. CHO

3 eq. SmI$_2$, THF ,-78°C

TBSO

MeO$_2$C

OH

TBSO H CO$_2$Me

HO

Enholm, E.J.; Trivellas, A. *Tetrahedron Lett.*, ***1994***, *35*, 1627

Ph

cat. Pd(OAc)$_2$, NaOAc , 1d
cat. SbCl$_3$, AcOH , 25°C

O

Ph

Ph O

88%

Cho, C.S.; Motofusa, S.; Uemura, S. *Tetrahedron Lett.*, ***1994***, *35*, 1739

Br

MeO$_2$C

CO$_2$Me

3 eq. Me$_3$SiSnBu$_3$, 2 eq. CsF
DMF , rt , 1h

Br

CO$_2$Me

93%

CO$_2$Me

Sato, H.; Isono, N.; Okamura, K.; Date, T.; Mori, M. *Tetrahedron Lett.*, ***1994***, *35*, 2035

MeTi(OiPr)$_4$MgCl , THF

Ni(acac)$_2$, -15°C → rt

O

O

85%

Flemming, S.; Kabbara, J.; Nickisch, K.; Neh, H.; Westermann, J. *Tetrahedron Lett.*, ***1994***, *35*, 6075

t-BuHgCl , Et_3SiH , DMSO

dark , hν (sunlamp) -35°C , 11h

CO_2Et → Et_3Si CO_2Et 93%

Russell, G.A.; Shi, B.Z. *Tetrahedron Lett.,* ***1994****, 35,* 3841

iPrCHO , $LiNO_3$, aq. MeCN

e^-

CO_2Me → CO_2Me 34%

Shono, T.; Soejima, T.; Takigawa, K.; Yamaguchi, Y.; Maekawa, H.; Kashimura, S. *Tetrahedron Lett.,* ***1994****, 35,* 4161

CO_2t-Bu CO_2t-Bu I

1. BuLi , THF -100°C → 30°C

2. H-2_O

CO_2t-Bu CO_2t-Bu 70% + other isomers

Cooke Jr., M.P.; Gopal, D. *Tetrahedron Lett.,* ***1994****, 35,* 2837

PhCHO , Ph_2CO

hν

Ph 69%

Kraus, G.A.; Liu, P. *Tetrahedron Lett.,* ***1994****, 35,* 7723

NO_2

MeMgCl , $CeCl_3$, THF

NO_2 Me 93%

Bartoli, G.; Bosco, M.; Sambri, L.; Marcantoni, E. *Tetrahedron Lett.,* ***1994****, 35,* 8651

hν , PPh_3 , DCA , DMF

(70 : 30) 55%

Pandey, G.; Hajra, S.; Ghorai, M.K. *Tetrahedron Lett.*, ***1994***, *35*, 7837

C_5H_{11} I CO_2Me Bu_3SnH , PhH MeO_2C C_5H_{11} C_5H_{11} CO_2Me C_5H_{11} C_5H_{11}

(1 : 1) 54%

Miura, K.; Itoh, D.; Hondo, T.; Hosomi, A. *Tetrahedron Lett.*, ***1994***, *35*, 9605

BuC≡CH , $PhC{\equiv}CSnEt_3$, Me_3SiCl

$Ni(acac)_2$, Dibal , THF , rt , 2h

80% (>95:5 *E:Z*)

Ikeda, S.; Sato, Y. *J. Am. Chem. Soc.*, ***1994***, *116*, 5975

CHO Bu $AlMe_3$/hexane/THF 5% CuBr , 0°C , 2h

94% (93:7 1,4:1,2)

Kabbara, J.; Flemming, S.; Nickisch, K.; Neh, H.; Westermann, J. *Synlett*, ***1994***, 679

Et_2Zn , MeCN , 1.5% $NiCl_2$, -30°C

30% chiral pyrrolidine

75% (82% ee , S)

Asami, M.; Usui, K.; Higuchi, S.; Inoue, S. *Chem. Lett.*, ***1994***, 297

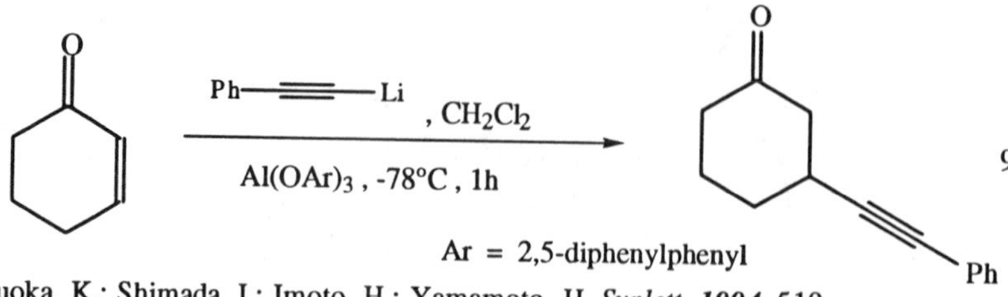

Ph—≡—Li , CH_2Cl_2

$Al(OAr)_3$, -78°C , 1h

O

Ph

99%

Ar = 2,5-diphenylphenyl

Maruoka, K.; Shimada, I.; Imoto, H.; Yamamoto, H. *Synlett*, ***1994***, 519

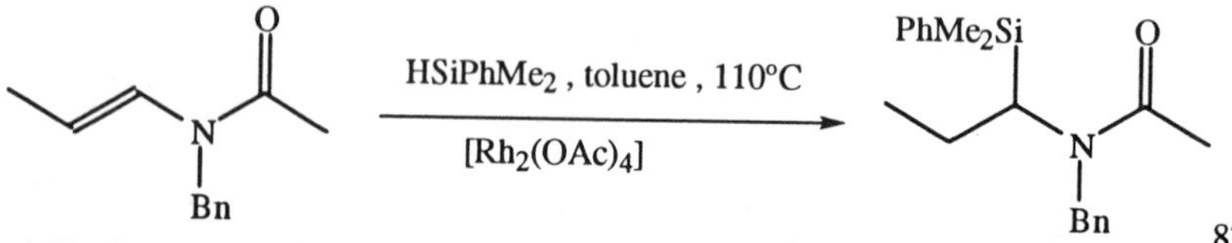

O

N

Bn

$HSiPhMe_2$, toluene , 110°C

$[Rh_2(OAc)_4]$

$PhMe_2Si$ O

N

Bn

87%

Murai, T.; Oda, T.; Kimura, F.; Onishi, H.; Kanda, T.; Kato, S. *J. Chem. Soc. Chem. Commun.*, ***1994***, 2143

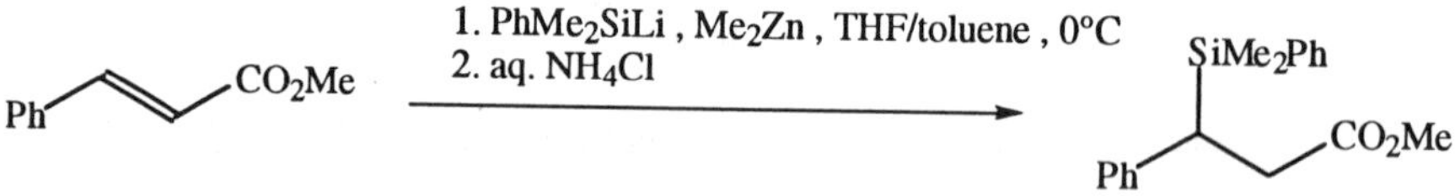

Ph CO_2Me

1. $PhMe_2SiLi$, Me_2Zn , THF/toluene , 0°C
2. aq. NH_4Cl

$SiMe_2Ph$

Ph CO_2Me

98%
73% using cuprate

Crump, R.A.N.C.; Fleming, I.; Urch, C.J. *J. Chem. Soc., Perkin Trans. 1.*, ***1994***, 701

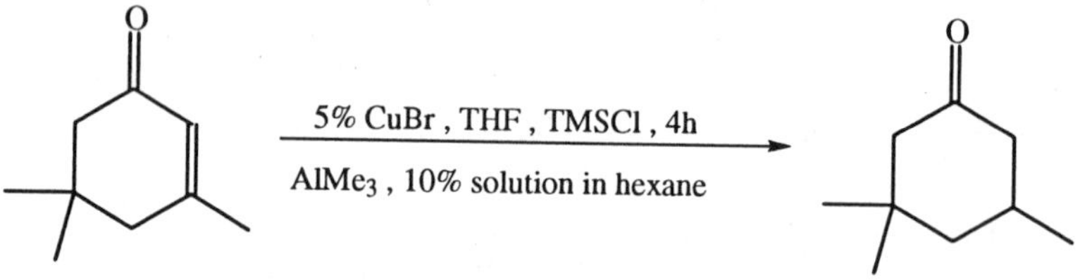

O

5% CuBr , THF , TMSCl , 4h

$AlMe_3$, 10% solution in hexane

O

96%

Kabbara, J.; Flemming, S.; Nickisch, K.; Neh, H.; Westermann, J. *Chem. Ber.*, ***1994***, *127*, 1489

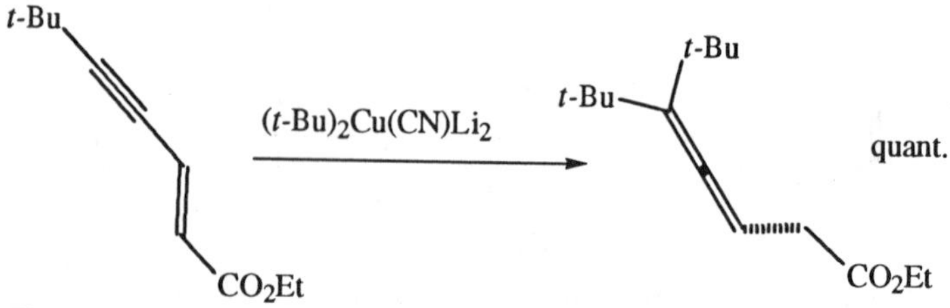

t-Bu

CO_2Et

$(t\text{-Bu})_2Cu(CN)Li_2$

t-Bu

t-Bu

CO_2Et

quant.

Gerold, A.; Krause, N. *Chem. Ber.*, ***1994***, *127*, 1547

, DME , 0.3h

LiOH , H_2O

CO_2Et → CO_2Et

96%

Bonadies, F.; Forcellese, M.L.; Locati, L.; Screttri, A.; Scolamiero, C. *Gazz. Chim. Ital.*, ***1994***, *124*, 467

CO_2Et — PhBr , $NiBr_2$, e^- / 3 H_2O , DMF-Py , 80°C / Bu_4NBr , Bu_4NI → Ph CO_2Et 63%

Condon-Guegnot, S.; Léonel, E.; Nédélec, J.-Y.; Périchon, J. *J. Org. Chem.*, ***1995***, *60*, 7684

$OSiMe_3$ — 5 M $LiClO_4$, ether → HO

70%

Sarawathy, V.G.; Sankararaman, S. *J. Org. Chem.*, ***1995***, *60*, 5024

t-BuZnBr , THF/pentane , -3°C / 2 TMSCl , 1.5 $BF_3•OEt_2$ → *t*-Bu 71%

Hanson, M.V.; Rieke, R.D. *J. Am. Chem. Soc.*, ***1995***, *117*, 10775

$CHPh_2$ — iPrI , Bu_3SnH , $Yb(OTf)_3$ / CH_2Cl_2/hexane/THF → $CHPh_2$ 89% (45:1)

Sibi, M.P.; Jasperse, C.P.; Ji, J. *J. Am. Chem. Soc.*, ***1995***, *117*, 10779

Ph — neat $Al(C_3H_7)_3$, 5% CuBr → C_3H_7 Ph

76%

Kabbara, J.; Flemming, S.; Nickisch, K.; Neh, H.; Westermann, J. *Tetrahedron*, ***1995***, *51*, 743

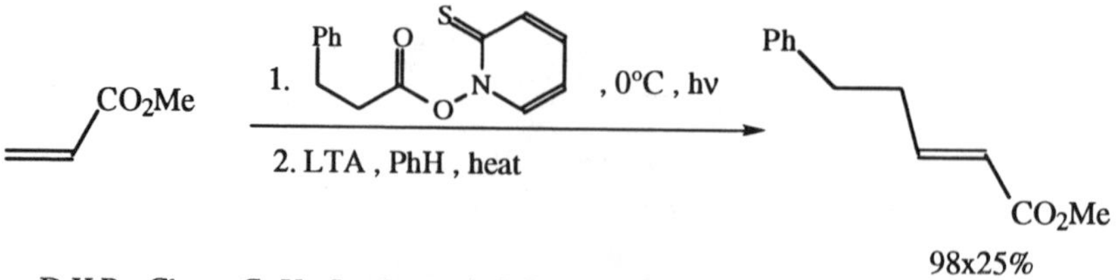

1. Ph...C(O)O–N(pyridine-2-thione), 0°C, hν
2. LTA, PhH, heat

CO_2Me → Ph...CO_2Me

98x25%

Barton, D.H.R.; Chern, C.-Y.; Jaszberenyi, J.Cs. *Tetrahedron*, ***1995***, *51*, 1867

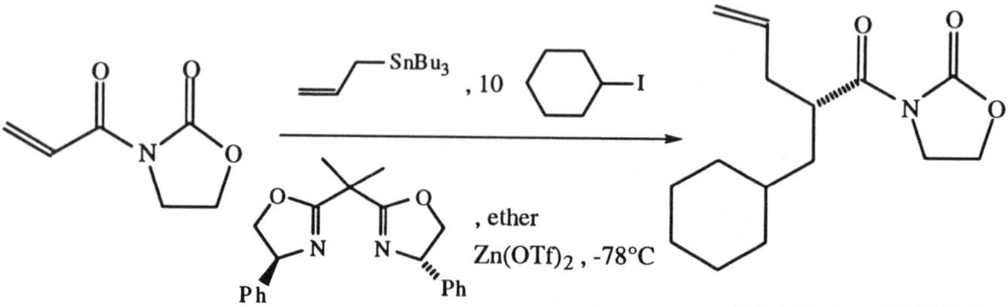

$SnBu_3$, 10 cyclohexyl–I, ether, $Zn(OTf)_2$, -78°C

61% (80% ee; 10:90 *R:S*)

Wu, J.H.; Radinov, R.; Porter, N.A. *J. Am. Chem. Soc.*, ***1995***, *117*, 11029

REVIEW:

"Intramolecular Michael and Anti-Michael Additions to Carbon-Carbon Triple Bonds," Rudorf, W.-D.; Schwarz, R. *Synlett*, ***1993***, 341

SECTION 74F: Cyclopropanations, including Halocyclopropanations

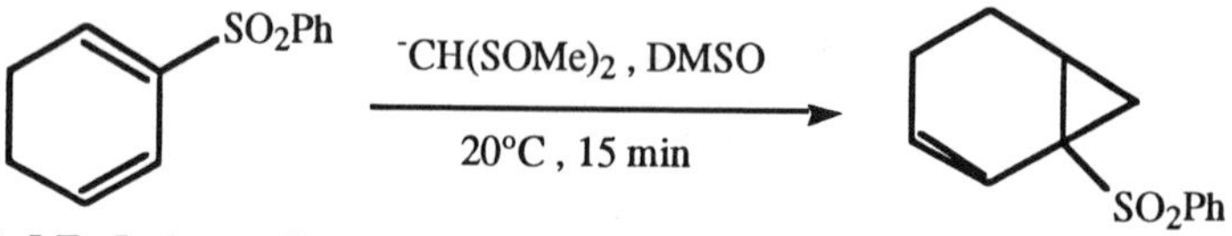

SO_2Ph — $^-CH(SOMe)_2$, DMSO, 20°C, 15 min → SO_2Ph 99%

Bäckvall, J-E.; Löfstöm, C.; Juntunen, S.K. *Tetrahedron Lett.*, ***1993***, *34*, 2007

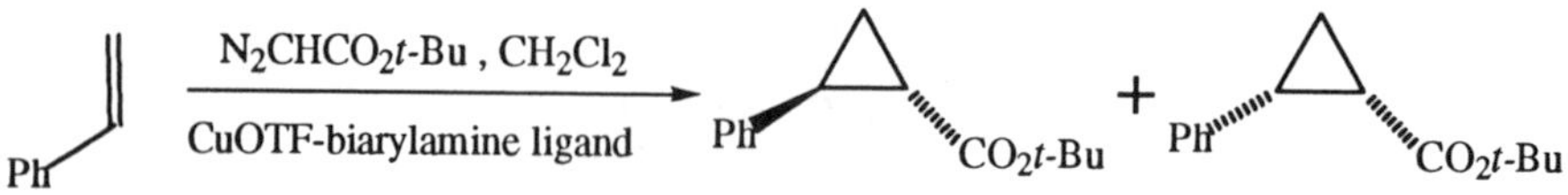

Ph — N_2CHCO_2t-Bu, CH_2Cl_2, CuOTF-biarylamine ligand → Ph ... CO_2t-Bu + Ph ... CO_2t-Bu

(86 : 14) 75%
92% ee

Ito, K.; Katsuki, T. *Tetrahedron Lett.*, ***1993***, *34*, 2661

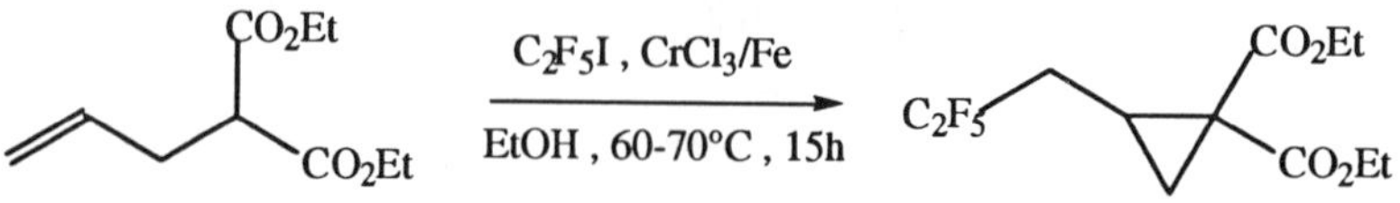

CO_2Et, CO_2Et — C_2F_5I, $CrCl_3$/Fe, EtOH, 60-70°C, 15h → C_2F_5 ... CO_2Et, CO_2Et 89%

Hu, C.-M.; Chen, J. *Tetrahedron Lett.*, ***1993***, *34*, 5957

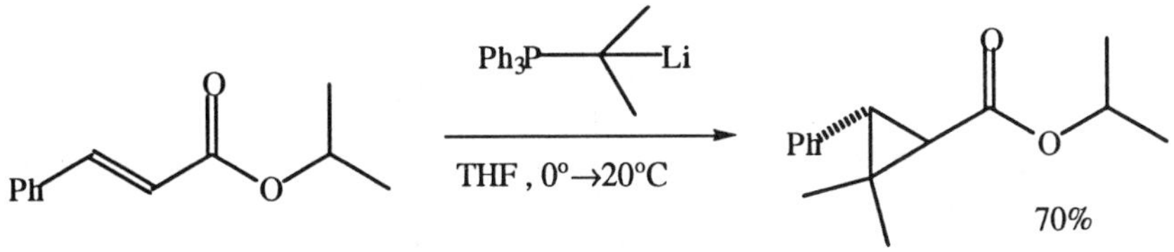

Ph₃P—C(Me)₂—Li

THF , 0°→20°C

Ph O 70%

Kreif, A.; Dubois, P. *Tetrahedron Lett.*, ***1993***, *34*, 2691

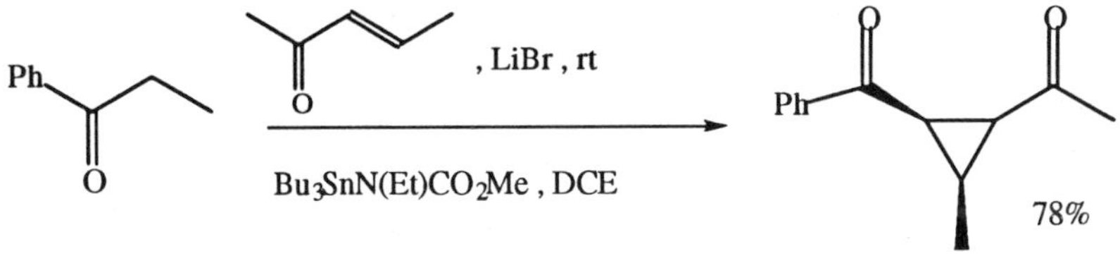

Ph O , LiBr , rt

$Bu_3SnN(Et)CO_2Me$, DCE

Ph O O 78%

(93:7 1,2-*trans:cis*)

Shibata, I; Mori, Y.; Yamasaki, H.; Baba, A.; Matsuda, H. *Tetrahedron Lett.*, ***1993***, *34*, 6567

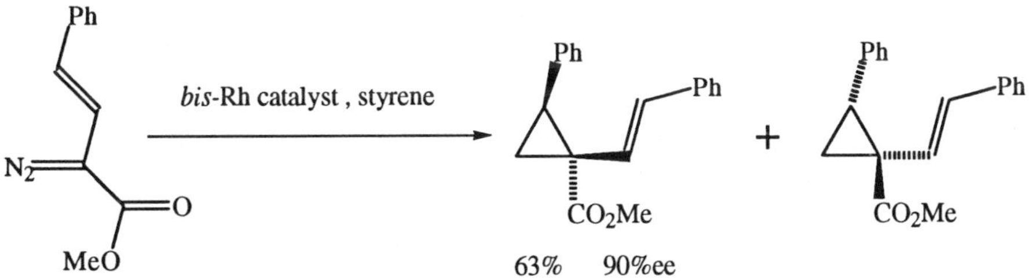

Ph N_2 O MeO

bis-Rh catalyst , styrene

Ph Ph CO_2Me + Ph Ph CO_2Me

63% 90%ee

Davies, H.M.L.; Hutcheson, D.K. *Tetrahedron Lett.*, ***1993***, *34*, 7243

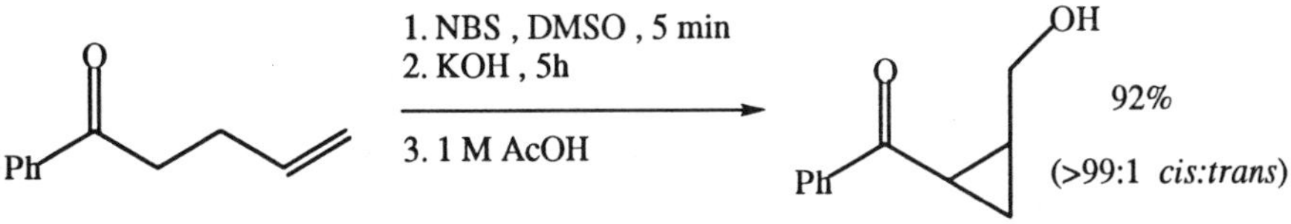

1. NBS , DMSO , 5 min
2. KOH , 5h
3. 1 M AcOH

Ph O OH 92%

(>99:1 *cis:trans*)

Dechoux, L.; Ebel, M.; Jung, L.; Stembach, J.F. *Tetrahedron Lett.*, ***1993***, *34*, 7405

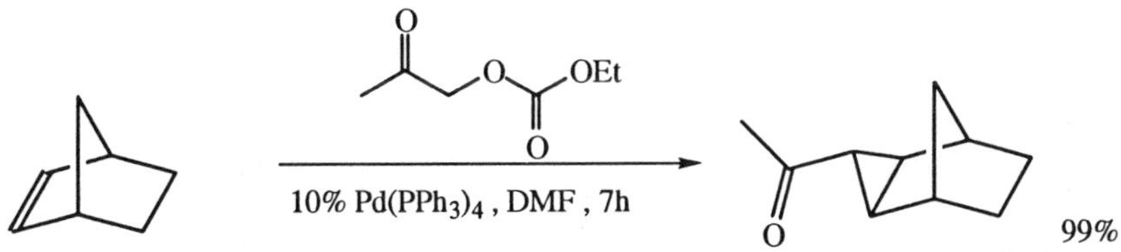

O OEt O

10% $Pd(PPh_3)_4$, DMF , 7h

O 99%

Ogoshi, S.; Morimoto, T.; Nishio, K.; Ohe, K.; Murai, S. *J. Org. Chem.*, ***1993***, *58*, 9

0°C

cat. $Rh_2(O_2CC_7H_{15})_4$

(97 : 3) 84%

Davies, H.M.L.; Huby, N.J.S.; Cantrell Jr., W.R.; Olive, J.L. *J. Am. Chem. Soc.*, ***1993***, *115*, 9468

$CrCl_2$ - DMF , 3h

32%

Montgomery, D.; Reynolds, K.; Stevenson, P. *J. Chem. Soc. Chem. Commun.*, ***1993***, 363

$LiCH(SPh)SO_2Ph$, THF

-78°C → 0°C

81%

Bailey, P.L.; Hewkin, C.T.; Clegg, W.; Jackson, R.F.W. *J. Chem. Soc., Perkin Trans. 1.*, ***1993***, 577

$Na^+ \ ^-CH(CO_2Me)_2$, THF-HMPA

-70°C → 0°C , 3h

63%

Watanabe, Y.; Ueno, Y.; Toru, T. *Bull. Chem. Soc. Jpn.*, ***1993***, *66*, 2042

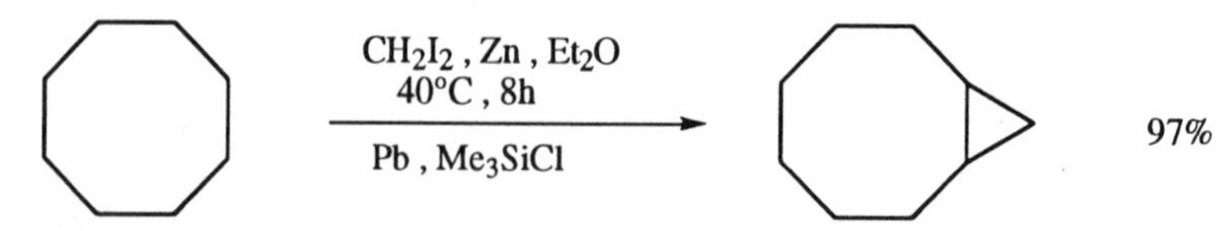

Takai, K.; Kakiuchi, T.; Utimoto, K. *J. Org. Chem.*, ***1994***, *59*, 2671

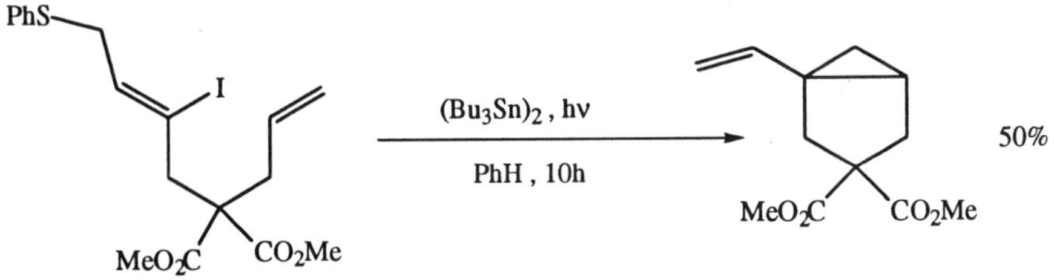

PhS, I, MeO$_2$C, CO$_2$Me

$(Bu_3Sn)_2$, hν

PhH , 10h

MeO$_2$C, CO$_2$Me

50%

Denis, R.C.; Gravel, D. *Tetrahedron Lett.*, ***1994***, *35*, 4531

(2.7

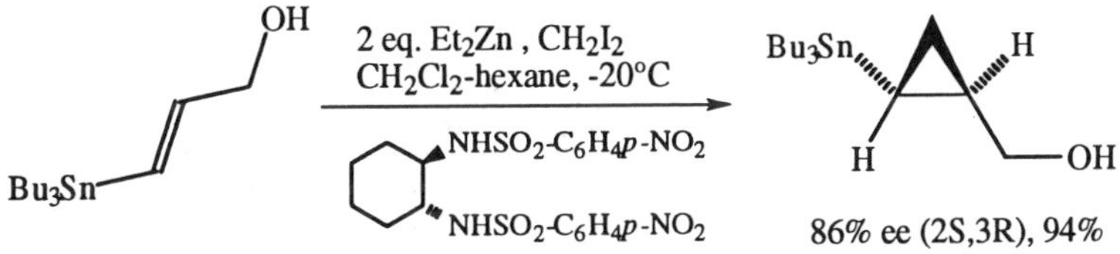

OH, Bu_3Sn

2 eq. Et_2Zn , CH_2I_2
CH_2Cl_2-hexane, -20°C

$NHSO_2$-C_6H_4p-NO_2

$NHSO_2$-C_6H_4p-NO_2

Bu_3Sn, H, H, OH

86% ee (2S,3R), 94%

Imai, N.; Sakamoto, K.; Takahashi, H.; Kobayashi, S. *Tetrahedron Lett.*, ***1994***, *35*, 7045

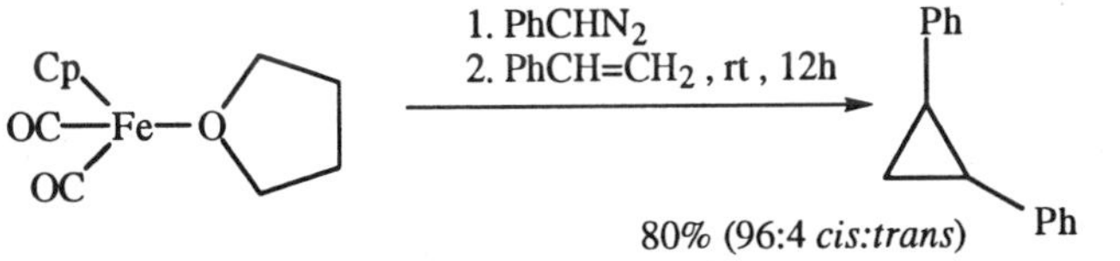

Cp, OC—Fe—O, OC

1. $PhCHN_2$
2. $PhCH{=}CH_2$, rt , 12h

Ph, Ph

80% (96:4 *cis:trans*)

Seitz, W.J.; Hossain, M.M. *Tetrahedron Lett.*, ***1994***, *35*, 7561

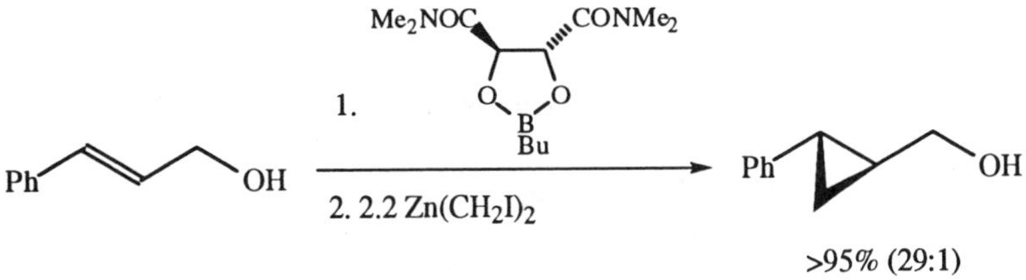

1. Me_2NOC, $CONMe_2$, O, O, B, Bu

Ph, OH

2. 2.2 $Zn(CH_2I)_2$

Ph, OH

>95% (29:1)

Charette, A.B.; Juteau, H. *J. Am. Chem. Soc.*, ***1994***, *116*, 2651

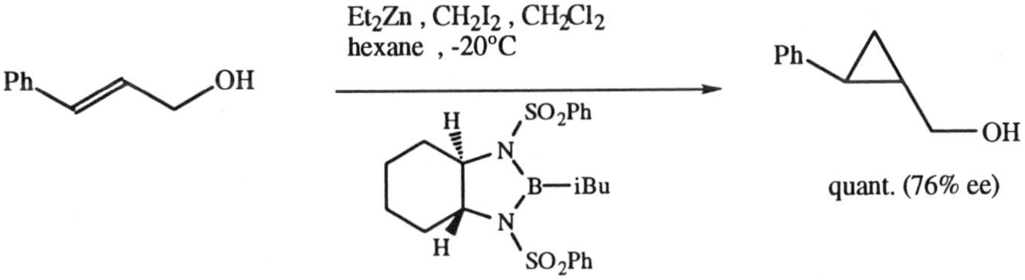

Ph, OH

Et_2Zn , CH_2I_2 , CH_2Cl_2
hexane , -20°C

H, SO_2Ph, N, B—iBu, N, H, SO_2Ph

Ph, OH

quant. (76% ee)

Imai, N.; Takahashi, H.; Kobayashi, S. *Chem. Lett.*, ***1994***, 177

Ph; $\overset{\ominus}{Br}$ $Bu_2\overset{\oplus}{Te}$ SiMe$_3$; Cs_2CO_3, 70°C; DME, trace water; Ph; O; Ph; Me$_3$Si; + ; O; Ph; Ph; Me$_3$Si

(92 : 8) 93%

Huang, Y.-Z.; Tang, Y.; Zhou, Z.-L.; Xia, W.; Shi, L.-P. *J. Chem. Soc., Perkin Trans. 1.*, ***1994***, 893

SO_2Ph; 3-pentanone/LDA, THF; 3d; O

62%

Ericsson, A.M.; Plobeck, N.A.; Bäckvall, J.-E. *Acta Chem. Scand. B.*, ***1994***, *48*, 252

CHO; Zn; $ClMe_2SiCH_2CH_2SiMe_2Cl$

57%

Motherwell, W.B.; Roberts, L.R. *Tetrahedron Lett.*, ***1995***, *36*, 1121

O; 2 $Ti(OiPr)_4$ 4 iPrMgBr; -45°C → 0°C; HO; OH

93% (58:42 *trans:cis*)

Kasatkin, A.; Sato, F. *Tetrahedron Lett.*, ***1995***, *36*, 6079

Ph; OH; 1. $Zn(CH_2I)_2$; 2. -78°C → -20°C; 3. $TiCl_4$; Ph; OH

90%

Charette, A.B.; Brochu, C. *J. Am. Chem. Soc.*, ***1995***, *117*, 11367

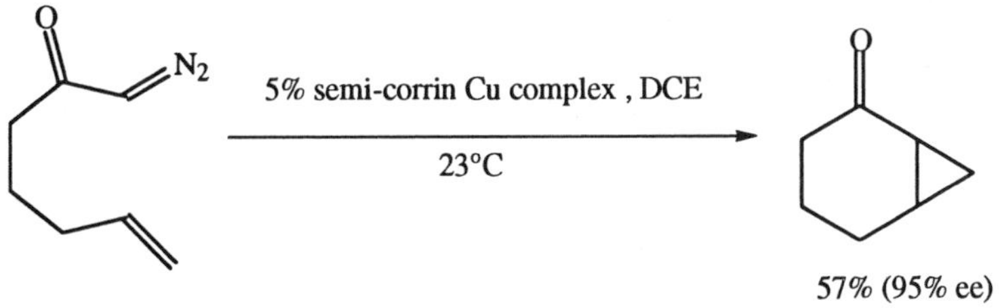

O

N_2

5% semi-corrin Cu complex , DCE

23°C

O

57% (95% ee)

Piqué, C.; Fähndrich, B.; Pfaltz, A. *Synlett*, ***1995***, 491

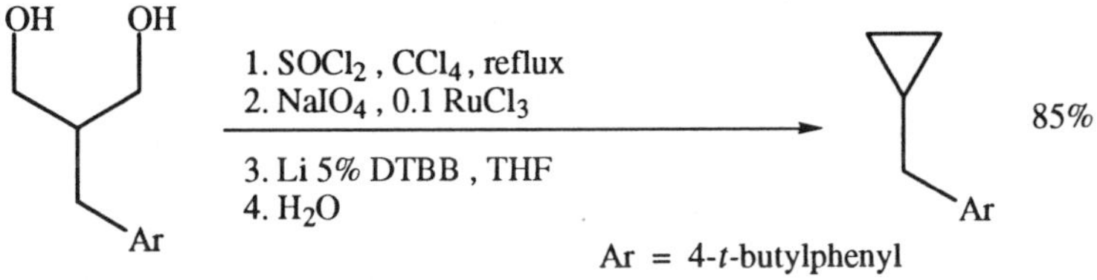

OH OH

1. $SOCl_2$, CCl_4, reflux
2. $NaIO_4$, 0.1 $RuCl_3$
3. Li 5% DTBB , THF
4. H_2O

Ar

Ar

85%

Ar = 4-*t*-butylphenyl

Guijarro, D.; Yus, M. *Tetrahedron*, ***1995***, *51*, 11445

REVIEW:

"The Asymmetric Cyclopropanantion of Acyclic Allylic Alcohols: Efficient Stereocontrol with Iodomethylzinc Reagents," Charette, A.B.; Marcoux, J.-F. *Synlett*, ***1995***, 1197

SECTION 75: ALKYLS, METHYLENES AND ARYLS FROM MISCELLANEOUS COMPOUNDS

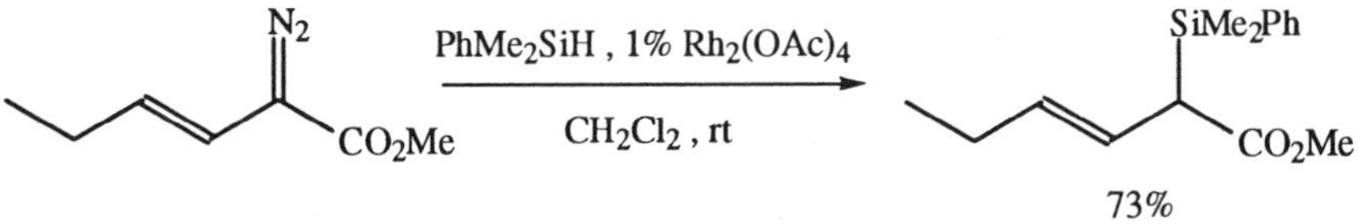

N_2

CO_2Me

$PhMe_2SiH$, 1% $Rh_2(OAc)_4$

CH_2Cl_2 , rt

$SiMe_2Ph$

CO_2Me

73%

Landais, Y.; Planchenault, D.; Weber, V. *Tetrahedron Lett.*, ***1994***, *35*, 9549

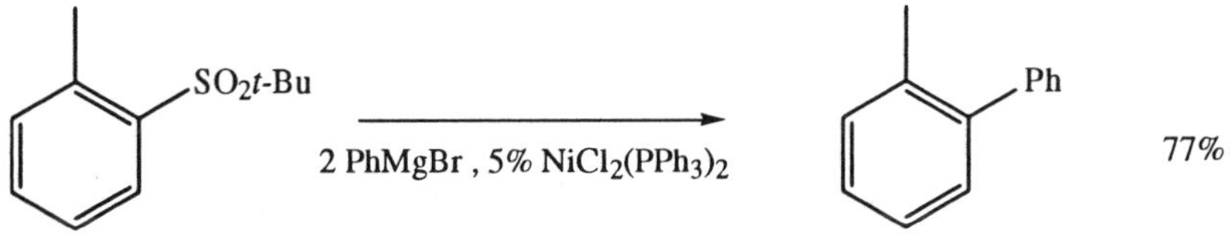

SO_2t-Bu

2 PhMgBr , 5% $NiCl_2(PPh_3)_2$

Ph

77%

Clayden, J.; Cooney, J.J.A.; Julia, M. *J. Chem. Soc., Perkin Trans. 1.*, ***1995***, 7

REVIEW:

"Organonickel Chemistry in Organic Synthesis. Some Applications of Alkyl and Metalacyclic Derivatives," Cámpora, J.; Paneque, M.; Poveda, M.L.; Carmona, E. *Synlett*, ***1994***, 465

CHAPTER 6
PREPARATION OF AMIDES

SECTION 76: AMIDES FROM ALKYNES

NH_2 —Ph

0.3 $Rh_6(CO)_{16}$, CO , H_2O
dioxane , NEt_3 , 175°C

H N O Ph

86%

Hirao, K.; Morii, N.; Joh, T.; Takahashi, S. *Tetrahedron Lett.*, ***1995***, *36*, 6243

SECTION 77: AMIDES FROM ACID DERIVATIVES

O Ph OH

1. $Sn[N(TMS)_2]_2$, THF , rt
2. N—H , 14h

O Ph N

89%

Burnell-Curty, C.; Roskamp, E.J. *Tetrahedron Lett.*, ***1993***, *34*, 5193

O Ph Cl

$TiCl_4$, Li , Me_3SiCl
CsF , N_2

O Ph NH_2

77%

\+

O O Ph N Ph H

11%

Kawaguchi, M.; Hamaoka, S.; Mori, M. *Tetrahedron Lett.*, ***1993***, *34*, 6907

$C_8H_{17}COOH$

$C_8H_{17}NH_2$, 180°C , 30 min

O C_8H_{17} NHC_8H_{17}

98%

Jursic, B.S.; Zdravkovski, Z. *Synth. Commun.*, ***1993***, *23*, 2761

NMM , PhCl , microwave

75°C , 2 min

(20 : 80) no yield

Bose, A.K.; Banik, B.K.; Manhas, M.S. *Tetrahedron Lett.*, ***1995***, *36*, 213

PhCH-NPh , toluene , Bu_3N

40% (5:95 *cis:trans*)

Browne, M.; Burnett, D.A.; Caplen, M.A.p; Chen, L.-Y.; Clader, J.W.; Domalski, M.; Dugar, S.; Pushpavanam, P.; Sher, R.; Vaccaro, W.; Viziano, M.; Zhao, H. *Tetrahedron Lett.*, ***1995***, *36*, 2555

1. NCS , PPh_3 ,CH_2Cl_2
2. $BnNH_2$

98%

Frøyen, P. *Synth. Commun.*, ***1995***, *25*, 959

PhCOOH

1. NsCl , NEt_3 , DMAP/MeCN , rt
2. $PhNH_2$, rt

95%

Lee, J.C.; Cho, Y.H.; Lee, H.K.; Cho, S.H. *Synth.Commun.*, ***1995***, *25*, 2877

SECTION 78: AMIDES FROM ALCOHOLS AND THIOLS

$CoCl_2$, Ac_2O , MeCN

70%

Mukhopadhyay, M.; Reddy, M.M.; Maikap, G.C.; Iqbal, J. *J. Org. Chem.*, ***1995***, *60*, 2670

MeCN , BF_3•OEt_2

92%

Firouzabadi, H.; Sardarian, A.R.; Badparva, H. *Synth. Commun.*, ***1994***, *24*, 601

SECTION 79: AMIDES FROM ALDEHYDES

NO ADDITIONAL EXAMPLES

SECTION 80: AMIDES FROM ALKYLS, METHYLENES AND ARYLS

PPA , AcOH , $NH_2OH•HCl$

110°C , 14h

80%

Cablewski, T.; Gurr, P.A.; Raner, K.D.; Strauss, C.R. *J. Org. Chem.*, ***1994***, *59*, 5814

SECTION 81: AMIDES FROM AMIDES

Conjugate reductions of unsaturated amides are listed in Section 74D (Alkyls from Alkenes).

$NHCO_2Me$

CO_2Et

1. Mg , MeOH , 0°C → rt
2. 2N HCl

O N H

91% (>99%ee)

Wei, Z.-Y.; Knaus, E.E. *Tetrahedron Lett.*, ***1993***, *34*, 4439

2.1 TBSOTf , 23°C
4.2 NEt_3 , 1.75h

Ph O N Me OMe

Ph O N Me H

75%

2.1 TBSOTf , 23°C

4.2 collidine , 2h

Ph O N Me CH_2OTBS

49%

Keck, G.E.; McHardey, S.F.; Murry, J.A. *Tetrahedron Lett.*, ***1993***, *34*, 6215

NH_2 O

$PhI(OAc)_2$, KOH

MeOH

$NHCO_2Me$

72%

Moriarty, R.M.; Chany II, C.JH.; Vaid, R.K.; Prakash, O.; Tuladhar, S.M. *J. Org. Chem.*, ***1993***, *58*, 2478

PhMgBr , THF

O N H OEt → O N H Ph 78%

Wei, Z.Y.; Knaus, E.E. *Org. Prep. Proceed. Int.*, ***1993***, *25*, 255

1. 2 eq. *s*-BuLi , (-)-sparteine
2. MeI

Ph NHMe O → H Me Ph NHMe O

84% (78% ee)

Beak, P.; Du, H. *J. Am. Chem. Soc.*, ***1993***, *115*, 2516

4.2 eq. $CuBr_2$•LiO*t*-Bu
rt , 6h

Ph NH$_2$ O → PhHN O*t*-Bu O 77%

Yamaguchi, J.; Hoshi, K.; Takeda, T. *Chem. Lett.*, ***1993***, 1273

, acetone , 0°C

N H O → N OH O 71%

Neset, S.M.; Benneche, T.; Undheim, K. *Acta Chem. Scand. B.*, ***1993***, *47*, 1141

CF_3SO_3H , CH_2Cl_2
0.75 h , 20°C

O NH$_2$ → O N—H

91%
(+ 4% lactone)

Marson, C.M.; Fallah, A. *Tetrahedron Lett.*, ***1994***, *35*, 293

HO Ph
DIAD , PPh_3 , THF

O N—H O → O N Ph O 52%

DIAD = diisopropylazodicarboxylate

Walker, M.A. *Tetrahedron Lett.*, ***1994***, *35*, 665

1. Br_2
2. NEt_3
3. $BnNH_2$

4. $Mg(OMe)_2$, MeOH

Bn N CO_2Me (S) H

86x79%

Garner, P.; Dogan, O.; Pillai, S. *Tetrahedron Lett.*, ***1994***, *35*, 1653

BnHN S OEt S

1. $(PhCO)_2O$, PhH , reflux
2. Bu_3SnH , AIBN , PhH

Bn N O

67%x"high yield"

Axon, J.; Boiteau, L.; Boivin, J.; Forbes, J.E.; Zard, S.Z. *Tetrahedron Lett.*, ***1994***, *35*, 1719

Ph N Me OTf

iPrOH , 80°C , 1h

Ph N CH_2OiPr H

85%

Hoffman, R.V.; Nayyar, N.K.; Shankweiler, J.M.; Klinekole III, B.W. *Tetrahedron Lett.*, ***1994***, *35*, 3231

N H O

$C_7H_{15}CHO$, H_2 , 40 bar
Pd/C-Na_2SO_4, EtOAc

100°C

N O C_8H_{17}

93%

Fache, F.; Jacquot, L.; LeMaire, M. *Tetrahedron Lett.*, ***1994***, *35*, 3313

Ph N Cl Boc

s-BuLi/TMEDA , THF

-78°C , 5h

Ph N Boc

82%

Beak, P.; Wu, S.; Yum, E.K.; Jun, Y.M. *J. Org. Chem.*, ***1994***, *59*, 276

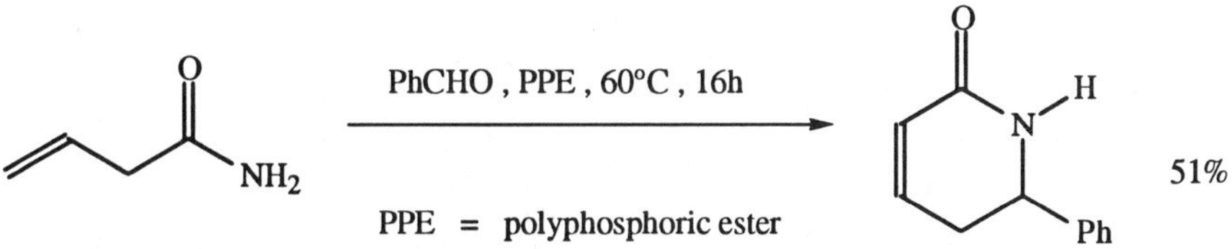

PhCHO , PPE , 60°C , 16h

NH2 → O, N, H, Ph 51%

PPE = polyphosphoric ester

Marson, C.M.; Grabowska, U.; Eallah, A.; Walsgrove, T.; Eggleston, D.S.; Baures, P.W. *J. Org. Chem.*, ***1994***, *59*, 291

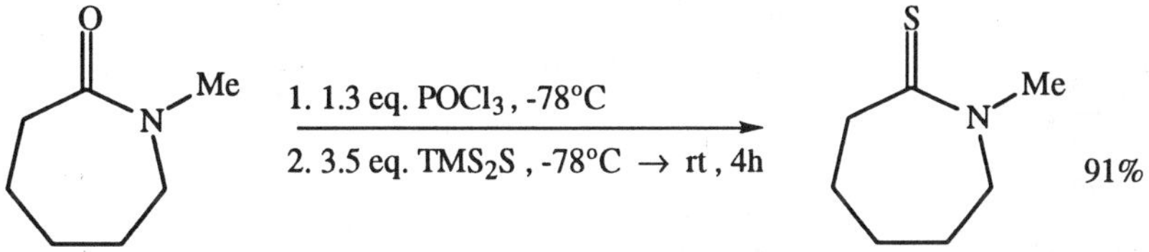

1. 1.3 eq. $POCl_3$, -78°C

2. 3.5 eq. TMS_2S , -78°C → rt , 4h

91%

Smith, D.C.; Lee, S.W.; Fuchs, P.L. *J. Org. Chem.*, ***1994***, *59*, 348

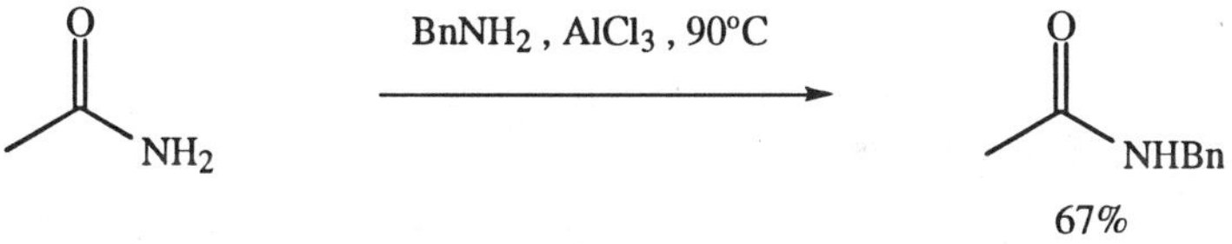

$BnNH_2$, $AlCl_3$, 90°C

NH_2 → NHBn 67%

Bon, E.; Bigg, D.C.H.; Bertrand, G. *J. Org. Chem.*, ***1994***, *59*, 4035

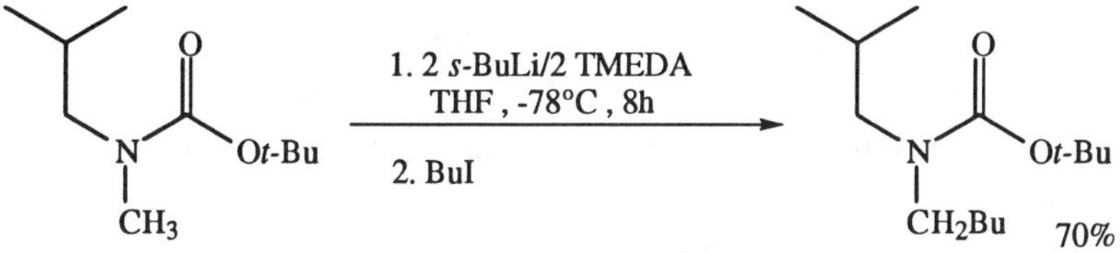

1. 2 *s*-BuLi/2 TMEDA THF , -78°C , 8h

2. BuI

O*t*-Bu, CH_3 → O*t*-Bu, CH_2Bu 70%

Snieckus, V.; Rogers-Evans, M.; Beak, P.; Lee, W.K.; Yum, E.K.; Freskos, J. *Tetrahedron Lett.*, ***1994***, *35*, 4067

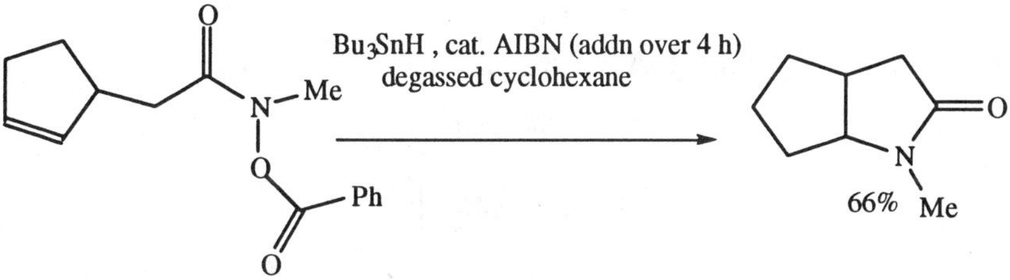

Bu_3SnH , cat. AIBN (addn over 4 h) degassed cyclohexane

66%

Callier, A.-C.; Quiclet-Sire, B.; Zard, S.Z. *Tetrahedron Lett.*, ***1994***, *35*, 6109

Bu_3SnH , AIBN

PhH (0.5M)

O I N O N

87%

Keusenkothen, P.F.; Smith, M.B. *J. Chem. Soc., Perkin Trans. 1.*, ***1994***, 2485

Ts N Ph

iPrMgCl , THF , 0°C , 6h

NHTs Ph + TsHN Ph

(45 : 57) 63%

Toshimitsu, A.; Abe, H.; Hirosawa, C.; Tamao, K. *J. Chem. Soc., Perkin Trans. 1.*, ***1994***, 3465

Me O N

1. 190°C , toluene , 17h
2. 215°C

O Me N

65% x low yield

Ciganek, E.; Wuonola, M.A.; Harlow, R.L. *J. Heterocyclic Chem.*, ***1994***, *31*, 1251

NH_2 O

1. $BtCH_2OH$

2. $CH_2{=}CHCH_2SiMe_3$ $BF_3{\bullet}OEt_2$

H N O 63%

Katritzky, A.R.; Ignatchenko, A.V.; Lang, H. *J. Org. Chem.*, ***1995***, *60*, 4002

OMe N $(CH_2)_3Br$ CO_2Me

Bu_3SnH , AIBN , PhH
reflux , 8h
(syringe pump , 7h)

OMe $(CH_2)_3NHCO_2Me$ + OMe N H CO_2Me

30% 36%

Lee, E.; Whang, H.S.; Chung, C.K. *Tetrahedron Lett.*, ***1995***, *36*, 913

(BnNEt$_3$)$_2$MoS$_4$, CH$_2$Cl$_2$

-78°C → rt , 30 min

82%

Ilankumaran, P.; Ramesha, A.R.; Chandrasekaran, S. *Tetrahedron Lett.*, ***1995***, *36*, 8311

1. 2.2 eq. BuLi , THF , -78°C → 0°C
2. PhCOCOPh
3. H$_3$O$^+$
4. Me$_3$SiCl , NaI , MeCN

75%x92%

Kiselyov, A.S. *Tetrahedron Lett.*, ***1995***, *36*, 493

2 SnCl$_2$, 2 KI
CH$_2$Cl$_2$, rt , 12h

53% + 10% benzophenone

Kataoka, T.; Iwama, T. *Tetrahedron Lett.*, ***1995***, *36*, 5559

HC≡CCH$_2$Br , DME/DMF , NaH

LiBr , 0°C → rt

87%

Liu, H.; Ko, S.-B.; Josien, H.; Curran, D.P. *Tetrahedron Lett.*, ***1995***, *36*, 8917

Mn(OAc)$_3$

AcOH

39%

D'Annibale, A.; Resta, S.; Trogolo, C. *Tetrahedron Lett.*, ***1995***, *36*, 9039

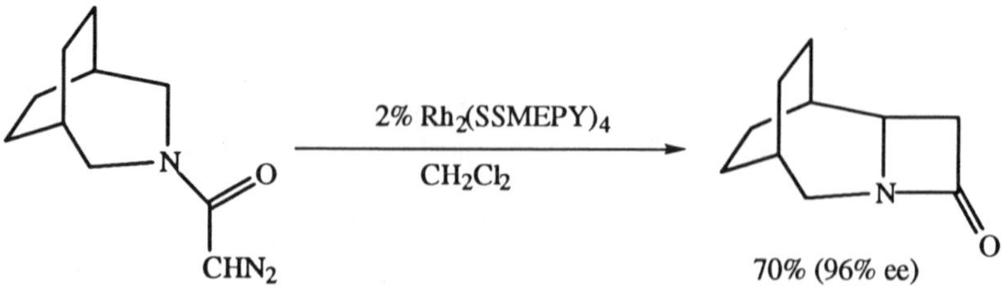

2% $Rh_2(SSMEPY)_4$

CH_2Cl_2

CHN_2

70% (96% ee)

Doyle, M.P.; Kalinin, A.V. *Synlett*, ***1995***, 1075

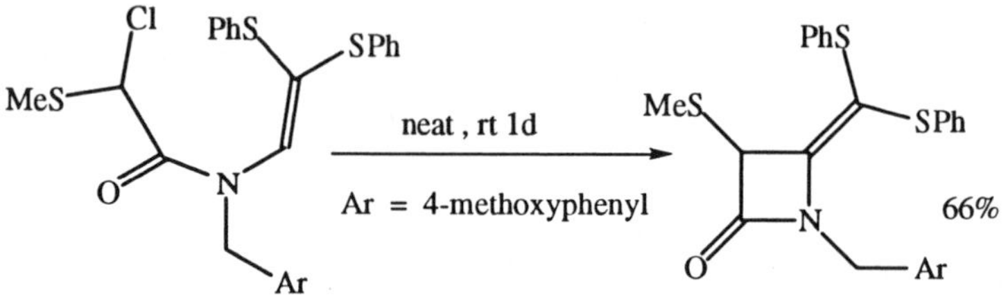

Cl PhS SPh MeS O N Ar

neat, rt 1d

Ar = 4-methoxyphenyl

PhS MeS SPh O N Ar 66%

Ishibashi, H.; Nakaharu, T.; Nishimura, M.; Nishikawa, A.; Kameoka, C.; Ikeda, M. *Tetrahedron*, ***1995***, *51*, 2929

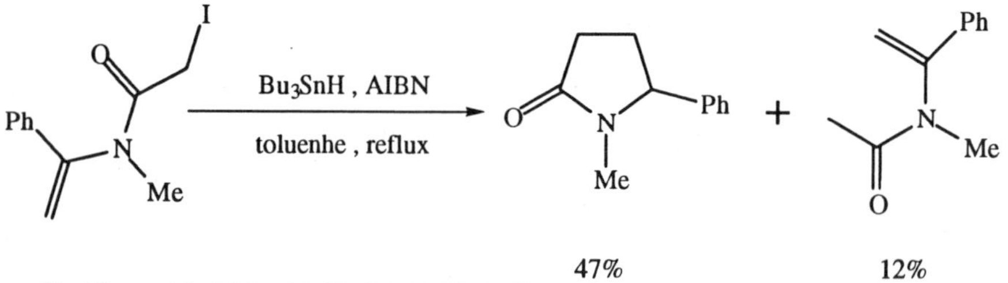

I O Ph N Me

Bu_3SnH, AIBN

toluenhe, reflux

O N Ph Me + Ph N Me O

47% 12%

Sato, T.; Chono, N.; Ishibashi, H.; Ikeda, M. *J. Chem. Soc., Perkin Trans. 1.*, ***1995***, 1115

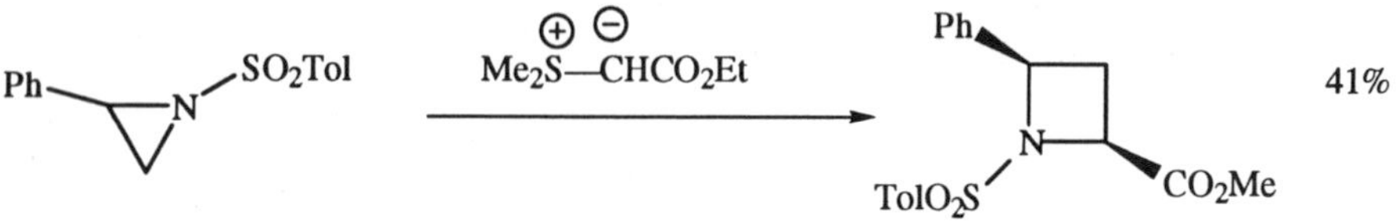

Ph N SO_2Tol

Me_2S^{+}—$CHCO_2Et^{-}$

Ph N $TolO_2S$ CO_2Me 41%

Nadir, U.K.; Arora, A. *J. Chem. Soc., Perkin Trans. 1.*, ***1995***, 2605

SECTION 82: AMIDES FROM AMINES

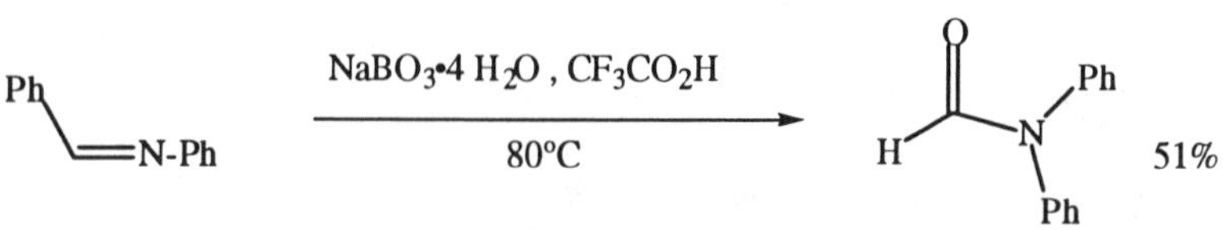

Ph =N-Ph

$NaBO_3 \cdot 4H_2O$, CF_3CO_2H

80°C

O H N Ph Ph 51%

Nongkunsarn, P.; Ramsden, C.A. *Tetrahedron Lett.*, ***1993***, *34*, 6773

NHBn, C_3H_7, CO_2Me → $Sn[N(TMS)_2]_2$, hexane , rt / pivalic acid → O, N, Bn, C_3H_7 89%

Wang, W.-B.; Roskamp, E.J. *J. Am. Chem. Soc.*, ***1993***, *115*, 9417

C_8H_{17}—NH_2 → $NCCH_2CHO$, rt , 6h → C_8H_{17}—NHCHO 78%

Duetsch, J.; Niclas, H.-J. *Synth. Commun.*, ***1993***, *23*, 1561

N—H → 1. N-cyclohexyl-N',N',N",N"-tetramethylguanidine , pyridine / toluene , CO_2 , -10°C 2. toluene , $SOCl_2$, -10°C → N, O, Cl 77%

McGhee, W.D.; Pan, Y.; Talley, J.J. *Tetrahedron Lett.*, ***1994***, *35*, 835

H_2N, OH → 1. 2 eq. Ph—P(O)(Ph)—Cl / 3 eq. NEt_3 , THF 2. excess NaH → O, P, Ph, Ph, N 67%

Osborn, H.M.I.; Cantrill, A.A.; Sweeney, J.B.; Howson, W. *Tetrahedron Lett.*, ***1994***, *35*, 3159

NHBn, OEt → xylene , 150°C , 3h → N, O, Bn 93%

MaGee, D.I.; Ramaseshan, M. *Synlett*, ***1994***, 743

NHBn, OTf → CO , MeCN , 65°C / $Pd(PPh_3)_4$, NBu_3 → N—Bn, O quant.

Crisp, G.T.; Meyer, A.G. *Tetrahedron*, ***1995***, *51*, 5585

C_3H_7–CH=CH–CH$_2$–NEt$_2$ → CO, Pd(OAc)$_2$, dppp → C_3H_7–CH=CH–CH$_2$–C(O)NEt$_2$ 78%

Murahashi, S.-I.; Imada, Y.; Nishimura, K. *Tetrahedron*, ***1994***, *50*, 453

SECTION 83: AMIDES FROM ESTERS

CH$_2$=CH–CH$_2$–OAc → H–N(OCO$_2$*t*-Bu)(CO$_2$*t*-Bu), K_2CO_3, rt, 1d → CH$_2$=CH–CH$_2$–N(OCO$_2$*t*-Bu)(CO$_2$*t*-Bu) 77%

Genet, J-P.; Thorimbert, S.; Touzin, A-M. *Tetrahedron Lett.*, ***1993***, *34*, 1159

Ph–CH$_2$–CO$_2$Me → 2 eq. Sn[NTMS$_2$]$_2$, 2 eq. Me$_2$NCH$_2$CMe$_2$OH, 2 eq. PhCHMeNH$_2$ → Ph–CH$_2$–C(O)–NH–CH(Ph)CH$_3$ 97%

Wang, W.-B.; Restituyo, J.A.; Roskamp, E.J. *Tetrahedron Lett.*, ***1993***, *34*, 7217

Ph–CH$_2$–C(O)OMe → 1.5 Sn[N(TMS)$_2$]$_2$, THF, 1.5 Et$_2$NH•HCl, rt, 18h → Ph–CH$_2$–C(O)NEt$_2$ 52%

Smith, L.A.; Wang, W.-B.; Burnell-Curty, C.; Roskamp, E.J. *Synlett*, ***1993***, 850

CH$_3$C(O)–O–CH=CH$_2$ → iPrNH$_2$, *t*-BuOH, 1.5d, 25°C → CH$_3$C(O)–NH–CH(CH$_3$)$_2$ 69%

Chen, S.-T.; Chen, S.-Y.; Chen, S.-J. *Tetrahedron Lett.*, ***1994***, *35*, 3583

C_3H_7–CH$_2$–C(O)OEt → NaEt$_2$Al(Bn)$_2$, toluene, reflux, 1h → C_3H_7–CH$_2$–C(O)NHBn 93%

Sim, T.B.; Yoon, N.M. *Synlett*, ***1994***, 827

1. $TiCl_4$, PySC(O)CH$_2$CH$_3$
2. NEt_3, CH_2Cl_2

PMP = 4-methoxyphenyl

(70 : 30) 99%

Annunziata, R.; Benaglia, M.; Cinquini, M.; Cozzi, F.; Ponzini, F.; Raimondi, L. *Tetrahedron*, ***1994***, *50*, 2939

1. BCl_3-SMe_2
2. Ph–CH=N–PMP

74% (78% ee)

PMP = N-4-methoxyphenyl

Annunziata, R.; Benaglia, M.; Cinquini, M.; Cozzi, F. *Tetrahedron Lett.*, ***1995***, *36*, 613

SECTION 84: AMIDES FROM ETHERS, EPOXIDES AND THIOETHERS

NO ADDITIONAL EXAMPLES

SECTION 85: AMIDES FROM HALIDES AND SULFONATES

1. Mg
2. PhNCO

83%

Cooke Jr., M.P.; Pollock, C.M. *J. Org. Chem.*, ***1993***, *58*, 7474

cat. $NiCl_2$(dppe), THF, rt

90%

Babudri, F.; Fiandanese, V.; Marchese, G.; Punzi, A. *Synlett*, ***1994***, 719

PhMgBr → 1. CS_2, THF; 2. $BtSO_2CF_3$; 3. $BuNH_2$ → Ph–C(=S)–NHBu 63%

Bt = benzotriazol-1'-yl

Katritzky, A.R.; Moutou, J.-L.; Yang, Z. *Synlett*, ***1995***, 99

SECTION 86: AMIDES FROM HYDRIDES

NO ADDITIONAL EXAMPLES

SECTION 87: AMIDES FROM KETONES

O O =O → MeO_2C~~~NH_2; TEA, $NaBH(OAc)_3$; DCE, rt, 45h → O O N O

aldehydes can also be used

92%

Abdel-Magid, A.F.; Harris, B.D.; Maryanoff, C.A. *Synlett*, ***1994***, 81

OTIPS → 1. 10 eq. $TMSN_3$, 2 eq. PPTS, CH_2Cl_2, rt, 2d; 2. UV, 0°C, 1h → O N—H

79x89%

Evans, P.A.; Modi, D.P. *J. Org. Chem.*, ***1995***, *60*, 6662

O N_3 → TFA, 0.75 h → O N 83%

Milligan, G.L.; Mossman, C.J.; Aubé, J. *J. Am. Chem. Soc.*, ***1995***, *117*, 10449

O → 1. NH_2OSO_3H, SiO_2; 2. H_2O, microwave → O N—H

86%

Laurent, A.; Jacquault, P.; Di Martino, J.-L.; Hamelin, J. *J. Chem. Soc. Chem. Commun.*, ***1995***, 1101

SECTION 88: AMIDES FROM NITRILES

PhCN → (UHP, cat. K_2CO_3, aq. acetone; rt, 1h) → $PhC(O)NH_2$ 85%

UHP = urea-hydrogen peroxide adduct

Balicki, R.; Kaczmarek, Ł. *Synth. Commun.*, ***1993***, *23*, 3149

→ (MnO_2, SiO_2, hexane; rt, 10d) → 92%

Breuilles, P.; Leclerc, R.; Uguen, D. *Tetrahedron Lett.*, ***1994***, *35*, 1401

→ (Al_2O_3 (neutral), 60°C, 1d) → 70%

Wilgus, C.P.; Downing, S.; Militor, E.; Bains, S.; Pagni, R.M.; Kabalka, G.W. *Tetrahedron Lett.*, ***1995***, *36*, 3469

SECTION 89: AMIDES FROM ALKENES

→ (TsN=IPh, Mn salen catalyst; pyridine N-oxide, MeCN, rt) → 34%

Noda, K.; Hosoya, N.; Irie, R.; Ito, Y.; Katsuki, T. *Synlett*, ***1993***, 469

→ (1. PhI=NTs, $Cu(OTf)_2$, MeCN; 2. 0.1 M NaOH) → 77%

Knight, J.G.; Muldowney, M.P. *Synlett*, ***1995***, 949

SECTION 90: AMIDES FROM MISCELLANEOUS COMPOUNDS

→ (BnN=C=O, hν (Pyrex filter)) → 42%

Rigby, J.H.; Ahmed, G.; Ferguson, M.D. *Tetrahedron Lett.*, ***1993***, *34*, 5397

N-OH; Bu_4NReO_4 , CF_3SO_3H , $MeNO_2$; 50% $H_2NOH{\bullet}HCl$; H; N; O; 84%

Narasaka, K.; Kusama, H.; Yamashita, Y.; Sato, H. *Chem. Lett.*, ***1993***, 489

O; N; TMSO; + N; Ph; 2.5 TMSOTf , CH_2Cl_2; O; O; N; H; aux; Ph

30% (>95:5 *endo:exo*)

Pouilhés, A.; Langlois, Y.; Nshimyumu Kiza, P.; Mbiya, K.; Ghosez, L. *Bull. Soc. Chim. Fr.*, ***1993***, *130*, 304

O; O; Ph; S; Me; 1. MeMgCl; 2. BEt_3; 3. H_2NOSO_3H; O; O; Ph; S; NH_2; 50%

Huang, H.-C.; Reinhard, E.J.; Reitz, D.B. *Tetrahedron Lett.*, ***1994***, *35*, 7201

O; N; hv , hexane , $PhNMe_2$; O; N

no yield

$\Phi_{max} = 0.77$

Post, A.J.; Nwaukwa, S.; Morrison, H. *J. Am. Chem. Soc.*, ***1994***, *116*, 6439

NO_2; O_2N; CO , MeOH , 7h; $Pd(dppp)Cl_2$, PhH; K_2CO_3 , 160°C; $NHCO_2Me$; MeO_2CHN; 59%

Reddy, N.P.; Masdeu, A.M.; El Ali, B.; Alper, H. *J. Chem. Soc. Chem. Commun.*, ***1994***, 863

Mn(tpp)Cl , MeCN

30°C

tpp = tetraphenylporphyrin

60%

Suda, K.; Sashima, M.; Izutsu, M.; Hino, F. *J. Chem. Soc. Chem. Commun.*, ***1994***, 949

Montmorillonite K-10 , microwave

7 min

91%

(17% with heating and no microwave)

Bosch, A.I.; de la Cruz, P.; Diez-Barra, E.; Loupy, A.; Langa, F. *Synlett*, ***1995***, 1259

DMF , $POCl_3$, 90°C ,. 5h

60%

Majo, V.J.; Venugopal, M.; Prince, A.A.M.; Perumal, P.T. *Synth. Commun.*, ***1995***, *25*, 3863

0.2 Bu_4NReO_4 , 0.2 CF_3SO_3H
$MeNO_2$, reflux , 0.5 $NH_2OH{\cdot}HCl$

91%

Kusama, H.; Yamashita, Y.; Narasaka, K. *Bull. Chem. Soc. Jpn.*, ***1995***, *68*, 373

SECTION 90A: PROTECTION OF AMIDES

2 $Mn(OAc)_3{\cdot}2\ H_2O$, EtOH
$Cu(OAc)_2{\cdot}H_2O$, 55°C

45%

Ghosh, A.; Miller, M.J. *Tetrahedron Lett.*, ***1993***, *34*, 83

N-OH

Ph Cl

1. $AlCl_3$, 60°C
2. PhOMe, 6h

O

PhHN

60%

OMe

(9:1 *p:o*)

Kim, J.N.; Ryu, E.K. *Tetrahedron Lett.*, ***1993***, *34*, 3567

O

Ph N Boc

Me

$Mg(ClO_4)_2$, MeCN

50°C, 20h

O

Ph NHMe

99%

Stafford, J.A.; Brackeen, M.F.; Karanewsky, D.S.; Valvano, N.L. *Tetrahedron Lett.*, ***1993***, *34*, 7873

Ph N O

CO_2t-Bu

$Mg(OMe)_2$, MeOH

Ph N O

H

90%

Wei, Z.-Y.; Knaus, E.E. *Tetrahedron Lett.*, ***1994***, *35*, 545

O

O N NHOMe

H O

5 eq. Ac_2O, 4% $Pd(PPh_3)_4$
2.2 eq. Bu_3SnH, CH_2Cl_2

rt, 10 min

AcHN NHOMe

58% O

Roos, E.C.; Bernabé, P.; Hiemstra, H.; Speckamp, W.N.; Kaptein, B.; Boesten, W.H.J. *J. Org. Chem.*, ***1995***, *60*, 1733

CHAPTER 7
PREPARATION OF AMINES

SECTION 91: AMINES FROM ALKYNES

NO ADDITIONAL EXAMPLES

SECTION 92: AMINES FROM ACID DERIVATIVES

NO ADDITIONAL EXAMPLES

SECTION 93: AMINES FROM ALCOHOLS AND THIOLS

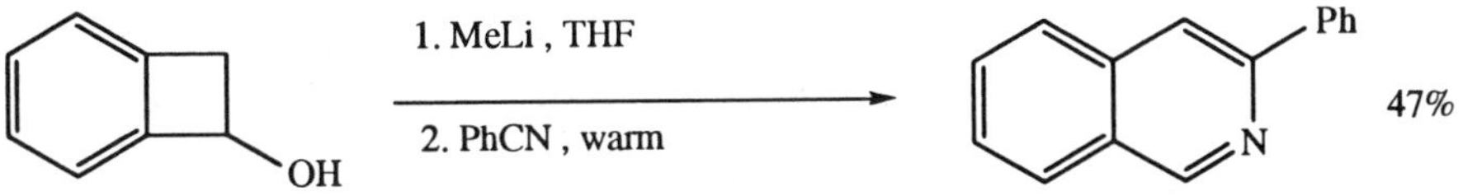

OH

1. MeLi , THF

2. PhCN , warm

Ph

N

47%

Fitzgerald, J.J.; Michael, F.E.; Olofson, R.A. *Tetrahedron Lett.*, ***1994***, *35*, 9191

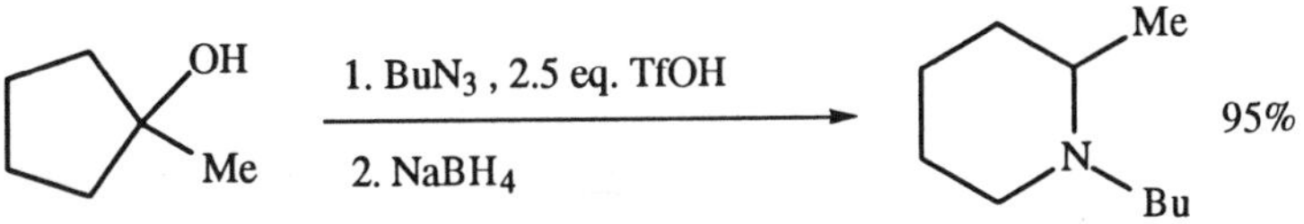

OH

Me

1. BuN_3 , 2.5 eq. TfOH

2. $NaBH_4$

Me

N

Bu

95%

Pearson, W.H.; Fang, W. *J. Org. Chem.*, ***1995***, *60*, 4960

$PhCH_2OH$

$PhCH_2NHTf$, DEAD , PPh_3

$(PhCH_2)_2N$—Ts

93%

Bell, K.E.; Knight, D.W.; Gravestock, M.B. *Tetrahedron Lett.*, ***1995***, *36*, 8681

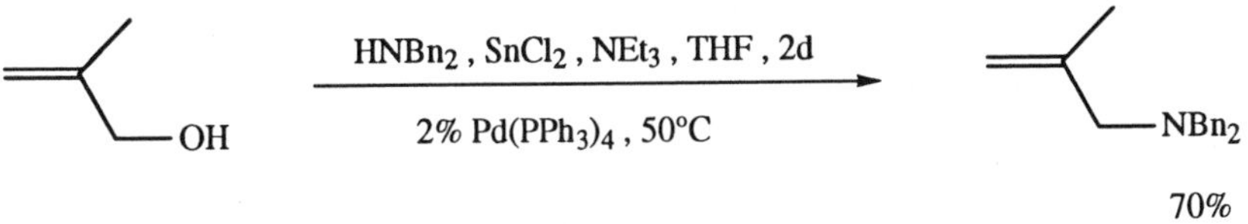

OH

$HNBn_2$, $SnCl_2$, NEt_3 , THF , 2d

2% $Pd(PPh_3)_4$, 50°C

NBn_2

70%

Masuyama, Y.; Kagawa, M.; Kurusu, Y. *Chem. Lett.*, ***1995***, 1121

SECTION 94: AMINES FROM ALDEHYDES

PhCHO → (Me_3Si–N-pyrrolidine; Me_3SiCl, 29h) → PhCH=N⁺(pyrrolidinium) Cl^- 92%

Jahn, U.; Schroth, W. *Tetrahedron Lett.*, ***1993***, *34*, 5863

Cyclohexyl-NH_2 → (1. $C_5H_{11}CHO$, BER, EtOH, Et_3NHCl; 2. BER, NaOH) → Cyclohexyl-$NHCH_2C_5H_{11}$

BER = borohydride exchange resin 89%

Yoon, N.M.; Kim, E.G.; Son, H.S.; Choi, J. *Synth. Commun.*, ***1993***, *23*, 1595

CHO → ($LiAl(NHBn)_4$, ether; 25°C, 7h) → =N-Bn 90%

Solladié-Cavallo, A.; Bencheqroun, M.; Bonne, F. *Synth. Commun.*, ***1993***, *23*, 1683

Ph, CHO → (1. Bzt-SNH_2, NaOH, EtOH; 2. $PhSN(TMS)_2$, cat. TBAF, THF; 3. Bu_3SnH, cat. AIBN, cyclohexane) → Ph, N 61x69%

Boivin, J.; Fouquet, E.; Zard, S.Z. *Tetrahedron*, ***1994***, *50*, 1745

Ph CHO → (1. $Me_2NH{\cdot}HCl$, $Ti(OiPr)_4$, NEt_3 EtOH, 25°C, 10h; 2. $NaBH_4$, 25°C, 10h) → Ph NMe_2 95%

Bhattacharyya, S. *J. Org. Chem.*, ***1995***, *60*, 4928

PhCHO → (1. $BnNH_2$, $Py{\cdot}BH_3$, MeOH MS 4Å, 16h; 2. 6 N HCl; 3. 8 N NaOH) → Ph N(H) Ph 87%

Bomann, M.D.; Guch, I.C.; DiMare, M. *J. Org. Chem.*, ***1995***, *60*, 5995

PhCHO → ($TsNH_2$, $Si(OEt)_4$, 160°C, 6h) → PhCH=NTs 68%

Love, B.E.; Raje, P.S.; Williams II, T.C. *Synlett*, ***1994***, 493

PhCHO —[$BnNH_2$, Bentonite, rt, 3h; ketones can also be used]→ PhCH=N-Bn 80%

Saoudi, A.; Benguedach, A.; Benhaoua, H. *Synth. Commun.*, ***1995***, *25*, 2349

Related Methods: Section 102 (Amines from Ketones)

SECTION 95: AMINES FROM ALKYLS, METHYLENES AND ARYLS

NO ADDITIONAL EXAMPLES

SECTION 96: AMINES FROM AMIDES

$C_6H_{13}CH_2C(=O)$-pyrrolidine —[$LiBH_3BN(iPr)_2$, THF, rt]→ $C_6H_{13}CH_2CH_2$-pyrrolidine 95%

Fisher, G.B.; Fuller, J.C.; Harrison, J.; Goralski, C.T.; Singaram, B. *Tetrahedron Lett.*, ***1993***, *34*, 1091

PhC(=O)$NHC_{10}H_{21}$ —[KH, THF, Cp_2ZrHCl, -20°C → rt]→ PhCH=N-$C_{10}H_{21}$ 86%

Schedler, D.J.A.; Godfrey, A.G.; Ganem, B. *Tetrahedron Lett.*, ***1993***, *34*, 5035

—[1. LiHMDS ; BnBr 2. LiHMDS ; MeI 3. $LiAlH_4$ 4. H_2, Pd-C]→ 89%

Westrum, L.J.; Meyers, A.I. *Tetrahedron Lett.*, ***1994***, *35*, 973

—[$TiCl_3$, 3 eq. C_8K]→ 90%

Furstner, A.; Hupperts, A.; Ptock, A.; Janssen, E. *J. Org. Chem.*, ***1995***, *60*, 5215

HBr , 163°C

NHPh•HBr

H—N N—Ph

88%

Poindexter, G.S.; Bruce, M.A.; LeBoulluec, K.L.; Monkovic, I. *Tetrahedron Lett.*, ***1994***, *35*, 7331

SO_2CF_3

t-BuLi , THF , -60°C , 3h

t-Bu

52%

Bozee-Ogor, S.; Salou-Guiziou, V.; Yaouanc, J.J.; Handel, H. *Tetrahedron Lett.*, ***1995***, *36*, 6063

3 Ti (powder) , THF , 21h
3 TMSCl , reflux

99%

Fürstner, A.; Hupperts, A. *J. Am. Chem. Soc.*, ***1995***, *117*, 4468

30% H_2O_2 , AcOH , 25°C

85%

Grivas, S.; Ronne, E. *Acta Chem. Scand. B.*, ***1995***, *49*, 225

Related Methods: Section 105A (Protection of Amines)

SECTION 97: AMINES FROM AMINES

1. 5 Li (powder) , THF
cat. $C_{10}H_8$, -78°C , 6h
2. H_2O

N—Ph → NHPh 93%

Almena, J.; Foubelo, F.; Yus, M. *Tetrahedron Lett.*, ***1993***, *34*, 1649

PhCHO , TFA

Et_3SiH , CH_2Cl_2 , 0°C

83%

Appleton, J.E.; Dack, K.N.; Green, A.D.; Steele, J. *Tetrahedron Lett.*, ***1993***, *34*, 1529

9-BBN , THF , 20h

56%

Ar = 3,4-dimethoxyphenyl

Gaul, M.D.; Fowler, K.W.; Grieco, P.A. *Tetrahedron Lett.*, ***1993***, *34*, 3099

1. Bu_2ZrCp_2

2. H_3O^+

70%

Maye, J.P.; Negishi, E. *Tetrahedron Lett.*, ***1993***, *34*, 3359

10 eq. MeLi , PhH , 50°C

>60% + 20%

Genin, M.J.; Biles, C.; Romero, D.L. *Tetrahedron Lett.*, ***1993***, *34*, 4301

1. *s*-BuLi , TMEDA
2. MeI
3. *s*-BuLi , TMEDA
4. MeI

62x41%

Beak, P.; Lee, W.K. *J. Org. Chem.*, ***1993***, *58*, 1109

t-Bu

1. 1.5 LDA, THF, 0°C

2. 1.2 Me_2Si N Si Me_2 Br

THF, 15h, 0°C → rt

3. 3 eq. K_2CO_3, MeOH heat, 3h

62%

DeKimpe, N.G.; Keppens, M.A.; Stevens, C.V. *Tetrahedron Lett., **1993**, 34*, 4693

2.5 eq. $LiBHEt_3$, THF, rt

84%

(1:1 *cis:trans*)

Blough, B.E.; Carroll, F.I. *Tetrahedron Lett., **1993**, 34*, 7239

1. Li, THF
2. PhBr, THF, rt, 6h
3. MeOH; 20% HCl

86%

Kanth, J.V.B.; Periasamy, M. *J. Org. Chem., **1993**, 58*, 3156

NH_2 I

5% $Pd(OAc)_2$, 5% PPh_3
100°C, 3.5d, Na_2CO_3

65%

Larock, R.C.; Berrios-Peña, N.G.; Fried, C.A.; Yum, E.K.; Tu, C.; Leong, W. *J. Org. Chem., **1993**, 58*, 4509

CO_2Me NH_2

1. $(CO)_5Cr$=C(ONMe$_2$)Me, Me_3COCl

2. Tol, 120°C, 20h

CO_2Me Me

65 x 60%

Söderberg, B.C.; Helton, E.S.; Austin, L.R.; Odens, H.H. *J. Org. Chem., **1993**, 58*, 5589

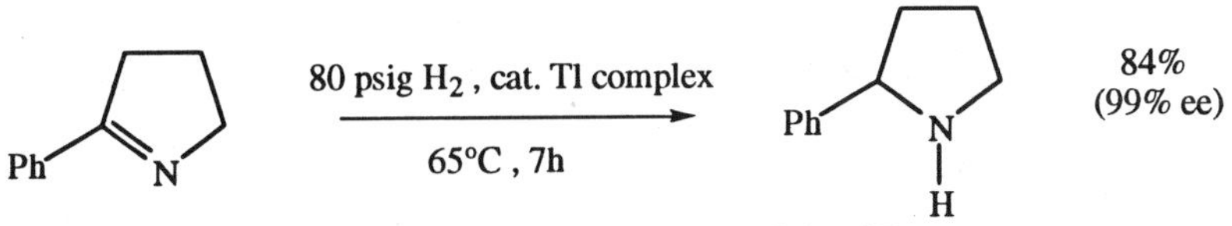

80 psig H_2, cat. Tl complex; 65°C, 7h — Ph N → Ph N H — 84% (99% ee)

Willoughby, C.A.; Buchwald, S.L. *J. Org. Chem.*, ***1993***, *58*, 7627

$PhNH_2$ —acetone, $NaBH_4$, 3M H_2SO_4, THF→ PhNHiPr 95%

Verardo, G.; Giumanini, A.G.; Strazzolini, P.; Poiana, M. *Synthesis*, ***1993***, 121

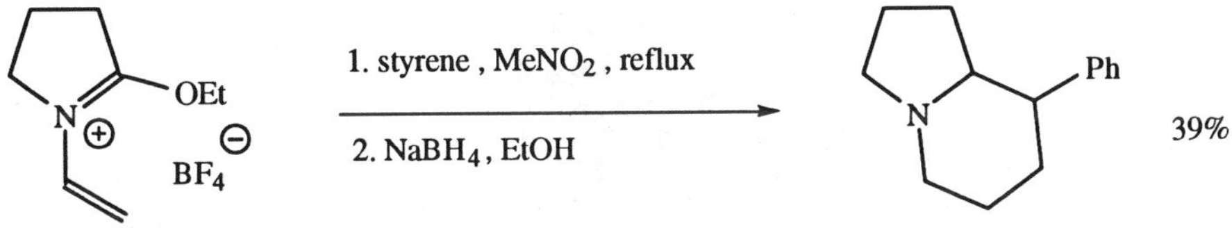

OEt N ⊕ ⊖ BF_4 — 1. styrene, $MeNO_2$, reflux; 2. $NaBH_4$, EtOH → N Ph — 39%

Sheu, J.; Smith, M.B.; Matsumoto, K. *Synth. Commun.*, ***1993***, *23*, 253

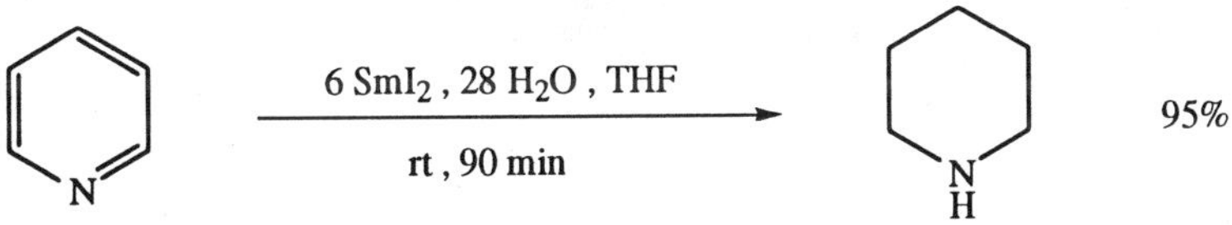

N — 6 SmI_2, 28 H_2O, THF; rt, 90 min → N H — 95%

Kamochi, Y.; Kudo, T. *Heterocycles*, ***1993***, *36*, 2383

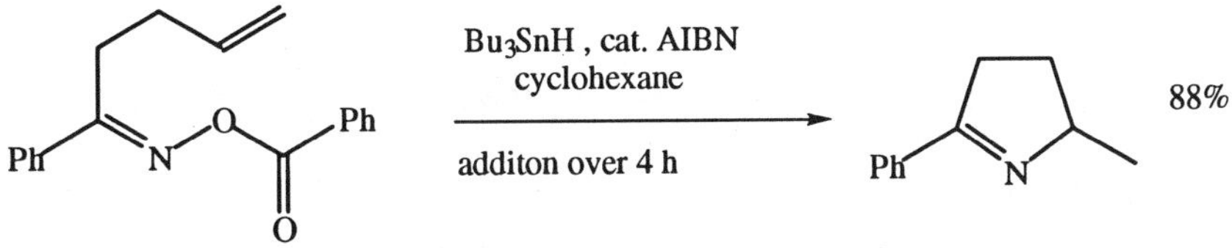

Ph N O Ph O — Bu_3SnH, cat. AIBN cyclohexane; additon over 4 h → Ph N — 88%

Boivin, J.; Schiano, A.-M.; Zard, S.Z. *Tetrahedron Lett.*, ***1994***, *35*, 249

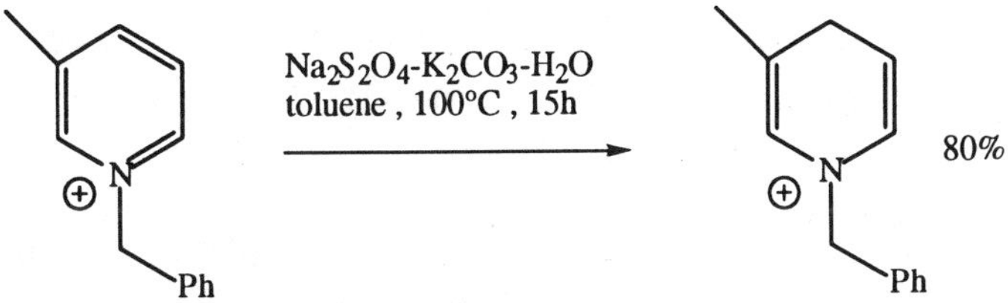

⊕ N Ph — $Na_2S_2O_4$-K_2CO_3-H_2O toluene, 100°C, 15h → ⊕ N Ph — 80%

the reaction fails for many N-R where R ≠ Bn

Wong, Y.-S.; Marazano, C.; Gnecco, D.; Das, B.C. *Tetrahedron Lett.*, ***1994***, *35*, 707

N-t-Bu

1. Br_2 , CH_2Cl_2 , 0°C
2. 2 eq. $LiAlH_4$, ether

N t-Bu 78%

DeKimpe, N.; Boelens, M.; Piqueur, J.; Baele, J. *Tetrahedron Lett.*, ***1994***, *35*, 1925

N—H

1. $(HCHO)_n$, $Ti(OiPr)_4$
2. $NaBH_4$

N—Me 95%

Bhattacharyya, S. *Tetrahedron Lett.*, ***1994***, *35*, 2401

Ph Ph N—H

1. MeMgCl , THF , rt
2. Cp_2ZrCl_2 , THF
3. hexane/dioxane, rt
4. MeMgCl , -30°C → rt
5. MeOH

H Ph N Ph 78%

Harris, M.C.J.; Whitby, R.J.; Blagg, J. *Tetrahedron Lett.*, ***1994***, *35*, 2431

N $SnBu_3$ MeO

1. Ph , BuLi
2. H_2O

Ph N + Ph N

(10 : 1) 88%

Pearson, W.H.; Stevens, E.P. *Tetrahedron Lett.*, ***1994***, *35*, 2641

NHt-Boc + $(iPr)_2CH$—N= F Cl

1. 2 t-BuLi
2. imine
3. TFA , CH_2Cl_2

N Cl 65%

Reuter, D.C.; Flippin, L.A.; McIntosh, J.; Caroon, J.M.; Hammaker, J. *Tetrahedron Lett.*, ***1994***, *35*, 4899

N N Ph Ph

PhMgCl
-78°C

Ph Ph *RHN NHR* + Ph Ph *RHN NHR*

(6 1) 47%

Bambridge, K.; Begley, M.J.; Simpkins, N.S. *Tetrahedron Lett.*, ***1994***, *35*, 3391

$MeCN$, $NaPF_6$; 2% $Pd(Ph_3)_4$ — 90%

Grellier, M.; Pfeffer, M.; van Koten, G. *Tetrahedron Lett.*, ***1994***, *35*, 2877

3 eq. LDA, THF, rt, 16h — 55%

Stevens, C.V.; De Kimpe, N.G.; Katritzky, A.R. *Tetrahedron Lett.*, ***1994***, *35*, 3763

$LiEt_2NBH_3$, THF, 0°C; 1h — 95% (66% de)

Fuller, J.C.; Belisle, C.M.; Goralski, C.T.; Singaram, B. *Tetrahedron Lett.*, ***1994***, *35*, 5389

Bu_3SnH, AIBN, PhMe; addn via syringe pump — 56%

Bowman, W.R.; Stephenson, P.T.; Terrett, N.K.; Young, A.R. *Tetrahedron Lett.*, ***1994***, *35*, 6369

0.05 $Pr_4N\ RuO_4$, 1.5 NMO; MeCN, rt, 1d — 68%

Goti, A.; Romani, M. *Tetrahedron Lett.*, ***1994***, *35*, 6567

$NaBH_4$, MeOH, heat — 89%

De Kimpe, N.; De Smaele, D. *Tetrahedron Lett.*, ***1994***, *35*, 8023

5% chiral Ti catalyst, H_2, THF, rt

75% (92% ee, R)

Lee, N.E.; Buchwald, S.L. *J. Am. Chem. Soc.*, ***1994***, *116*, 5985

2 MeLi, toluene, -78°C

90% (70% ee, R)

Denmark, S.E.; Nakajima, N.; Nicaise, O.J.-C. *J. Am. Chem. Soc.*, ***1994***, *116*, 8797

1. Cp_2ZrBu_2, THF, -78°C
2. aq. MeOH

70%

Kemp, M.I.; Whitby, R.J.; Coote, S.J. *Synlett*, ***1994***, 451

1. $PhS(=O)(=NH)CH_2CPh_2OH$ (10%), $BH_3 \cdot SMe_2$
2. workup

68% (50% ee, S)

Bolm, C.; Felder, M. *Synlett*, ***1994***, 655

1. Bu_3SnH, AIBN, THF syringe pump, reflux
2. 2M HCl
3. 5M NaOH

49% (2.8 : 1.2 : 1 (minor product)

Bowman, W.R.; Clark, D.N.; Marmon, R.J. *Tetrahedron*, ***1994***, *50*, 1295

1. DCE , $ClCO_2Et$
2. $PhNEt_2{\bullet}BI_3$

N—Bn → N—H

80%

Kanth, J.V.B.; Reddy, Ch.K.; Periasamy, M. *Synth. Commun.*, ***1994***, *24*, 313

1. $(HCHO)_n$, $ZnCl_2$, THF
2. $Zn(BH_4)_2$, THF , 25°C , 10h
3. aq. NH_3

NH_2 → NHMe

70%

Bhattacharyya, S.; Chatterjee, A.; Duttachowdhury, S.K.
J. Chem. Soc., Perkin Trans. 1., ***1994***, 1

1. NCS , PhH
2. Bu_3SnH , AIBN , PhH , reflux

Ph, NHMe, Me → Ph, N, Me, Me

42%

Tokuda, M.; Fujita, H.; Suginome, H. *J. Chem. Soc., Perkin Trans. 1.*, ***1994***, 777

, 48% HBr
dioxane , sealed tube
100°C , 4d

NH_2, Cl → NH*t*-Bu, Cl

78%

Gage, J.R.; Wagner, J.M. *J. Org. Chem.*, ***1995***, *60*, 2613

MeO , 100°C , 9 h
PPTs , NEt_3 , sealed tube

NH_2, N, Cl → $N{=}CMe_2$, N, Cl

PPTs = pyridinium *p*-toluenesulfonate

quant.

Morris, J.; Wishka, D.G. *J. Org. Chem.*, ***1995***, *60*, 2642

OEt, $OSiMe_3$, AcOH
$NaCNBH_4$, MeOH

NHMe → N, Me

75%

Gillaspy, M.L.; Lefker, B.A.; Hada, W.A.; Hoover, D.J. *Tetrahedron Lett.*, ***1995***, *36*, 7399

Bn N SnMe$_3$ → 0.2 MeLi, SnMe$_4$, THF → SnMe$_3$ N Bn 70%

Coldham, I.; Hufton, R. *Tetrahedron Lett.*, ***1995***, *36*, 2157

N Boc → 1. *sec*-BuLi, TMEDA 2. PhI, [P(*o*-Tol)$_3$]$_4$Pd-CuCN 75°C, 18h, toluene → N Boc Ph 54%

Dieter, R.K.; Li, S. *Tetrahedron Lett.*, ***1995***, *36*, 3613

N → Br CO$_2$Me In, THF, 15h → N CO$_2$Me 50%

Bossard, F.; Dambrin, V.; Lintanf, V.; Beuchet, P.; Mosset, P. *Tetrahedron Lett.*, ***1995***, *36*, 6055

Ph N Me MeO → BuLi, ether, 0°C LiBr, 5h → Ph Bu N H Me OMe

80% (98:2 *anti:syn*)

Hashimoto, Y.; Kobayashi, N.; Kai, A.; Saigo, K. *Synlett*, ***1995***, 961

I N Ar → Bu$_3$SnH, AIBN, PhH, 80°C → N Ar 71%

Ar = 4-chlorophenyl

Gioanola, M.; Leardini, R.; Nanni, D.; Pareschi, P.; Zanardi, G. *Tetrahedron*, ***1995***, *51*, 2039

Bn_2NH → [1. $(HCHO)_n$, $ZnCl_2$; 2. $ZnCl_2$, $NaBH_4$, CH_2Cl_2, rt] → Bn_2NMe 90%

Bhattacharyya, S. *Synth. Commun.*, ***1995***, *25*, 2061

NH_2 O_2N → [styrene (Ph), HCHO; MeCN, TFA, reflux] → H N O_2N Ph 90%

Mellor, J.M.; Merriman, G.D. *Tetrahedron*, ***1995***, *51*, 6115

Ph Ph N Ph → [1. K_2CO_3, KOH, BnBr, CH_2Cl_2 chiral ammonium ligand; 2. H_3O^+] → Ph Ph H_2N 70% (91% ee)

Eddine, J.J.; Cherqaoui, M. *Tetrahedron Asymmetry*, ***1995***, *6*, 1225

Ph N Me → [$Pd(dba)_2$, dppb; 20°C, 25h] → Ph N H Me quant.

Lemaire-Audoire, S.; Savignac, M.; Dupuis, C.; Genêt, J.-P. *Bull. Soc. Chim. Fr.*, ***1995***, *132*, 1157

SECTION 98: AMINES FROM ESTERS

OAc → [$NHEt_2$, $Ni(dppe)_2$, THF; 80°C] → NEt_2 no yield

Bricout, H.; Carpentier, J.-F.; Mortreux, A. *J. Chem. Soc. Chem. Commun.*, ***1995***, 1863

SECTION 99: AMINES FROM ETHERS, EPOXIDES AND THIOETHERS

OMe OMe → [[pyrrolidine N–H / BuLi]; THF, reflux, overnight] → OMe N 70%

ten Hoeve, W.; Kruse, C.G.; Luteyn, J.M.; Thiecke, J.R.G.; ten Hoeve, W.; Kruse, C.G.; Luteyn, J.M.; Thiecke, J.R.G.; Wynberg, H. *J. Org. Chem.*, ***1993***, *58*, 5101

SECTION 100: AMINES FROM HALIDES AND SULFONATES

Br

Ph_2N—N

H

SmI_2 , THF , HMPA

-42°C

NHNPh2 + NHNPh2

(7 : 1) 88%

Sturino, C.F.; Fallis, A.G. *J. Am. Chem. Soc.*, ***1994***, *116*, 7447

Br

Me

1. [Bu_3Sn—NEt_2 , MeNHBn]

5% $PdCl_2(Po\text{-}Tol_3)_2$

toluene, 105°C , 4h

2. HCl

3. MeOH

NMeBn

Me

79%

Guram, A.S.; Buchwald, S.L. *J. Am. Chem. Soc.*, ***1994***, *116*, 7901

Ph Br

iAmONO , Bu_6Sn_2 , hν (UV lamp)

4h

Ph N—OH 84%

Fletcher, R.J.; Kiril, M.; Murphy, J.A. *Tetrahedron Lett.*, ***1995***, *36*, 323

Bu Br

P((*o*-Tol)3 , $LiNTMS_2$, 2h

piperidine , 100°C

Bu N(TMS)2

84%

Louie, J.; Hartwig, J.F. *Tetrahedron Lett.*, ***1995***, *36*, 3689

C_3H_7

OMs

$BuNH_2$, DMSO , 80°C

2h

C_3H_7

NHBu

46%

Hénin, F.; Mahuet, E.; Muller, C.; Muzart, J. *Synth. Commun.*, ***1995***, *25*, 1331

Ph Br

NHMePh , $Pd(dba)_2$/2 P(*o*-Tol)3

65°C , toluene

Ph NMePh

88%

Guram, A.S.; Rennels, R.A.; Buchwald, S.L. *Angew. Chem. Int. Ed. Engl.*, ***1995***, *34*, 1348

SECTION 101: AMINES FROM HYDRIDES

OMe

1. 65°C , 24h

$Cl_3CH_2CO_2C\text{-}N{=}N\text{-}CO_2CH_2CCl_3$

2. Zn , AcOH

OMe

H_2N

99 x 71%

Zaltsgendler, I.; Leblanc, Y.; Bernstein, M.A. *Tetrahedron Lett.*, ***1993***, *34*, 2441

SECTION 102: AMINES FROM KETONES

N

O

$=NMe_2$ ClO_4

DMF , NH_4OAc

N

N

N

53%

Westerwelle, U.; Risch, N. *Tetrahedron Lett.*, ***1993***, *34*, 1775

O

Ph Ph

1. LDA , nitroethylene , $ZnCl_2$, THF , -30°C

2. Ni(R) , EtOH , H_2 (50 psi)

3. $NaBH_3CN$, AcOH , EtOH , 0°C

Ph

Ph

N H

77x95x100%

Pal, K.; Behnke, M.L.; Tong, L. *Tetrahedron Lett.*, ***1993***, *34*, 6205

O

H^+ , $Si(OEt)_4$, H_2NCHPh_2

N

Ph

Ph

72%

Love, B.E.; Ren, J. *J. Org. Chem.*, ***1993***, *58*, 5556

Ph

O

NO_2

1,10-phenanthroline , CO

$Ru_3(CO)_{12}$, p-dioxane

140°C , 16h

Ph N

91%

Watanabe, Y.; Yamamoto, J.; Akazome, M.; Kondo, T.; Mitsudo, T. *J. Org. Chem.*, ***1995***, *60*, 8328

Related Methods: Section 94 (Amines from Aldehydes)

SECTION 103: AMINES FROM NITRILES

$PhMgCl$, CH_2Cl_2
5h, -78°C → rt

N—OH, Ph

96%

Kim, J.N.; Kim, H.R.; Ryu, E.K. *Tetrahedron Lett.*, ***1993***, *34*, 5117

$Cr(CO)_5$, OMe, Ph

t-BuCN, *t*-BuOMe
0°C (1h), reflux (2d)

OMe, NH*t*-Bu

71%

Merlic, C.A.; Burns, E.E. *Tetrahedron Lett.*, ***1993***, *34*, 5401

Br, CN

BuMgCl, PhH/ether, rt

N, Bu

84%

Fry, D.F.; Fowler, C.B.; Dieter, R.K. *Synlett*, ***1994***, 836

Ph—CN

2 eq. HC≡CH, hν, H_2O, 3h
[$Co(\eta^5-C_5H_5)(cod)$]

N, Ph

41%

Heller, B.; Oehme, G. *J. Chem. Soc. Chem. Commun.*, ***1995***, 179

SECTION 104: AMINES FROM ALKENES

Ph

$PhNH_2$, PhNHLi, 12d
$[RhCl(PEt_3)_2]_2$, 70°C

Ph, N-Ph

65% (+ 30%N-phenyl phenethylamine
and 5% N-phenyl-2-phenylethane)

Brunet, J.-J.; Neibecker, D.; Philippot, K. *Tetrahedron Lett.*, ***1993***, *34*, 3877

$NsONHCO_2Et$, CH_2Cl_2
K_2CO_3 , 1h , rt

62%

EtO_2C

Fioravanti, S.; Loreto, M.A.; Pellacani, L.; Tardella, P.A. *Tetrahedron Lett.*, ***1993***, *34*, 4353

Ph

1. O_3 , CH_2Cl_2 , -78°C
2. 2 eq. N—H
4Å MS , CH_2Cl_2 , reflux

Ph N

85%

Hon, Y.-S.; Lu, L. *Tetrahedron Lett.*, ***1993***, *34*, 5309

Ph

PhNHOH , Fe(Pc)

Pc = phthalocyanine

NHPh

Ph

76%

+ 22% $PhNH_2$

Johannsen, M.; Jørgensen, K.A. *J. Org. Chem.*, ***1994***, *59*, 214

Bu

$HZr(Cl)Cp_2$:$RONH_2$

THF

Bu NH_2

R = $Me_3C_6H_2OSO_2$-

77%

Zheng, B.; Srebnik, M. *J. Org. Chem.*, ***1995***, *60*, 1912

SECTION 105: AMINES FROM MISCELLANEOUS COMPOUNDS

N O N Ph

1. BuLi , BF_3•OEt_2 , toluene
2, $LiAlH_4$

NH_2

Bu

70% (60% de)

Dieter, R.K.; Datar, R. *Can. J. Chem.*, ***1993***, *71*, 814

O P Ph Ph Ph

$(EtO)_3SiH$, $Ti(OiPr)_4$

67°C

P Ph Ph Ph

85%

Coumbe, T.; Lawrence, N.J.; Muhammad, F. *Tetrahedron Lett.*, ***1994***, *35*, 625

1. Bu_3SnH, AIBN, MeCN, 100°C
2. PhBr, NEt_3, 100°C, 5% $Pd(PPh_3)_4$

Fukuyama, T.; Chen, X.; Peng, G. *J. Am. Chem. Soc.*, ***1994***, *116*, 3127

toluene, 75°C, 14h

Ar = 3,4-dimethoxyphenyl 76%

Rodriques, J.A.R.; Leiva, G.C.; de Sousa, J.D.F. *Tetrahedron Lett.*, ***1995***, *36*, 59

$(BnNEt_3)_2MoS_4$, MeCN, rt 86%

Ilankumaran, P.; Chandrasekaran, S. *Tetrahedron Lett.*, ***1995***, *36*, 4881

$C_8H_{17}CH_2N=C=O$ → $C_8H_{17}CH_2NH_2$ quant.

1. $Hg(OAc)_2$, aq. THF
2. $NaBH_4$, OH^-

Malanga, C.; Urso, A.; Lardicci, L. *Tetrahedron Lett.*, ***1995***, *36*, 8859

Et_3SiH, $Ni(OAc)_2 \cdot 4\ H_2O$ 95%

Vetter, A.H.; Berkessel, A. *Synthesis*, ***1995***, 419

10% $Yb(OTf)_3$, MeCN 74%

Makioka, Y.; Shindo, T.; Taniguchi, Y.; Takaki, K.; Fujiwara, Y. *Synthesis*, ***1995***, 801

$$\text{PhCH=N-OH} \xrightarrow[\text{BER = borohydride exchange resin}]{\text{BER-Ni(OAc)}_2\text{, MeOH, 75°C}} \text{PhCH}_2\text{NH}_2 \quad 81\%$$

Bandgar, B.P.; Nikat, S.M.; Wadgaonkar, P.P. *Synth. Commun.*, ***1995***, *25*, 863

$$C_6H_{13}(Me)C\text{=N-OH} \xrightarrow[2.\ NaBH_4,\ I_2]{1.\ Ac_2O} C_6H_{13}(Me)CH\text{-NH}_2 \quad 86\%$$

Barbry, D.; Champagne, P. *Synth. Commun.*, ***1995***, *25*, 3503

$$\text{cyclohexylidene=N-Ph} \xrightarrow{Cp_2MoH_2\text{, toluene}} \text{cyclohexyl-NH-Ph} \quad 89\%$$

Minato, M.; Fujiwara, Y.; Ito, T. *Chem. Lett.*, ***1995***, 647

$$Ph(Me)C\text{=N-Bn} \xrightarrow[\text{MeOH}]{BnNH_2,\ H_2,\ Ir(R\text{-BINAP})} Ph(Me)CH\text{-N-Bn}$$

quant. (66% ee , S)

Tani, K.; Onouchi, J.; Yamagata, T.; Kataoka, Y. *Chem. Lett.*, ***1995***, 955

AMINES FROM AZIDES

$$Bu(CH_2)_6N_3 \xrightarrow[\text{rt, 30 min}]{\text{BER-Ni(OAc)}_2\text{, MeOH}} Bu(CH_2)_6NH_2 \quad 96\%$$

BER = borohydride exchange resin

Yoon, N.M.; Choi, J.; Shon, Y.S. *Synth. Commun.*, ***1993***, *23*, 3047

$$BuN_3 \xrightarrow{NaBH_4\text{, cat. }CuSO_4\text{, MeOH}} BuNH_2 \quad 95\%$$

Rao, H.S.P.; Siva, P. *Synth. Commun.*, ***1994***, *24*, 549

$$\text{2-azidobiphenyl} \xrightarrow{SmI_2\text{, THF}} \text{2-aminobiphenyl } (H_2N) \quad 96\%$$

Benati, L.; Montevecchi, P.C.; Nanni, D.; Spagnolo, P.; Volta, M. *Tetrahedron Lett.*, ***1995***, *36*, 7313

$LiMe_2NBH_3$, THF , 0°C — cyclopentyl azide → cyclopentylamine 90%

Alvarez, S.G.; Fisher, G.B.; Singaram, B. *Tetrahedron Lett.*, ***1995***, *36*, 2567

$C_{10}H_{21}$—N_3 → (2.5 SmI_2 , THF , rt , 1h) → $C_{10}H_{21}$—NH_2 85%

Goulaouic-Dubois, C.; Hesse, M. *Tetrahedron Lett.*, ***1995***, *36*, 7427

C_5H_{11} N_3 → ($BHCl_2$•SMe_2 , CH_2Cl_2) → C_5H_{11} NH_2 95%

Salunkhe, A.M.; Brown, H.C. *Tetrahedron Lett.*, ***1995***, *36*, 7987

AMINES FROM NITRO COMPOUNDS

Ph, N, NO_2 → (1. 7 eq. SmI_2 , THF/MeOH rt , 1d; 2. oxalic acid/water) → Ph, N, NH_2

95%
(6:1 *trans:cis*)

Sturgess, M.A.; Yarberry, D.J. *Tetrahedron Lett.*, ***1993***, *34*, 4743

NO_2, O, O → (hν , TiO_2; EtOH) → NH_2, O, O 79%

Mahdavi, F.; Bruton, T.C.; Li, Y. *J. Org. Chem.*, ***1993***, *58*, 744

NO_2 → (borohydride exchange resin; $Ni(OAc)_2$, MeOH , rt , 30 min) → NH_2 94%

Yoon, N.M.; Choi, J. *Synlett*, ***1993***, 135

NO_2, Br → (Baker's yeast , NaOH aq. MeOH , 80°C , 4h) → NH_2, Br

Baik, W.; Han, J.L.; Lee, K.C.; Lee, N.H.; Kim, B.H.; Hahn, J.-T. *Tetrahedron Lett.*, ***1994***, *35*, 3965

$$PhNO_2 \xrightarrow[\text{2. MeOH , 10°C}]{\text{1. CuBr•SMe}_2\text{ , NaBH}_4\text{ , THF}} PhNH_2 \quad 81\%$$

Patel, H.V.; Vyas, K.A. *Org. Prep. Proceed. Int.*, ***1995***, *27*, 81

$$C_6H_5\text{-}NO_2 \xrightarrow{\text{H}_2\text{N-NMe}_2\text{ , FeCl}_3\text{ , MeOH , 18h}} C_6H_5\text{-}NH_2 \quad 89\%$$

Boothroud, S.R.; Kerr, M.A. *Tetrahedron Lett.*, ***1995***, *36*, 2411

$$PhNO_2 \xrightarrow{\text{NaBH}_4\text{-BiCl}_3\text{/EtOH , 2h}} PhNH_2 \quad 86\%$$

Ren, P.-D.; Pan, S.-F.; Dong, T.-W.; Wu, S.H. *Synth. Commun.*, ***1995***, *25*, 3799

$$Cl\text{-}C_6H_4\text{-}NO_2 \xrightarrow[\text{EtOH , 30 min}]{\text{NaBH}_4\text{-(NH}_4\text{)}_2\text{SO}_4} Cl\text{-}C_6H_4\text{-}NH_2 \quad 90\%$$

Gohain, S.; Prajapati, D.; Sandhu, J.S. *Chem. Lett.*, ***1995***, 725

REVIEWS:

"Asymmetric Reductions of C-N Double Bonds. A Review," Zhu, Q.-C.; Hutchins, R.O. *Org. Prep. Proceed. Int.*, ***1994***, *26*, 193

"Reduction of Nitro-Substituted Tertiary Alkanes," Weis, C.D.; Newkome, G.R. *Synthesis*, ***1995***, 1053

SECTION 105A: PROTECTION OF AMINES

(FMoc)NH–CH(CH(Ot-Bu)CH₃)–C(=O)–NH–CH(CO₂Me)–CH(CH₃)CH₂CH₃ → (t-Boc)NH–CH(CH(Ot-Bu)CH₃)–C(=O)–NH–CH(CO₂Me)–CH(CH₃)CH₂CH₃

KF , NEt_3 , DMF / Boc_2O , rt , 10h — 85%

Li, W-R.; Jiang, J.; Joullié, M.M. *Tetrahedron Lett.*, ***1993***, *34*, 1413

$$PhCH_2NH\text{-}C(=O)\text{-}OCH_2CH{=}CH_2 \xrightarrow[\text{aq. MeCN , 9 min}]{\text{Pd(OAc)}_2\text{ + TPPTS , Et}_2\text{NH}} PhCH_2NH_2 \quad 72\%$$

also with carbonates to give alcohols

Genêt, J.P.; Blart, E.; Savignac, M.; Lemeune, S.; Paris, J.-M. *Tetrahedron Lett.*, ***1993***, *34*, 4189

Ph O O N O N Cl CO_2H

1. TMSI, HMDS, MeCN
2. DABCO
3. HCl, H_2O

^+H_3N O N Cl CO_2^- 70%

Fisher, J.W.; Dunigan, J.M.; Hatfield, L.D.; Hoying, R.C.; Ray, J.E.; Thomas, K.L. *Tetrahedron Lett.*, ***1993***, *34*, 4755

Ph N → Ph NH_2

$Pd(PPh_3)_4$, NDMBA, 30°C
CH_2Cl_2, 1.5h

quant

NDMBA = N,N-dimethylbarbituric acid

Garro-Helion, F.; Merzouk, A.; Guibé, F. *J. Org. Chem.*, ***1993***, *58*, 6109

MeO_2C CO_2t-Bu NHBoc → MeO_2C CO_2t-Bu $NH_3^+ Cl^-$

1M HCl, EtOAc
rt, 5h

quant.

Gibson, F.S.; Bergmeier, S.C.; Rapoport, H. *J. Org. Chem.*, ***1994***, *59*, 3216

$Bn_2N—SO_2Ph$ → $Bn_2N—H$ 92%

SmI_2, THF, DMPU, reflux
4.5h

Vedejs, E.; Lin, S. *J. Org. Chem.*, ***1994***, *59*, 1602

Ph N Me → Ph N(H) Me

$Pd(dba)_2$, dppb, 20°C, 25 min
$2\text{-}(SH)C_6H_4COOH$, THF

quant.

Lemaire-Audoire, S.; Savignac, M.; Genêt, J.P.; Bernard, J.-M. *Tetrahedron Lett.*, ***1995***, *36*, 1267

MeHN NH_2 → MeHN NHTFA

CF_3CO_2Et, THF, 0°C

selective for 1° amines

>95%

Xu, D.; Prasad, K.; Repic, O.; Blacklock, T.J. *Tetrahedron Lett.*, ***1995***, *36*, 7357

1. $ClC(O)OCH(Cl)CH_3$, DCE , reflux
2. MeOH , reflux

N-Bn 2-Me tetrahydropyridine → N-H 2-Me tetrahydropyridine 74%

Yang, B.V.; O'Rourke, D.; Li, J. *Synlett*, ***1993***, 195

Me_2NBn —($TiCl_3$-Li , THF , reflux , 22h)→ Me_2NH 55%

Talukdar, S.; Banerji, A. *Synth. Commun.*, ***1995***, *25*, 813

CHAPTER 8

PREPARATION OF ESTERS

SECTION 106: ESTERS FROM ALKYNES

xylenes , 150°C

72%

Liang, L.; Ramaseshan, M.; MaGee, D.I. *Tetrahedron*, ***1993***, *49*, 2159

SECTION 107: ESTERS FROM ACID DERIVATIVES

The following types of reactions are found in this section:

1. Esters from the reaction of alcohols with carboxylic acids, acid halides and anhydrides.
2. Lactones from hydroxy acids
3. Esters from carboxylic acids and halides, sulfoxides and miscellaneous compounds

Li [Binol-AlHOEt] , THF

-78°C → rt

69%

(84% ee)

Matsuki, K.; Inoue, H.; Takeda, M. *Tetrahedron Lett.*, ***1993***, *34*, 1167

$PhCH_2OH$, 2,6-lutidine

, CH_2Cl_2

-78°C → rt

CO_2H → CO_2CH_2Ph

81%

Folmer, J.J.; Weinreb, S.M. *Tetrahedron Lett.*, ***1993***, *34*, 2737

N_2; H, $Si(iPr)_3$; 90°C → 93%

Soderquist, J.A.; Miranda, E.I. *Tetrahedron Lett.*, ***1993***, *34*, 4905

CO_2H → CO_2Et; 3 eq. TEOA, 24h; 110°C, Tol; 92%

TEOA = triethyl orthoacetate

Trujillo, J.I.; Gopalan, A.S. *Tetrahedron Lett.*, ***1993***, *34*, 7355

$PhCO_2H$ → ($C_8H_{17}OH$, TsOH, neat; microwave, 3 min) → $PhCO_2C_8H_{17}$ 97%

Loupy, A.; Petit, A.; Ramdani, M.; Yvanaeff, C.; Majoub, M.; Labiad, B.; Villemin, D. *Can. J. Chem.*, ***1993***, *71*, 90

Ph CO_2H → (1. NBS CH_2Cl_2, MS 4Å 0°C, PPh_3; 2. $PhCH_2OH$, Py, 1h) → Ph, O, Ph, O 98%

Sucheta, K.; Reddy, G.S.R.; Ravi, D.; Ramo Rao, N. *Tetrahedron Lett.*, ***1994***, *35*, 4415

Ph CO_2H → (Ph, OH; 20% $TiCl_4$ + 2 $AgClO_4$; $(ArCO)_2O$, MS3Å) → O, O, Ph, Ph 90%

Ar = 3,5-trifluoromethylphenyl

Shiina, I.; Miyoshi, So.; Miyashita, M.; Mukaiyama, T. *Chem. Lett.*, ***1994***, 515

HO $(CH_2)_{10}$ OH, O → (5% $TiCL_2(OTf)_2$, $(ArCO)_2O$; 3 TMSCl, CH_2Cl_2 (2 mM); reflux (slow addition over 5 h)) → O, O, $(CH_2)_{10}$

83% + 10% diolide

Ar = 4-trifluoromethylphenyl

Shiina, I.; Mukaiyama, T. *Chem. Lett.*, ***1994***, 677

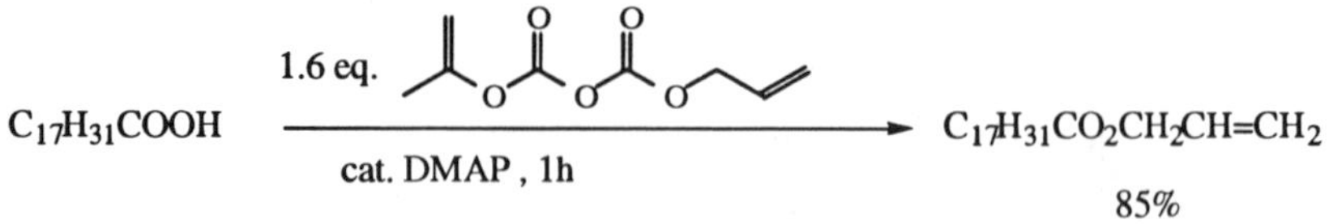

Takeda, K.; Akiyama, A.; Konda, Y.; Takayanagi, H.; Harigaya, Y. *Tetrahedron Lett.*, ***1995***, *36*, 113

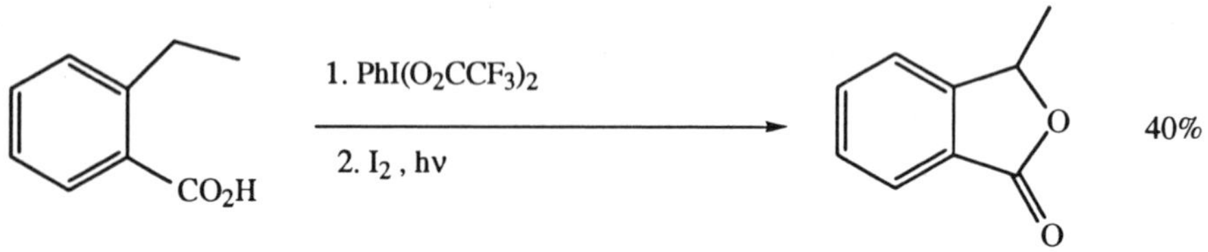

Togo, H.; Muraki, T.; Yokoyama, M. *Tetrahedron Lett.*, ***1995***, *36*, 7089

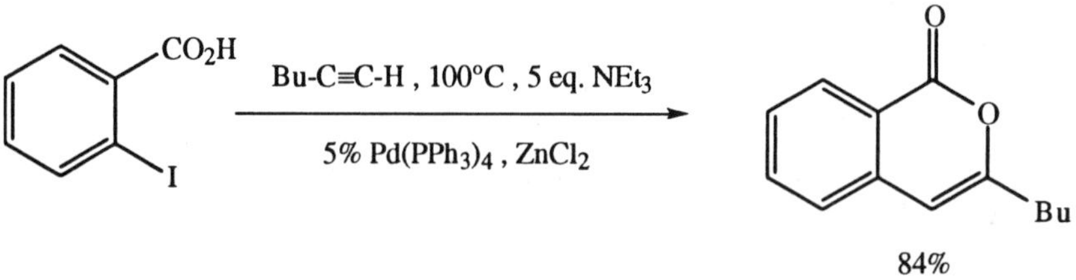

Liao, H.-Y.; Cheng, E.H. *J. Org. Chem.*, ***1995***, *60*, 3711

Further examples of the reaction RCO_2H + R'OH → RCO_2R' are included in Section 108 (Esters from Alcohols and Phenols) and in Section 30A (Protection of Carboxylic Acids).

SECTION 108: ESTERS FROM ALCOHOLS AND THIOLS

Further examples of the reaction ROH → RCO_2R' are included in Section 107 (Esters from Acid Derivatives) and in Section 45A (Protection of Alcohols and Phenols).

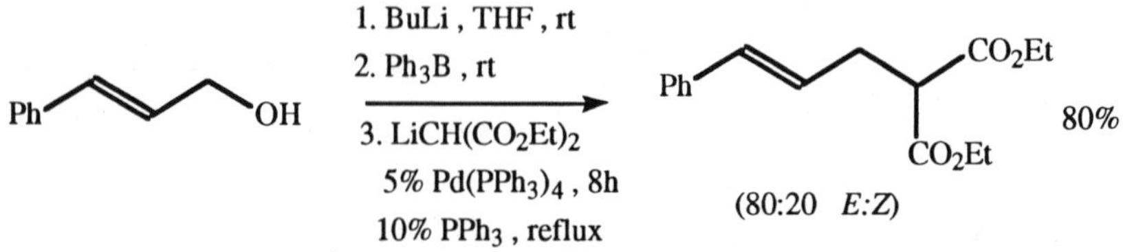

Starý, I,; Stará, I.G.; Kočovský, P. *Tetrahedron Lett.*, ***1993***, *34*, 179

3.5 $Ca(OCl)_2$, 2 AcOH; 32 MeOH , dark , 2d; MeCN , rt — CO_2Me 89%

McDonald, C.E.; Nice, L.E.; Shaw, A.W.; Nestor, N.B. *Tetrahedron Lett.*, ***1993***, *34*, 2741

$C_{13}H_{27}CH_2OH$ → 1. $SOCl_2$, DMF; 2. H_2O → $C_{13}H_{27}CH_2OCHO$ 95%

Fernández, I.; Garcia, B.; Muñoz, S.; Pedro, J.R.; de la Salud, R. *Synlett*, ***1993***, 489

OH → KNO_3/BF_3•1.25 H_2O; CH_2Cl_2 , 0°C → rt → ONO_2 95%

Olah, G.A.; Wang, Q.; Li, X.; Prakash, G.K.S. *Synthesis*, ***1993***, 207

OH → 1. $Cr(CO)_5$•THF , THF , 55°C; 2. CAN → O O 81%

Quayle, P.; Rahman, S.; Ward, E.L.M. *Tetrahedron Lett.*, ***1994***, *35*, 3801

C_3H_7 OH → $Pb(OAc)_4$, PhH , CO; 40°C , 1d → C_3H_7 O O 51%

Tsunoi, S.; Ryu, I.; Sonoda, N. *J. Am. Chem. Soc.*, ***1994***, *116*, 5473

HO OH → $NaBrO_2$, Al_2O_3 , CH_2Cl_2 → O O 57%

Morimoto, T.; Hirano, M.; Iwasaki, K.; Ishikawa, T. *Chem. Lett.*, ***1994***, 53

OH → $MeC(OEt)_3$, $EtCO_2H$, DMF; microwave , 12 min → CO_2Et 63%

Srikrishna, A.; Nagaraju, S.; Kondaiah, P. *Tetrahedron*, ***1995***, *51*, 1809

OH

TMAD , DEAD , Bu_3P , PhH
4-$(MeO)C_6H_4COOH$, 60°C , 1d

O_2CAr

TMAD = $Me_2NCON{=}NCONMe_2$
A Mitsunobu reaction for sterically hindered alcohols 98%

Tsudoda, T.; Yamamiya, Y.; Kawamura, Y.; Itô, S. *Tetrahedron Lett.*, ***1995***, *36*, 2529

OH

OH

3 eq. NCS , CH_2Cl_2 , 20°C , 5h

O

O

88%

Kondo, S.; Kawasoe, S.; Kunisada, H.; Yuki, Y. *Synth. Commun.*, ***1995***, *25*, 719

C_3H_7 OH

$NaBrO_3$, $NaHSO_3$, H_2O

rt , 2h

O

O C_3H_7

76%

Takase, K.; Masuda, H.; Kai, O.; Nishiyama, Y.; Sakaguchi, S.; Ishii, Y. *Chem. Lett.*, ***1995***, 871

SECTION 109: ESTERS FROM ALDEHYDES

Bu CHO

1. Bu_3SnLi
2. MOMCl , iPr_2NEt
3. O_3 , CH_2Cl_2, -78°C
(quant.)

O

Bu O O

Linderman, R.J.; Jaber, M. *Tetrahedron Lett.*, ***1994***, *35*, 5993

Et CHO

O N

OMe

mcpba , CH_2Cl_2 , rt , 2.5h

Et OCHO

O N

OMe

94%

Alcaide, B.; Aly, M.F.; Sierra, M.A. *Tetrahedron Lett.*, ***1995***, *36*, 3401

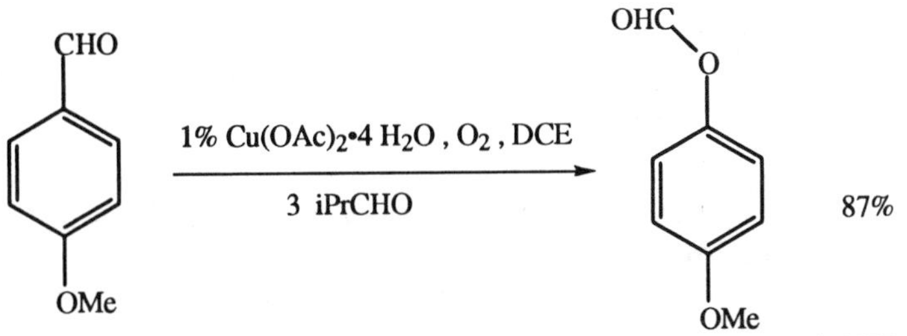

CHO / OMe → (1% $Cu(OAc)_2 \cdot 4\ H_2O$, O_2 , DCE; 3 iPrCHO) → OHC–O / OMe 87%

Anoune, N.; Lantéri, P.; Longeray, R.; Arnaud, C. *Tetrahedron Lett.*, ***1995***, *36*, 6679

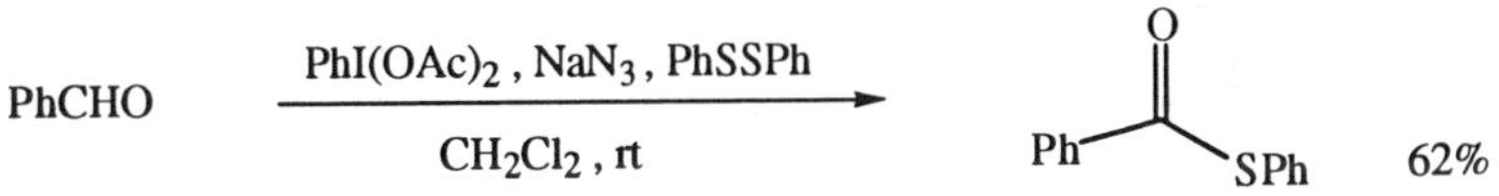

PhCHO → ($PhI(OAc)_2$, NaN_3 , PhSSPh; CH_2Cl_2 , rt) → PhC(O)SPh 62%

Tingoli, M.; Temperini, A.; Testaferri, L.; Tiecco, M. *Synlett*, ***1995***, 1129

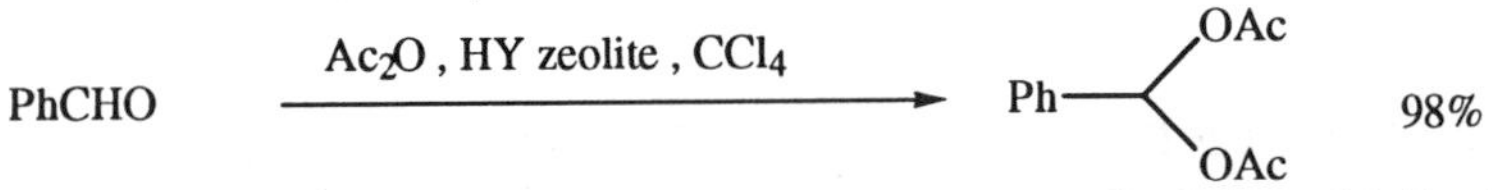

PhCHO → (Ac_2O , HY zeolite , CCl_4) → PhCH(OAc)OAc 98%

Pereira, C.; Gigante, B.; Marcelo-Curto, M.J.; Carreyre, H.; Pérot, G.; Guisnet, M. *Synthesis*, ***1995***, 1077

Related Methods: Section 117 (Esters from Ketones)

SECTION 110: ESTERS FROM ALKYLS, METHYLENES AND ARYLS

No examples of the reaction R-R → RCO_2R' or $R'CO_2R$ (R,R' = alkyl, aryl, etc.) occur in the literature. For the reaction R-H → RCO_2R' or $R'CO_2R$, see Section 116 (Esters from Hydrides).

NO ADDITIONAL EXAMPLES

SECTION 111: ESTERS FROM AMIDES

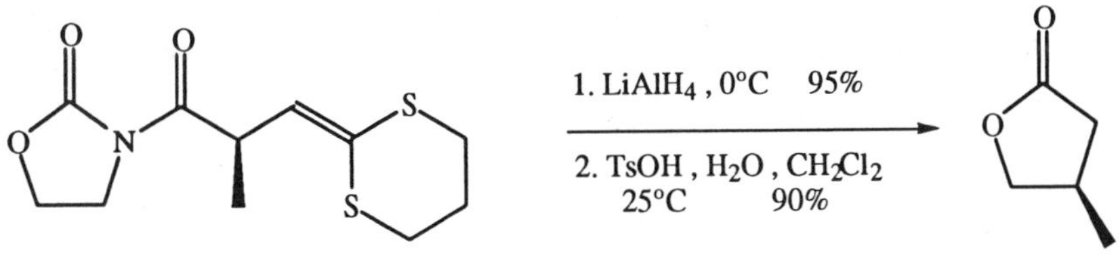

(1. $LiAlH_4$, 0°C 95%; 2. TsOH , H_2O , CH_2Cl_2 25°C 90%)

Canan Koch, S.S.; Chamberlin, A.R. *J. Org. Chem.*, ***1993***, *58*, 2725

Ph–CH$_2$–C(=O)–NHOMe —[$TiCl_4$, MeOH , 3h]→ Ph–CH$_2$–C(=O)–OMe

86%

Fisher, L.E.; Caroon, J.M.; Stabler, S.R.; Lundberg, S.; Zaidi, S.; Sorensen, C.M.; Sparacino, M.L.; Muchowski, J.M. *Can. J. Chem.*, ***1994***, *72*, 142

N—Ph (hexahydrophthalimide) —[1. BDMAB , Zn complex , -10°C; 2. $NaBH_4$]→ lactone

69% (84% ee , 1R2S)

Kang, J.; Lee, J.W.; Kim, J.I.; Pyun, C. *Tetrahedron Lett.*, ***1995***, *36*, 4265

SECTION 112: ESTERS FROM AMINES

NO ADDITIONAL EXAMPLES

SECTION 113: ESTERS FROM ESTERS

Conjugate reductions and conjugate alkylations of unsaturated esters are found in Section 74 (Alkyls from Alkenes).

Me_2SiClH —[1. N_2CHCO_2Et , CH_2Cl_2 , $Rh_2(OAc)_4$; 2. iPrOH , NEt_3]→ (iPrO)$Me_2SiCH_2CO_2Et$ 74%

Andrey, O.; Landais, Y.; Panchenault, D. *Tetrahedron Lett.*, ***1993***, *34*, 2927

Me_3SiO–CH$_2$–$(CH_2)_{10}$–C(=O)–$OSiMe_3$ —[10% $TiCl_4$, 2 $AgClO_4$, CH_2Cl_2; $(ArCO_2)_2O$, rt , 3 hr addition]→ lactone $(CH_2)_{10}$

Ar - 4-trifluoromethylphenyl

75%

Mukaiyama, T.; Izumi, J.; Miyashita, M.; Shiina, I. *Chem. Lett.*, ***1993***, 907

Ph–CH$_2$–C(=O)–OMe —[1. LDA , Me_3siCl; 2. H_2Se]→ Ph–CH$_2$–C(=Se)–OMe 74%

Wright, S.W. *Tetrahedron Lett.*, ***1994***, *35*, 1331

$(H_2C)_{15}$ OH, OCH_2CF_3, O — 0.1 eq. Bu_3SnOMe , decalin , heat → $(H_2C)_{15}$ lactone — 81% + 14% diolide

[with BuSnH/AIBN → 70% lactone + 0% diolide]

White, J.D.; Green, N.J.; Fleming, F.F. *Tetrahedron Lett.*, ***1993***, *34*, 3515

OAc, Ph, Ph — 2.5% $[Pd(\eta^3\text{-}C_3H_5)Cl]_2$, CH_2Cl_2 , 36h , $CH_2(CO_2Me)_2$, BSA , AcOK , rt, 6% (ligand) → MeO_2C, CO_2Me, Ph, Ph — 84% (89% ee,S)

BSA = N,O-*bis*-(trimethylsilyl)acetamide

Kubota, H.; Nakajima, M.; Koga, K. *Tetrahedron Lett.*, ***1993***, *34*, 8135

CO_2Et — TMSCl , NaI , MeCN → O, O — 82%

Piva, O. *Tetrahedron*, ***1994***, *50*, 13687

I—$(H_2C)_{10}$—O—C(=O)—CH=CH$_2$ — hν , $NaBH_3CN$, MeOH, 3h , rt → O, O, $(CH_2)_{10}$ — 83x74%

Abe, M.; Hayashikoshi, T.; Kurata, T. *Chem. Lett.*, ***1994***, 1789

N_2, O, O — $[Rh_2(OAc)_4]$ → O, O — 46% (40:60 *cis:trans*)

Müller, P.; Polleux, P. *Helv. Chim. Acta*, ***1994***, *77*, 645

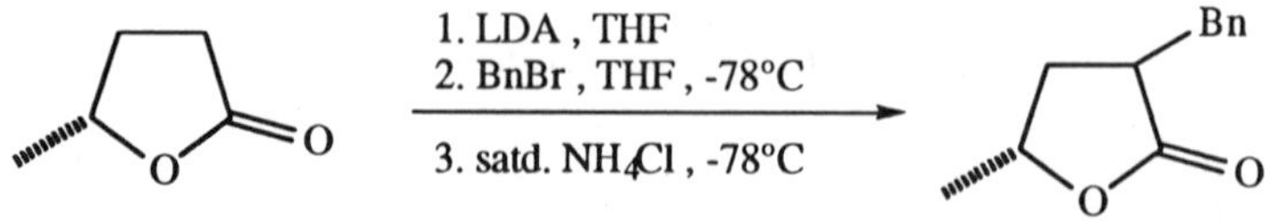
1. LDA , THF
2. BnBr , THF , -78°C
3. satd. NH_4Cl , -78°C
Bn
O
O

83% (10:1 *anti:syn*)

Anceau, C.; Dauphin, G.; Coudert, G.; Guillaumet, G. *Bull. Soc. Chim. Fr.*, ***1994***, *131*, 291

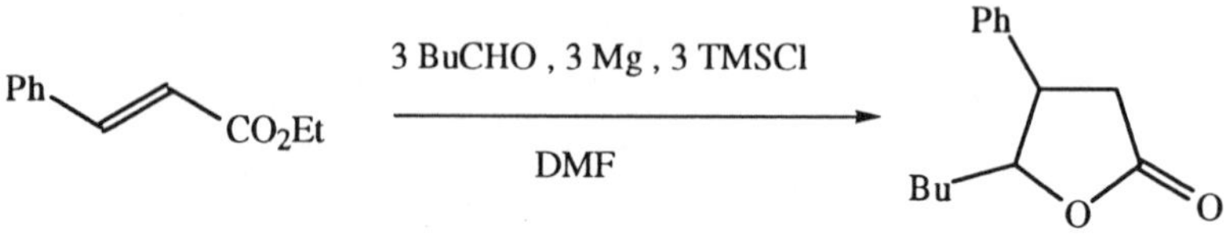
Ph
CO_2Et
3 BuCHO , 3 Mg , 3 TMSCl
DMF
Ph
Bu
O
O

74%

Ohno, T.; Ishino, Y.; Tsumagari, Y.; Nishiguchi, I. *J. Org. Chem.*, ***1995***, *60*, 458

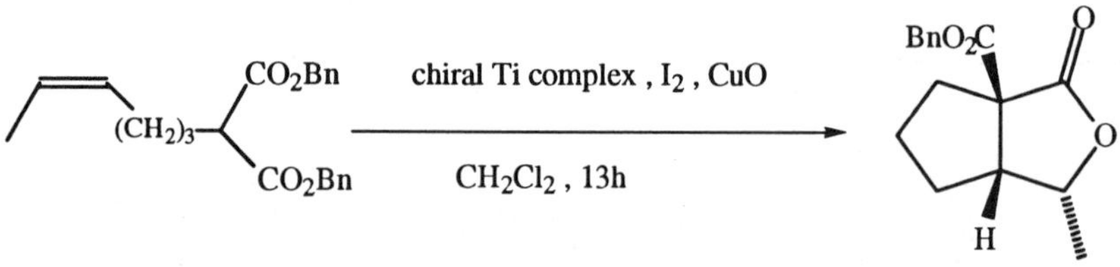
CO_2Bn
$(CH_2)_3$
CO_2Bn
chiral Ti complex , I_2 , CuO
CH_2Cl_2 , 13h
BnO_2C
O
O
H

quant
(76% ee)

Inoue, T.; Kitagawa, O.; Kurumizawa, S.; Ochiai, O.; Taguchi, T. *Tetrahedron Lett.*, ***1995***, *36*, 1479

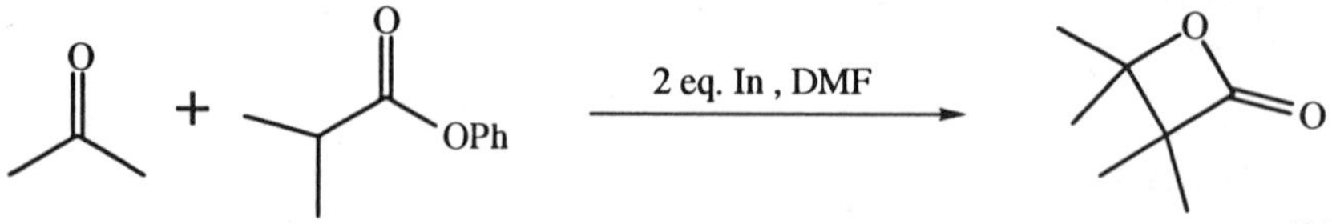
O
+
O
OPh
2 eq. In , DMF
O
O

62%

Schick, H.; Ludwig, R.; Kleiner, K.; Kunath, A. *Tetrahedron*, ***1995***, *51*, 2939

SECTION 114: ESTERS FROM ETHERS, EPOXIDES AND THIOETHERS

Ph
OTMS
1.5 AcBr , 5% $SnBr_2$
Ph
OAc

9%

Oriyama, T.; Oda, M.; Gono, J.; Koga, G. *Tetrahedron Lett.*, ***1994***, *35*, 2027

TS-1 , H_2O_2

TS-1 = titanium silicate

55%

Sasidharan, M.; Suresh, S.; Sudalai, A. *Tetrahedron Lett.*, ***1995***, *36*, 9071

$PhCH_2CH_2OCH_2Ph$ → ($SnBr_2$-AcBr , CH_2Cl_2; rt , 1h) → $PhCH_2CH_2OAc$ 92%

Oriyama, T.; Kimura, M.; Oda, M.; Koga, G. *Synlett*, ***1993***, 437

NdBr$_3$, AcOH , microwave; 1.7 min

OBu Ph → OAc Ph 82%

Yulin, J.; Yuncheng, Y. *Synth. Commun.*, ***1994***, *24*, 1045

$PdCl_2$-$CuCl_2$, O_2 , acetone; CO (60 atm) , 2d

16%

Miyamoto, M.; Minami, Y.; Ukaji, Y.; Kinoshita, H.; Inomata, K. *Chem. Lett.*, ***1994***, 1149

67 atm CO , 180°C , PhH , 2d; $[Rh(cod)Cl]_2$

OEt → CO_2Et 73%

Khumtaveeporn, K.; Alper, H. *J. Chem. Soc. Chem. Commun.*, ***1995***, 917

SECTION 115: ESTERS FROM HALIDES AND SULFONATES

BuBr → (1. $(MeS)_3CLi$, THF , -78°C; 2. 35% aq. HBF_4 , THF , 70°C) → BuCOSMe 91x84%

Barbero, M.; Cadamuro, S.; Degani, I.; Dughera, S.; Fochi, R. *J. Chem. Soc., Perkin Trans. 1.*, ***1993***, 2075

$BrZnCH_2CO_2Et$; CH_2Cl_2 , 0°C → 20°C

Cl Ph → CO_2Et Ph 69%

Bott, K. *Tetrahedron Lett.*, ***1994***, *35*, 555

cat. Bu_4NNO_3 , $NaNO_3$, PhH/H_2O; sealed tube (135°C)

OTs Ph → ONO_2 Ph 81%

Hwu, J.R.; Yyas, K.A.; Patel, H.V.; Lin, C.-H.; Yang, J.-C. *Synthesis*, ***1994***, 471

Br

ArOK , CO , $PdCl_2$(dppp) , PhH

100°C

Ar = 2,5-di-*t*-butyl-4-methylphenyl

84%

Kubota, Y.; Hanaoka, T.-a.; Takeuchi, K.; Sugi, Y. *J. Chem. Soc. Chem. Commun.*, ***1994***, 1553

OMs

$PhCH_2H$, CsF , DMF , 60°C

17h

C_6H_{13} Me

OBz

50% (100% ee)

Sato, T.; Otera, J. *Synlett*, ***1995***, 336

Related Methods: Section 25 (Acid Derivatives from Halides).

SECTION 116: ESTERS FROM HYDRIDES

This section contains examples of the reaction R-H → RCO_2R' or $R'CO_2R$ (R = alkyl, aryl, etc.).

5% $Pd(OAc)_2$, 1.5 H_2O_2

10% O= =O , AcOH 2h , 50°C

OAc

77%

Åkermark, B.; Larsson, E.M.; Oslob, J.D. *J. Org. Chem.*, ***1994***, *59*, 5729

$PhCO_3t$-Bu , 5% CuOTf , 23°C MeCN , 4d

8%

O Ph O

57% conversion
74% (74% ee)

Gokhale, A.S.; Minidis, A.B.E.; Pfaltz, A. *Tetrahedron Lett.*, ***1995***, *36*, 1831

$RuCl_3 \bullet n\ H_2O$, rt

2 eq. $MeCO_3H$
TFA:CH_2Cl_2

O_2CCF_3 + =O

69% 12%

Murahashi, S.; Oda, Y.; Komiya, N.; Naota, T. *Tetrahedron Lett.*, ***1994***, *35*, 7953

5% chiral bis-oxazoline Cu complex
$PhCO_2t$-Bu , MeCN , -20°C

OBz

49 % (81% ee)

Andrus, M.B.; Argade, A.B.; Chen, X.; Pamment, M.G. *Tetrahedron Lett.*, ***1995***, *36*, 2945

$Cu(OTf)_2$(chiral amine)$_3$, $PhCO_3t$-Bu

rt , 40h

OBz

68% (74% ee , S)

Kawasaki, K.; Tsumura, S.; Katsuki, T. *Synlett*, ***1995***, 1245

2 eq. $PhCO_3t$-Bu , 3 eq. $PhCO_3H$

PhH , 4h , 0.1 Cu_2O , reflux , air

OBz

59% (45% ee , S)

Levina, A.; Muzart, J. *Tetrahedron Asymmetry*, ***1995***, *6*, 147

$Cu(OAc)_2$, S-proline , $EtCO_2H$

$PhCO_3t$-Bu , 20°C , 16h

O

O

70% (57% ee)

Rispens, M.T.; Zondervan, C.; Rispens, M.T.; Zondervan, C.; Feringa, B.L. *Tetrahedron Asymmetry*, ***1995***, *6*, 661

Also via: Section 26 (Acid Derivatives) and Section 41 (Alcohols).

SECTION 117: ESTERS FROM KETONES

O

O_2 (1 atm) , 3 eq. PhCHO , 21h
0.2 $Cu(OAc)_2 \bullet H_2O$, DCE

dark

O

O

94%

Bolm, C.; Schlingloff, G.; Weickhardt, K. *Tetrahedron Lett.*, ***1993***, *34*, 3405

THF , 25°C

H—Nb Cp Cp Ph Ph

CO_2Me +

72% 28%

Fermin, M.C.; Bruno, J.W. *Tetrahedron Lett.*, ***1993***, *34*, 7545

Cl CO_2Et , DMF

e⁻ , $SmCl_3$, Mg anode

60%

Hebri, H.; Duñach, E.; Périchon, J. *J. Chem. Soc. Chem. Commun.*, ***1993***, 499

Ph

$MeOCO_2Me$, 195°C , K_2CO_3

6.4h

Ph OMe

71%

Selva, M.; Marques, C.A.; Tundo, P. *Gazz. Chim. Ital.*, ***1993***, *123*, 515

3 PhCHO , CCl_4

O_2 , 40°C

90%

Kaneda, K.; Ueno, S.; Imanaka, T.; Shimotsuma, E.; Nishiyama, Y.; Ishii, Y. *J. Org. Chem.*, ***1994***, *59*, 2915

NBS , DME/H_2O

0°C , 30 min

82%

Marotta, E.; Piombi, B.; Righi, P.; Rosini, G. *J. Org. Chem.*, ***1994***, *59*, 7526

PhCHO , 5% MnO_2

DCE , O_2 , rt , 1d

68%

Inokuchi, T.; Kanazaki, M.; Sugimoto, T.; Torii, S. *Synlett*, ***1994***, 1037

PhCHO, O_2, $Mg_{10}Al_2(OH)_{24}CO_3$ / DCE, 40°C, 5h — 90%

Kaneda, K.; Ueno, S.; Imanaka, T. *J. Chem. Soc. Chem. Commun.*, ***1994***, 797

3 eq. mCPBA, rt / 3 eq. $KHCO_3$ — OBn, OMe — 48%

Chida, N.; Tobe, T.; Ogawa, S. *Tetrahedron Lett.*, ***1994***, *35*, 7249

O_2, *t*-BuCHO, Cu complex — Ph — 41% (65% ee)

Bolm, C.; Schlingloff, G.; Weickhardt, K. *Angew. Chem. Int. Ed. Engl.*, ***1994***, *33*, 1848

O_2, 1% Cu complex / *t*-BuCHO — (3 [67% ee] : 1 [92% ee]) 61%

Bolm, C.; Schlingloff, G. *J. Chem. Soc. Chem. Commun.*, ***1995***, 1247

Also via: Section 27 (Acid Derivatives).

SECTION 118: ESTERS FROM NITRILES

$C_5H_{11}CN$ — cat. $RuH_2(PPh_3)_4$, H_2O / HO–(isoamyl alcohol) — C_5H_{11} ester — 72%

Naota, T.; Shichijo, Y.; Murahashi, S.-I. *J. Chem. Soc. Chem. Commun.*, ***1994***, 1359

SECTION 119: ESTERS FROM ALKENES

OMe O_3, -78°C, MeOH/CH_2Cl_2; 5 eq. NaOH; C_6H_{13}; OMe; C_6H_{13}; CO_2Me; 63%

Marshall, J.A.; Garofalo, A.W. *J. Org. Chem.*, ***1993***, *58*, 3675

$CH_2(CO_2Me)_2$, $Mn(OAc)_3$; ultrasound; CO_2Me; O; O; 78%

Allegretti, M.; D'Annibale, A.; Trogolo, C. *Tetrahedron*, ***1993***, *49*, 10705

C_7H_{15}; V_2O_5, TFA, 70°C, 5h; C_7H_{15}; O_2CCF3; 86%

Choudary, B.M.; Reddy, P.N. *J. Chem. Soc. Chem. Commun.*, ***1993***, 405

$(CH_2)_7$ CO_2Me; ICH_2CO_2Et, Cu, 130°C; O; O; $(CH_2)_7$ CO_2Me; 70%

Metzger, J.O.; Mahler, R. *Angew. Chem. Int. Ed. Engl.*, ***1995***, *34*, 902

Also via: Section 44 (Alcohols).

SECTION 120: ESTERS FROM MISCELLANEOUS COMPOUNDS

$OSiMe_2t$-Bu; SO_2Ph; 1. BuLi; 2. $CNCO_2Me$ 44%; 3. TBAF, THF 65%; 4. Al/Hg, aq. THF 60%; O; O

Robin, S.; Huet, F. *Tetrahedron Lett.*, ***1993***, *34*, 2945

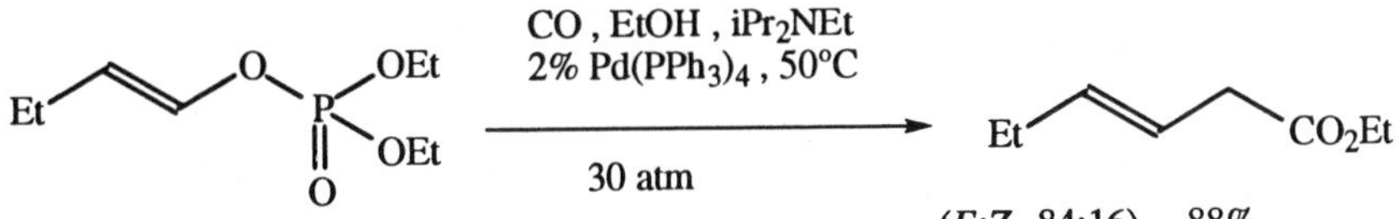

Et–CH=CH–O–P(=O)(OEt)(OEt) → Et–CH=CH–CH$_2$–CO_2Et

CO , EtOH , iPr_2NEt
2% $Pd(PPh_3)_4$, 50°C
30 atm

(E:Z 84:16) 88%

also works with allylic acetates

Murahashi, S.-I.; Imada, Y.; Taniguchi, Y.; Higashiura, S. *J. Org. Chem.*, ***1993***, *58*, 1538

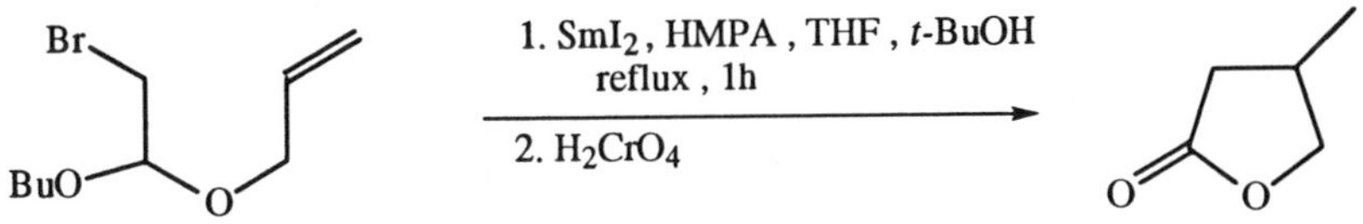

1. SmI_2 , HMPA , THF , t-BuOH
reflux , 1h
2. H_2CrO_4

Br, BuO, O → O, O

99%

Fukuzawa, S.; Tsuchimoto, T. *Synlett*, ***1993***, 803

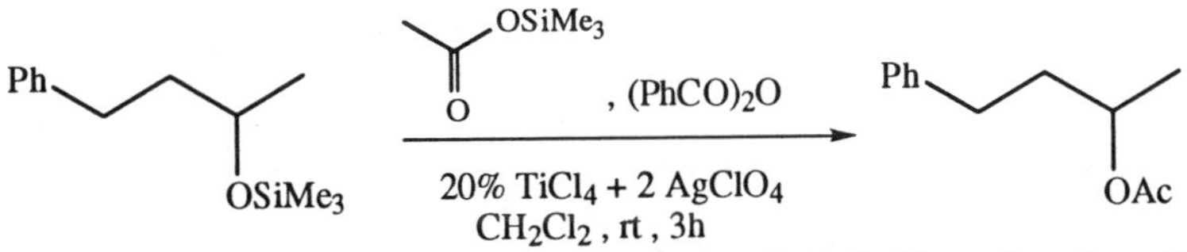

OSiMe$_3$, O , $(PhCO)_2O$

Ph, $OSiMe_3$ → Ph, OAc 90%

20% $TiCl_4$ + 2 $AgClO_4$
CH_2Cl_2 , rt , 3h

Miyashita, M.; Shiina, I.; Miyoshi, S.; Mukaiyama, T. *Bull. Chem. Soc. Jpn.*, ***1993***, *66*, 1516

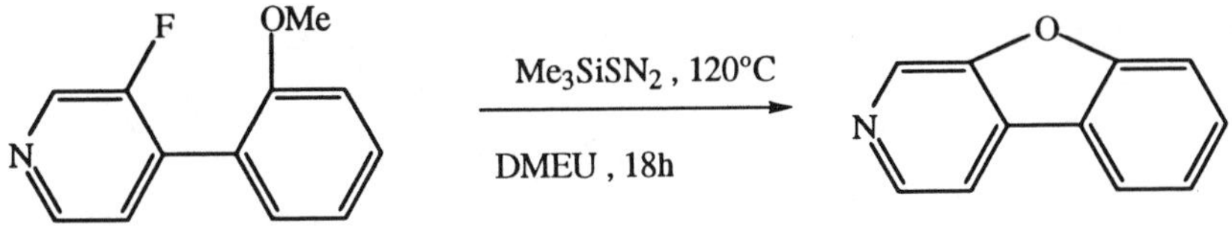

F, OMe, N → O, N

Me_3SiSN_2 , 120°C
DMEU , 18h

DMEU = 1,3-dimethyl-2-imidazolidinone 63%

Lai, L.-L.; Lin, R.Y.; Huang, W.-H.; Shiao, M.-J. *Tetrahedron Lett.*, ***1994***, *35*, 3545

CHAPTER 9

PREPARATION OF ETHERS, EPOXIDES AND THIOETHERS

SECTION 121: ETHERS, EPOXIDES AND THIOETHERS FROM ALKYNES

1. Me_3Si—≡—Bu, MeCN, rt
2. NaOPh, MeOH

62%

Kitamura, T.; Zheng, L.; Taniguchi, H.; Sakurai, M.; Tanaka, R. *Tetrahedron Lett.*, ***1993***, *34*, 4055

SECTION 122: ETHERS, EPOXIDES AND THIOETHERS FROM ACID DERIVATIVES

NO ADDITIONAL EXAMPLES

SECTION 123: ETHERS, EPOXIDES AND THIOETHERS FROM ALCOHOLS AND THIOLS

$Te(iBu)_2^{\oplus}$ $Br^{\ominus}$

NaOH, THF, rt, 5h

85%

Xu, C.; Lu, S.; Huang, X. *Synth. Commun.*, ***1993***, *23*, 2527

1. $Me_3SiSSiMe_3$
2. $BF_3 \cdot OEt_2$, CH_2Cl_2
3. aq. $NaHCO_3$

75%

Tsay, S.-C.; Yep, G.L.; Chen, B.-L.; Lin, L.C.; Hwu, J.R. *Tetrahedron*, ***1993***, *49*, 8969

MeCOSH , hν , 1h

(2 : 1) 57%

Kirpichenko, S.V.; Tolstikova, L.L; Suslova, E.N.; Voronkov, M.G. *Tetrahedron Lett.*, ***1993***, *34*, 3889

1. $NaCNBH_3$, Bu_3SnCl AIBN , EtOH , reflux
2. NaOH , ether

67% (96%ee)

Corey, E.J.; Helal, C.J. *Tetrahedron Lett.*, ***1993***, *34*, 5227

CO (100 bar) , MeOH
150°C

98%

Bott, K. *Chem. Ber.*, ***1993***, *126*, 1955

$PhCH_2OH$ — $PhCH_2Cl$, cetylMe_3NBr , 5 min; microwave → $PhCH_2OCH_2Ph$ 85%

Yuncheng, Y.; Yulin, J.; Jun, P.; Xiaohui, Z.; Conggui, Y. *Gazz. Chim. Ital.*, ***1993***, *123*, 519

1. $AlMe_3$, CH_2I_2 (67% — 6:1 *anti:syn*)
2. *p*-TsOH (>30:1 (2,3-*anti:syn*)

Panek, J.S.; Garbaccio, R.M.; Jain, N.F. *Tetrahedron Lett.*, ***1994***, *35*, 6453

BuOH , 0.2 M CAN
reflux , 75 min

92%

Iranpoor, N.; Mothaghineghad, E. *Tetrahedron*, ***1994***, *50*, 1859

OH → 1. ,6-di-*t*-butylpyridine , AgOTf CH_2Cl_2 , 0°C 2. MeI → OMe 95%

Burk, R.M.; Gac, T.S.; Roof, M.T.S. *Tetrahedron Lett.*, ***1994***, *35*, 8111

SPDBTO, OH → 1. $Hg(OAc)_2$, CH_2Cl_2 rt , 18h 2. aq. NaCl 3. Bu_3SnH , THF → SPDBTO, O + SPDBTO, O (7 : 1) 93%

Garavelas, A.; Mavropoulos, I.; Perlmutter, P.; Westman, G. *Tetrahedron Lett.*, ***1995***, *36*, 463

OH → 5% $Pd(OAc)_2$, DMSO , O_2 23°C , 1d → H, H, O 90%

Rönn, M.; Bäckvall, J.-E.; Andersson, P.G. *Tetrahedron Lett.*, ***1995***, *36*, 7749

OH → BnBr , Cs_2CO_3 , MeCN 80°C → OBn 92%

Lee, J.C.; Yuk, J.Y.; Cho, S.H. *Synth. Commun.*, ***1995***, *25*, 1367

OH → 1. Ac_2O 2. $(Cp)_2TiMe_2$ 3. H_2 , Pd/C → OiPr 80x80%

Le Diguarher, T.; Billington, D.C.; Dorey, G. *Synth. Commun.*, ***1995***, *25*, 1633

SECTION 124: ETHERS, EPOXIDES AND THIOETHERS FROM ALDEHYDES

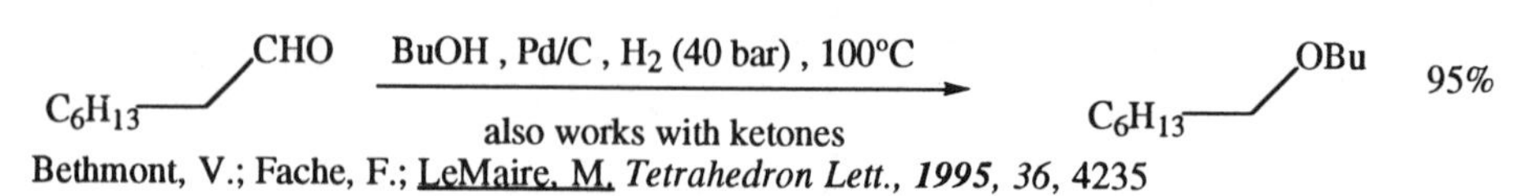

Bethmont, V.; Fache, F.; LeMaire, M. *Tetrahedron Lett.*, ***1995***, *36*, 4235

PhCHO $\xrightarrow[Et_3SiH\ ,\ TMSOTf\ ,\ CH_2Cl_2\ ,\ -30°C\ \rightarrow\ 0°C]{Me_3SiO\text{-allyl}}$ Ph–CH$_2$–O–CH$_2$CH=CH$_2$ 99%

Hatakeyama, S.; Mori, H.; Kitano, K.; Yamada, H.; Nishizawa, M. *Tetrahedron Lett.*, ***1994***, *35*, 4367

cyclohexyl–CHO $\xrightarrow[Al(OAr)_3\ ,\ -78°C\ ,\ 1h]{CH_2N_2}$ cyclohexyloxirane

Ar = 2,5-diphenylphenyl 83%

Maruoka, K.; Concepcion, A.B.; Yamamoto, H. *Synlett*, ***1994***, 521

PhCHO $\xrightarrow[-5°C\ ,\ 1d]{Ar-S(=N^{\ominus}Ts)-Me\ ,\ NaH\ ,\ THF}$ styrene oxide (H, Ph) 63% (70% ee)

Baird, C.P.; Taylor, P.C. *J. Chem. Soc. Chem. Commun.*, ***1995***, 893

SECTION 125: ETHERS, EPOXIDES AND THIOETHERS FROM ALKYLS, METHYLENES AND ARYLS

NO ADDITIONAL EXAMPLES

SECTION 126: ETHERS, EPOXIDES AND THIOETHERS FROM AMIDES

NO ADDITIONAL EXAMPLES

SECTION 127: ETHERS, EPOXIDES AND THIOETHERS FROM AMINES

NO ADDITIONAL EXAMPLES

SECTION 128: ETHERS, EPOXIDES AND THIOETHERS FROM ESTERS

Ph–CH$_2$–CO$_2$Me $\xrightarrow[2.\ TMSOTf\ ,\ Py]{1.\ Dibal\ ,\ -78°C}$ Ph–CH$_2$–CH(OMe)–OSiMe$_3$ 91%

Kiyooka, S.; Shirouchi, M.; Kaneko, Y. *Tetrahedron Lett.*, ***1993***, *34*, 1491

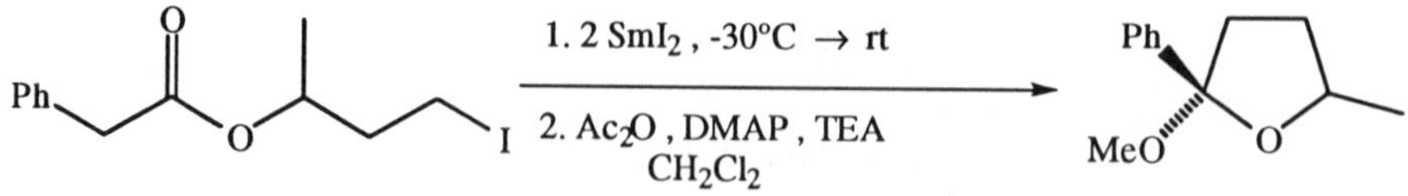

Ph
O
O
I
1. 2 SmI2 , -30°C → rt
2. Ac2O , DMAP , TEA
CH2Cl2
Ph
MeO
O

84% (1.5:1)

Molander, G.A.; McKie, J.A. *J. Am. Chem. Soc.*, ***1993***, *115*, 5821

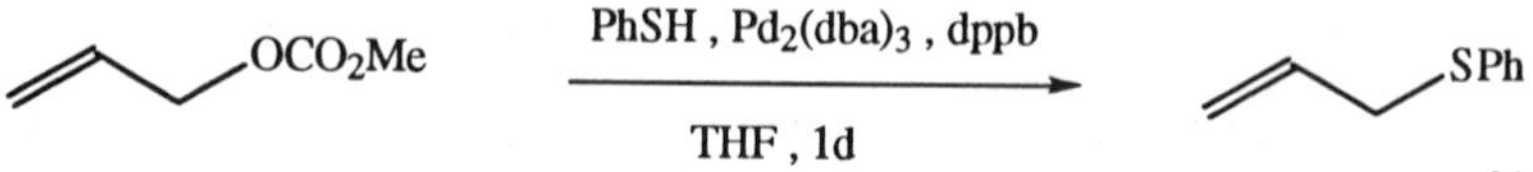

OCO2Me
PhSH , Pd2(dba)3 , dppb
THF , 1d
SPh

83%

Goux, C.; Lhoste, P.; Sinou, D. *Tetrahedron*, ***1994***, *50*, 10321

SECTION 129: ETHERS, EPOXIDES AND THIOETHERS FROM ETHERS, EPOXIDES AND THIOETHERS

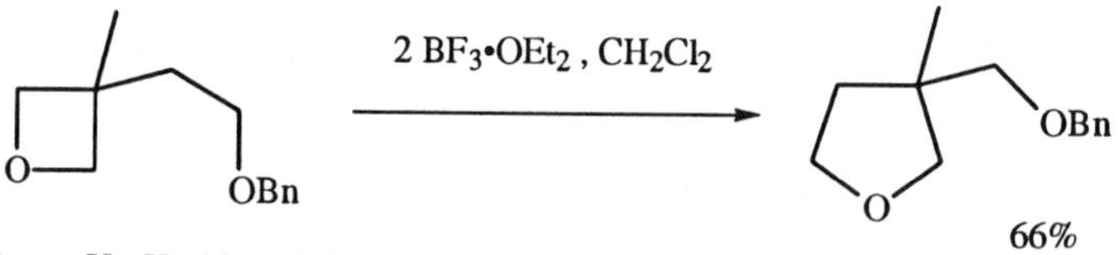

O
OBn
2 BF3•OEt2 , CH2Cl2
OBn
O

66%

Itoh, A.; Hirose, Y.; Kashiwagi, H.; Masaki, Y. *Heterocycles*, ***1994***, *38*, 2165

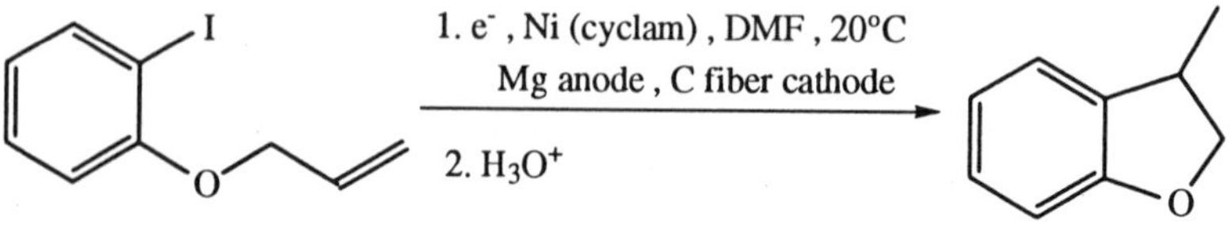

I
O
1. e⁻ , Ni (cyclam) , DMF , 20°C
Mg anode , C fiber cathode
2. H3O+
O

90%

Olivero, S.; Clinet, J.C.; Duñach, E. *Tetrahedron Lett.*, ***1995***, *36*, 4429

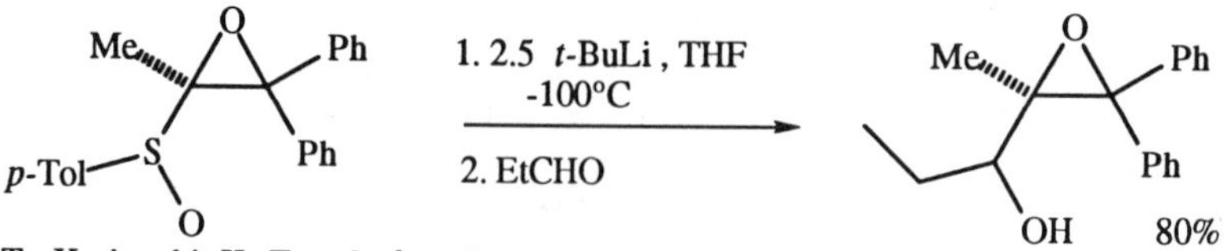

Me
O
Ph
Ph
p-Tol
S
O
1. 2.5 t-BuLi , THF
-100°C
2. EtCHO
Me
O
Ph
Ph
OH

80%

Satoh, T.; Horiguchi, K. *Tetrahedron Lett.*, ***1995***, *36*, 8235

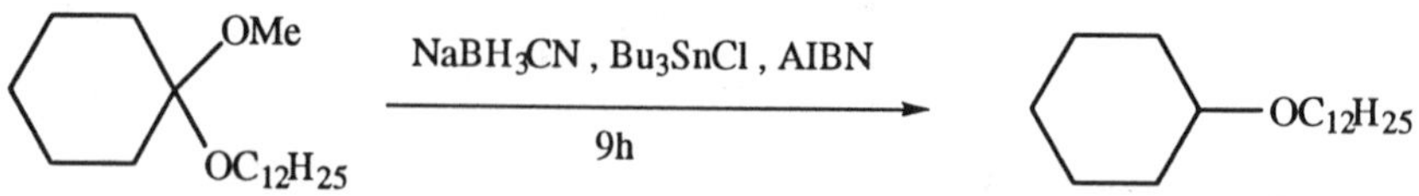

OMe
OC12H25
NaBH3CN , Bu3SnCl , AIBN
9h
OC12H25

62%

Srikrishna, A.; Viswajanani, R. *Synlett*, ***1995***, 95

PhI , 2% $Pd(OAc)_2$, 3 NEt_3 6% PPh_3 , DMF , 70°C

Ph Ph (>98 : 2) 66%

Hillers, S.; Rieger, O. *Synlett*, ***1995***, 153

SECTION 130: ETHERS, EPOXIDES AND THIOETHERS FROM HALIDES AND SULFONATES

$C_8H_{17}Br$ —BER-PhSSPh , MeOH , reflux→ C_8H_{17}—S—Ph 96%

Yoon, N.M.; Choi, J.; Ahn, J.H. *J. Org. Chem.*, ***1994***, *59*, 3490

MeSCH$_2$Cl , $AlCl_3$ / CH_2Cl_2 , 0°C

SMe 87%

Olah, G.A.; Wang, Q.; Neyer, G. *Synthesis*, ***1994***, 276

S I —PhSH , rt→ S SPh 88%

Lee, S.B.; Hong, J.-I. *Tetrahedron Lett.*, ***1995***, *36*, 8439

Related Methods: Section 123 (Ethers from Alcohols).

SECTION 131: ETHERS, EPOXIDES AND THIOETHERS FROM HYDRIDES

OMe —PhSH , $(CF_3)_2CHOH$ / PIFA , rt→ SPh OMe 62%

PIFA = phenyliodine(III) *bis*(trifluoroacetate)

Kita, Y.; Takada, T.; Mihara, S.; Tohma, H. *Synlett*, ***1995***, 211

SECTION 132: ETHERS, EPOXIDES AND THIOETHERS FROM KETONES

O Br Ph$_2$MeSi —Bu_3SnH , AIBN / PhH (0.05 M)→ OSiMePh$_2$ Me 76%

Tsai, Y-M.; Tang, K-H.; Jiaang, W-T. *Tetrahedron Lett.*, ***1993***, *34*, 1303

$SiMe_3$

TBAF

Ph Ph

S

Ph Ph

50%

Capperucci, A.; Degl'Innocenti, A.; Ferrara, M.C.; Bonini, B.F.; Mazzanti, G.; Zanti, P.; Ricci, A. *Tetrahedron Lett.*, ***1994***, *35*, 161

N_2

OAc , $Rh_2(OAc)_4$

PhF , 10 h

69%

Pirrung, M.C.; Lee, Y.R. *Tetrahedron Lett.*, ***1994***, *35*, 6231

Related Methods: Section 124 (Epoxides from Aldehydes).

SECTION 133: ETHERS, EPOXIDES AND THIOETHERS FROM NITRILES

NO ADDITIONAL EXAMPLES

SECTION 134: ETHERS, EPOXIDES AND THIOETHERS FROM ALKENES

Asymmetric Epoxidation

CO_2Et

Ph–(pyridine N-oxide) , NaOCl

Mn-salen derivative
0°C , pH 11.3

CO_2Et

58%
(77% ee)

Chang, S.; Lee, N.H.; Jacobsen, E.N. *J. Org. Chem.*, ***1993***, *58*, 6939

Mn (salen) catalyst , 1h

1% H_2O_2 , 0°C , CH_2Cl_2

77%
(64%ee)

Schwenkreis, T.; Berkessel, A. *Tetrahedron Lett.*, ***1993***, *34*, 4785

threitol-strapped Mn-porphyrin , CH_2Cl_2

iodosylbenzene , 1,5-dicyclohexylimidazole

86% (69% ee, R+)

Collman, J.P.; Lee, V.J.; Zhang, X.; Ibers, J.A.; Brauman, J.I. *J. Am. Chem. Soc.*, ***1993***, *115*, 3834

chloroperoxidase , acetone

H_2O_2 , pH 5

78% (96% ee , 2R3S)

Allain, E.J.; Hager, L.P.; Deng, L.; Jacobsen, E.N. *J. Am. Chem. Soc.*, ***1993***, *115*, 4415

t-BuOOH

35% (22% ee , S,S)

Manoury, E.; Mouloud, H.A.H.; Balavoine, G.G.A. *Tetrahedron Asymmetry*, ***1993***, *4*, 2339

Mn salen catalyst , O_2

PhH , rt

51% (52% ee)

Mukaiyama, T.; Yamada, T.; Nagata, T.; Imagawa, K. *Chem. Lett.*, ***1993***, 327

cat. Mn salen , 0°C ,
buffered bleach , pH 11.3

69% (93% ee, S,S)

Brandes, B.D.; Jacobsen, E.N. *J. Org. Chem.*, ***1994***, *59*, 4378

cat. Mn(salen) , CH_2Cl_2

49%

(42%ee , 1S,2S)

Pietikäinen, P. *Tetrahedron Lett.*, ***1994***, *35*, 941

Mn salen catalyst , PhIO , 0°C

O

72% (98% ee)

Sasaki, H.; Irie, R.; Katsuki, T. *Synlett*, ***1994***, 356

Ph Ph

2 PhIO , Mn salen catalyst

rt

Ph Ph O

27% (15% ee , 1R2R)

Sasaki,H.; Irie, R.; Hamada, T.; Suzuki, K.; Katsuki, T. *Tetrahedron*, ***1994***, *50*, 11827

Ph

Mn salen catalyst , *t*-BuCHO

O_2 , rt

O Ph

28% (63:37 *cis:trans*)
80% ee for cis

Nagata, T.; Imagawa, K.; Yamada, T.; Mukaiyama, T. *Chem. Lett.*, ***1994***, 1259

O_2 , Mn salen complex , rt
NMI , pivaladehyde , PhF

O

78% (63% ee)

Yamada, T.; Imagawa, K.; Nagata, T.; Mukaiyama, T. *Bull. Chem. Soc. Jpn.*, ***1994***, *67*, 2248

Ph

Mn (salen) catalyst , Bu_4NIO_4

imidazole , rt , 3h

O Ph

53% (47% ee , 1R,2S (-))

Pietikäinen, P. *Tetrahedron Lett.*, ***1995***, *36*, 319

20% 4-Ph-pyridine N-oxide , 0°C
5% Mn salen complex , CH_2Cl_2

O

37%
35% ee

Brandes, B.D.; Jacobsen, E.N. *Tetrahedron Lett.*, ***1995***, *36*, 5123

8% Mn salen complex , mcpba , NMO

CH_2Cl_2 , -78°C

89% (86% ee)

Palucki, M.; McCormick, G.J.; Jacobsen, E.N. *Tetrahedron Lett.*, ***1995***, *36*, 5457

chloroperoxidase , KOAc , H_2O_2

antifoam A , aq. acetone , pH 5.2
sodium citrate

89% (49% ee)

Dexter, A.F.; Lakner, F.J.; Campbell, R.A.; Hager, L.P. *J. Am. Chem. Soc.*, ***1995***, *117*, 6412

O_2 , Mn salen catalyst , *t*-BuCHO

70% (64% ee)

Nagata, T.; Imagawa, K.; Yamada, T.; Mukaiyama, T. *Bull. Chem. Soc. Jpn.*, ***1995***, *68*, 1455

Non-Asymmetric Epoxidation

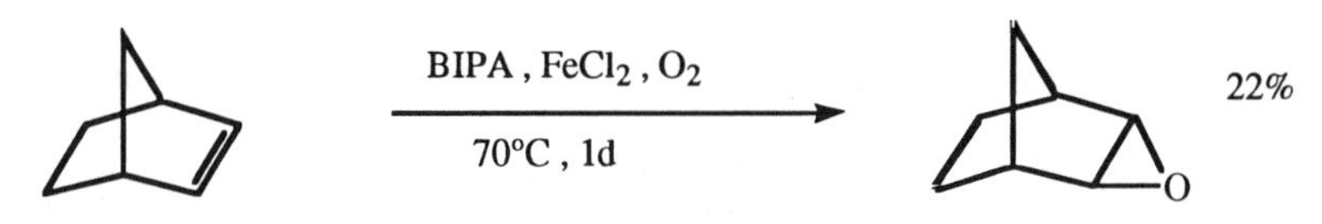

BIPA = N,N'-bis{2-(4-imidazoloyl)ethyl}-2,6-pyridinecarboxamide

Hirao, T.; Moriuchi, T.; Mikami, S.; Ikeda, I.; Ohshiro, Y. *Tetrahedron Lett.*, ***1993***, *34*, 1031

cat. [K-10-montmorillonite-
Ni(acac)]

iBuCHO , CH_2Cl_2 , rt
autoclave , 10 bar

67%

Bouhlel, E.; Laszlo, P.; Levart, M.; Montaufier, M-T.; Singh, G.P.
Tetrahedron Lett., ***1993***, *34*, 1123

Laszlo, P.; Levart, M. *Tetrahedron Lett.*, ***1993***, *34*, 1127

[$CoCl_2$/salicylaldehyde-L-serine methyl ester]

O_2 , MeCN , rt , 25h

64%

Punniyamurthy, T.; Bhatia, B.; Iqbal, J. *Tetrahedron Lett.*, ***1993***, *34*, 4657

Fe cat. , O_2 , iPrCHO , rt

Ph Ph Ph Ph O 96%

Saalfrank, R.W.; Reihs, S.; Hug, M. *Tetrahedron Lett.*, ***1993***, *34*, 6033

10% $(NH_4)_5H_4PV_6W_6O_{40}$•6 H_2O
O_2 , DCE , 25°C , 4h
2 eq. iPrCHO

C_5H_{11} O C_5H_{11} 77%

Hamamoto, M.; Nakayama, K.; Nishiyama, Y.; Ishii, Y. *J. Org. Chem.*, ***1993***, *58*, 6421

$Co(mac)_2$, $EtCH(OEt)_2$
O_2 , MS4Å , 45°C

C_7H_{15} C_7H_{15} O 97%

mac = 3-methyl-2,4-pentanedione

Mukaiyama, T.; Yorozu, K.; Yakai, T.; Yamada, T. *Chem. Lett.*, ***1993***, 439

$Cu(OH)_2$, CH_3CHO , CH_2Cl_2
O_2 , rt , 17h

O 79%

Murahashi, S.-I.; Oda, Y.; Naota, T.; Komiya, N. *J. Chem. Soc. Chem. Commun.*, ***1993***, 139

urea-H_2O_2 complex/maleic acid

Bu Bu O 71%

Astudillo, L.; Galindo, A.; González, A.G.; Mansilla, H. *Heterocycles*, ***1993***, *36*, 1075

1. $[Mn^{III}(TPP)(Cl)]$, CH_2Cl_2
2. imidazole , Bu_4NIO_4 , 5h

O

Mohajer, D.; Tangestaninejad, S. *Tetrahedron Lett.*, ***1994***, *35*, 945

O , $KHSO_5$
H_2O/CH_2Cl_2 (pH 11)
18-crown-6 , 5°C

O + O

(9 : 91) 80%

Kurihara, M.; Ito, S.; Tsutsumi, N.; Miyata, N. *Tetrahedron Lett.*, ***1994***, *35*, 1577

AgOAc/bipy/anode
ACN (0.1 M $LiClO_4$)
H_2O

87%

Kandzia, C.; Steckhan, E. *Tetrahedron Lett.*, ***1994***, *35*, 3695

iPrCHO , $Fe(AAEMA)_3$, O_2
rt , 10h

AAEMA =

95%

Mastrorilli, P.; Nobile, C.F. *Tetrahedron Lett.*, ***1994***, *35*, 4193

$Ph_2P(O)$–O–$P(O)Ph_2$
H_2O_2

Ph Ph

87%

Kende, A.S.; Delair, P.; Blass, B.E. *Tetrahedron Lett.*, ***1994***, *35*, 8123

1. $(EtO)_2P(S)SBn$
2. TBAF•3 H_2O

82%

Capozzi, G.; Mecichetti, S.; Neri, S.; Skowronska, A. *Synlett*, ***1994***, 267

NaOCl , 4-methylpyridine
porphyrin catalyst/Mn

Ph Ph O

88%

Gonsalves, A.Md'A.R.; Pereira, M.M.; Serra, A.C.; Johnstone, R.A.W.; Nunes, M.L.P.G. *J. Chem. Soc., Perkin Trans. 1.*, ***1994***, 2053

C_7H_{15}

O_2 (1 atm) , 2% $Co(acac)_2$
$EtCH(OEt)_2$, 45°C , 10h

C_7H_{15}
O

48%

Yorozu, K.; Takai, T.; Yamada, T.; Mukaiyama, T. *Bull. Chem. Soc. Jpn.*, ***1994***, *67*, 2195

Oxone , 0°C , Bu_4NHSO_4

acetone , CH_2CL_2 , buffer

OBn

O

OBn

99%

Denmark, S.E.; Forbes, D.C.; Hays, D.S.; DePue, J.S.; Wilde, R.G. *J. Org. Chem.*, ***1995***, *60*, 1391

Ph

Me

Me, O, O, F_3C , MeCN , 0°C

15 min

Ph O Me 96%

Yang, D.; Wong, M.-K.; Yip, Y.-C. *J. Org. Chem.*, ***1995***, *60*, 3887

CO_2H

O

hν , methylene blue

PhH , H_2O

O

50%

Das, S.; Thanulingam, T.L.; Rajesh, C.S.; George, M.V. *Tetrahedron Lett.*, ***1995***, *36*, 1337

OH

H_2O_2 , DMF , pH 4-9

OH

O

98%

Chen,Y.; Reymond, J.-L. *Tetrahedron Lett.*, ***1995***, *36*, 4015

t-BuOOH , Ti(OiPr)$_4$, SiO_2

rt , 1d

O

73%

Fraile, J.M.; García, J.I.; Mayoral, J.A.; de Mènorval, L.C.; Rachdi, F. *J. Chem. Soc. Chem. Commun.*, ***1995***, 539

PhIO , NaOCl , Mn salen catalyst

O

92%

Rasmussen, K.G.; Thomsen, D.S.; Jørgensen, K.A. *J. Chem. Soc., Perkin Trans. 1.*, ***1995***, 2009

t-BuOOH , MeOH

CrS-2- molecular sieve

85%

Joseph, R.; Sasidharan, M.; Kumar, R.; Sudalai, A.; Ravindranathan, T. *J. Chem. Soc. Chem. Commun.*, ***1995***, 1341

Formation of Other Ethers

PhSeSePh , CAN

MeOH , 30 min

SePh

OMe

93%

Bosman, C.; D'Annibale, A.; Resta, S.; Trogolo, C. *Tetrahedron Lett.*, ***1994***, *35*, 6525

C_7H_{15} C_5H_{11} OH

$AgBF_4$, acetone , rt

2h

C_7H_{15} O C_5H_{11}

97%

Marshall, J.A.; Sehon, C.A. *J. Org. Chem.*, ***1995***, *60*, 5966

REVIEW:

"Rational Design of Manganese-Salen Epoxidation Catalyts; Preliminary Results," Hosoya, N.; Hatayama, A.; Irie, R.; Sasaki, H.; Katsuki, T. *Tetrahedron*, ***1994***, *50*, 4311

SECTION 135: ETHERS, EPOXIDES AND THIOETHERS FROM MISCELLANEOUS COMPOUNDS

OMe

PhSSPh , $SbCl_5$, DCE

reflux , 3h

PhS OMe

76%

Mukaiyama, T.; Suzuki, K. *Chem. Lett.*, ***1993***, 1

$PhSO_2Me$ $\xrightarrow{LiAlH_4 , TiCl_4}$ PhSMe 93%

Akgün, E.; Mahmoud, K.; Mathis, C.A. *J. Chem. Soc. Chem. Commun.*, ***1994***, 761

$[SiO_2/SOCl_2]$, 10 min — 73%

Mohanazadeh, F.; Momeni, A.R.; Ranjbar, Y. *Tetrahedron Lett.*, ***1994***, *35*, 6127

Cp_2TiCl_2, Sm, THF, rt — Ph–S–Me 77-88%

Zhang, Y.; Yu, Y.; Bao, W. *Synth. Commun.*, ***1995***, *25*, 1825

$TiCl_4$, Sm, THF, rt, 30 min — Ph–S–Ph

Wang, J.Q.; Zhang, Y.M. *Synth. Commun.*, ***1995***, *25*, 3545

CHAPTER 10

PREPARATION OF HALIDES AND SULFONATES

SECTION 136: HALIDES AND SULFONATES FROM ALKYNES

NO ADDITIONAL EXAMPLES

SECTION 137: HALIDES AND SULFONATES FROM ACID DERIVATIVES

Ph–CH=CH–CO_2H —(PhIO , NBS , MeCN , 60°C)→ Ph–CH=CH–Br

73% (96:6 *cis:trans*)

Graven, A.; Jørgensen, K.S.; Dahl, S.; Stanczak, A. *J. Org. Chem.*, ***1994***, *59*, 3543

SECTION 138: HALIDES AND SULFONATES FROM ALCOHOLS AND THIOLS

$PhCH_2OH$ —(PPh_3 , $(Cl_3COC)_2O$)→ $PhCH_2Cl$ 95%

Rivero, I.A.; Somanathan, R.; Hellberg, L.H. *Synth. Commun.*, ***1993***, *23*, 711

1-butylcyclohexan-1-ol (OH) —(BBr_3 , CH_2Cl_2 , 0°C)→ 1-bromo-1-butylcyclohexane (Br) quant.

Pelletier, J.D.; Poirier, D. *Tetrahedron Lett.*, ***1994***, *35*, 1051

C_8H_{17}—OH —(TMSCl , DMSO)→ C_8H_{17}—Cl 95%

Snyder, D.C. *J. Org. Chem.*, ***1995***, *60*, 2638

Cyclopentanol (–OH) → I_2 , pet ether , reflux , 1.5h → cyclopentyl–I 42%

Joseph, R.; Pallan, P.S.; Sudalai, A.; Ravindranathan, T. *Tetrahedron Lett.*, ***1995***, *36*, 609

t-Butyl–OH → Me_3SiCl , $BiCl_3$ / rt , 90 min → t-Butyl–Cl quant.

Labrouillère, M.; Le Roux, C.; Oussaid, A.; Gaspard-Iloughmane, H.; Dubac, J. *Bull. Soc. Chim. Fr.*, ***1995***, *132*, 522

REVIEWS:

"An Alternative Synthesis of Aryl and Heteroaryl Bromides from Activated Hydroxy Compounds," Katritzky, A.R.; Li, J.; Stevens, C.V.; Ager, D.J. *Org. Prep. Proceed. Int.*, ***1994***, *26*, 439

SECTION 139: HALIDES AND SULFONATES FROM ALDEHYDES

NO ADDITIONAL EXAMPLES

SECTION 140: HALIDES AND SULFONATES FROM ALKYLS, METHYLENES AND ARYLS

For the conversion R-H → R-Halogen, see Section 146 (Halides from Hydrides).

Ph–CH=CH–CH$_2$–SiMe$_3$ → 2 eq. $VO(OEt)Cl_2$, N_2 / MeCN , 0°C → rt → Ph–CH=CH–CH$_2$–Cl 60%

Fujii, T.; Hirao, T.; Ohshiro, Y. *Tetrahedron Lett.*, ***1993***, *34*, 5601

SECTION 141: HALIDES AND SULFONATES FROM AMIDES

NO ADDITIONAL EXAMPLES

SECTION 142: HALIDES AND SULFONATES FROM AMINES

Ph–CH$_2$CH$_2$–NH$_2$ → 1. $TolSO_2Cl$ 2. base / 3. NBS , THF , hν → Ph–CH$_2$CH$_2$–Br 86%

Collazo, L.R.; Guziec Jr., F.S.; Hu, W.-X.; Pankayatselvan, R. *Tetrahedron Lett.*, ***1994***, *35*, 7911

SECTION 143: HALIDES AND SULFONATES FROM ESTERS

NO ADDITIONAL EXAMPLES

SECTION 144: HALIDES AND SULFONATES FROM ETHERS, EPOXIDES AND THIOETHERS

NO ADDITIONAL EXAMPLES

SECTION 145: HALIDES AND SULFONATES FROM HALIDES AND SULFONATES

$ClCF_2CO_2Me$ —[n-$C_8H_{17}I$, KF, CuI / CdI_2, HMPA, 120°C, 8h]→ $C_8H_{17}CF_3$ + CO_2 + MeCl

77%

Chen, Q.-Y.; Duan, J.-X. *Tetrahedron Lett.*, ***1993***, *34*, 4241

SECTION 146: HALIDES AND SULFONATES FROM HYDRIDES

α-Halogenations of aldehydes, ketones and acids are found in Sections 338 (Halide-Aldehyde), 369 (Halide-Ketone), 359 (Halide-Esters) and 319 (Halide-Acids).

NO_2 —[BrF_3, Br_2, 0*C]→ NO_2 Br

Rozen, S.; Lerman, O. *J. Org. Chem.*, ***1993***, *58*, 239

OMe —[N N N N ⊕ HBr_3 / CH_2Cl_2, MeOH]→ Br OMe

90%

Bisarya, S.C.; Rao, R. *Synth. Commun.*, ***1993***, *23*, 779

NH_2 —[PyH Br_3, THF]→ Br NH_2

87% (19:68 *o:p*)

Reeves, W.P.; King II, R.M. *Synth. Commun.*, ***1993***, *23*, 855

OMe → NBS/cat. HZSM-5, CCl_4 → Br–C6H4–OMe 70% (p:o = 28)

Paul, V.; Sudalai, A.; Daniel, T.; Srinivasan, K.V. *Tetrahedron Lett.*, ***1994***, *35*, 7055

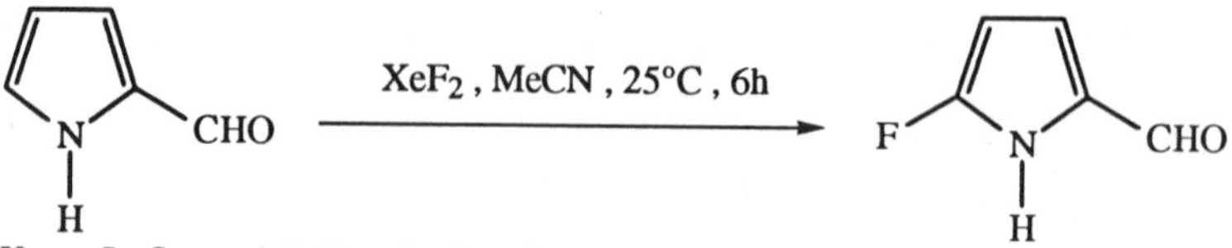

CHO → XeF_2, MeCN, 25°C, 6h → F, CHO 35%

Wang, J.; Scott, A.I. *Tetrahedron Lett.*, ***1994***, *35*, 3679

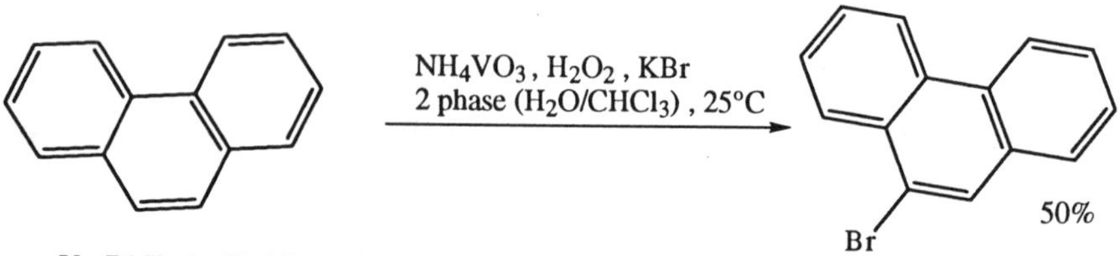

NH_4VO_3, H_2O_2, KBr
2 phase ($H_2O/CHCl_3$), 25°C → Br 50%

Conte, V.; Di Furia, F.; Moro, S. *Tetrahedron Lett.*, ***1994***, *35*, 7429

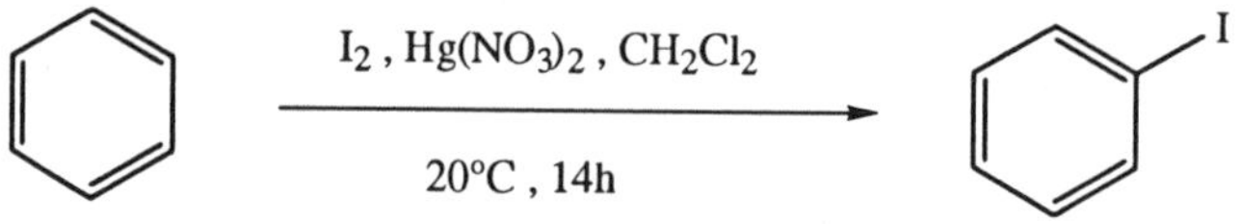

I_2, $Hg(NO_3)_2$, CH_2Cl_2, 20°C, 14h → I 52%

Bachki, A.; Foubelo, F.; Yus, M. *Tetrahedron*, ***1994***, *50*, 5139

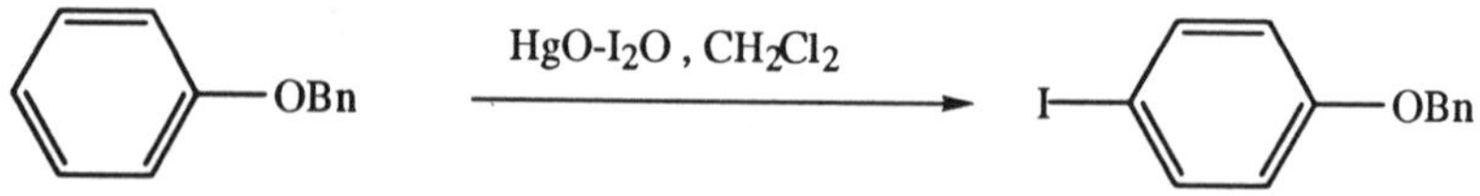

OBn → HgO-I_2O, CH_2Cl_2 → I–C6H4–OBn 87%

Orito, K.; Hatakeyama, T.; Takeo, M.; Suginome, H. *Synthesis*, ***1995***, 1273

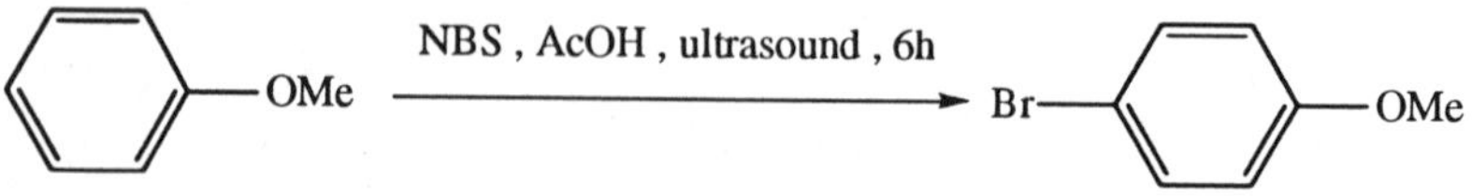

OMe → NBS, AcOH, ultrasound, 6h → Br–C6H4–OMe 92%

Paul, V.; Sudalai, A.; Daniel, T.; Srinivasan, K.V. *Synth. Commun.*, ***1995***, *25*, 2401

SECTION 147: HALIDES AND SULFONATES FROM KETONES

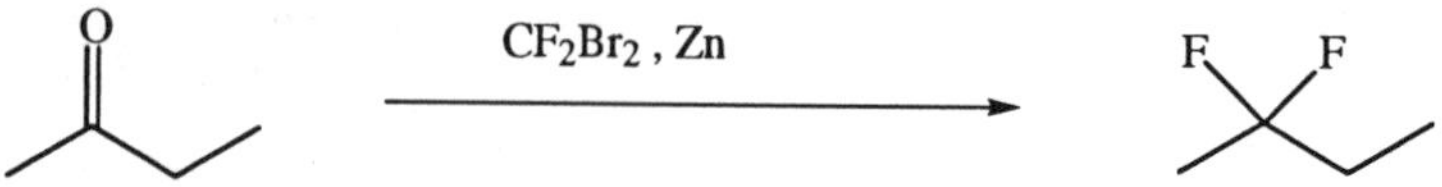

O → CF_2Br_2, Zn → F F 58%

Hu, C.-M.; Qing, F.-L.; Shen, C.-X. *J. Chem. Soc., Perkin Trans. 1.*, ***1993***, 335

SECTION 148: HALIDES AND SULFONATES FROM NITRILES

NO ADDITIONAL EXAMPLES

SECTION 149: HALIDES AND SULFONATES FROM ALKENES

For halocyclopropanations, see Section 74E (Alkyls from Alkenes).

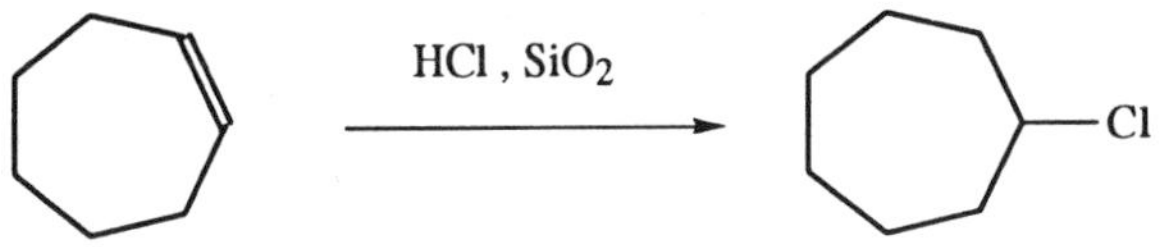

HCl, SiO_2

Cl

62%

Krupp, P.J.; Daus, K.A.; Tubergen, M.W.; Kepler, K.D.; Wilson, V.P.; Craig, S.L.; Baillargeon, M.M.; Breton, G.W. *J. Am. Chem. Soc.*, ***1993***, *115*, 3071

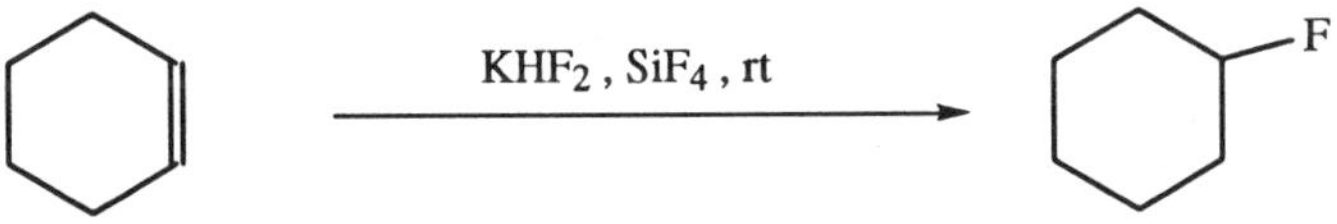

KHF_2, SiF_4, rt

F

79%

Tamura, M.; Shibakami, M.; Kurosawa, S.; Arimura, T.; Sekiya, A. *J. Chem. Soc. Chem. Commun.*, ***1995***, 1891

SECTION 150: HALIDES AND SULFONATES FROM MISCELLANEOUS COMPOUNDS

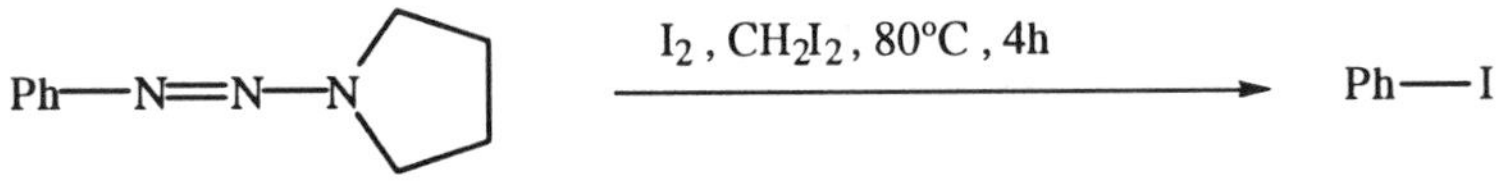

Ph—N=N—N

I_2, CH_2I_2, 80°C, 4h

Ph—I >98%

Wu, Z.; Moore, J.S. *Tetrahedron Lett.*, ***1994***, *35*, 5539

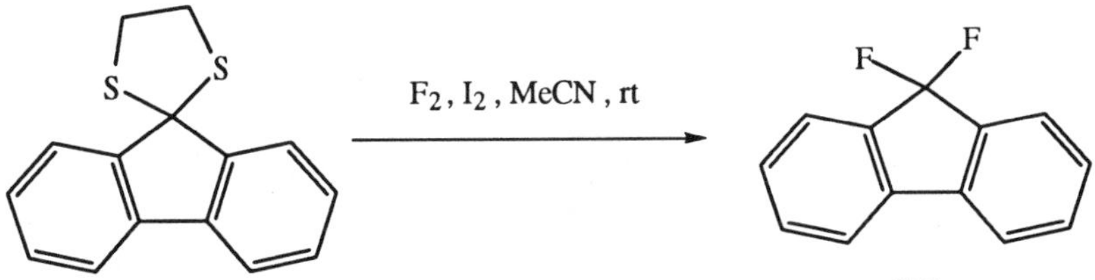

S S

F_2, I_2, MeCN, rt

F F

52%

Chambers, R.D.; Sandford, G.; Atherton, M. *J. Chem. Soc. Chem. Commun.*, ***1995***, 177

$C_{12}H_{25}SO_3Na$ —($SOCl_2$, DMF, 100°C)→ $C_{12}H_{25}Cl$ 81%

Carlsen, P.H.J.; Rist, Ø.; Lund, T.; Helland, I. *Acta Chem. Scand. B.*, ***1995***, *49*, 701

CHAPTER 11

PREPARATION OF HYDRIDES

This chapter lists hydrogenolysis and related reactions by which functional groups are replaced by hydrogen: e.g. $RCH_2X \rightarrow RCH_2\text{-}H$ or R-H.

SECTION 151: HYDRIDES FROM ALKYNES

NO ADDITIONAL EXAMPLES

SECTION 152: HYDRIDES FROM ACID DERIVATIVES

This section lists examples of decarboxylations ($RCO_2H \rightarrow R\text{-}H$) and related reactions.

NO ADDITIONAL EXAMPLES

SECTION 153: HYDRIDES FROM ALCOHOLS AND THIOLS

This section lists examples of the hydrogenolysis of alcohols and phenols ($ROH \rightarrow R\text{-}H$).

$(CH_2)_6$ OH

1. $(Bu_3Sn)_2O$, Tol , reflux , 1d
2. 2 eq. EtNCO (90%)
3. Et_3SiH , PhH , 140°C
0.05M , benzoyl peroxide (70%)

$(CH_2)_6$

Nishiyama, K.; Oba, M.; Ishimi, M.; Sugawara, T.; Ueno, R. *Tetrahedron Lett.*, ***1993***, *34*, 3745

Ph OH

1. $[PPh_3/Tf_2O]$, CH_2Cl_2
2. $NaBH_4$, rt
3. 1N HCl

Ph

87%

Hendrickson, J.B.; Singer, M.; Hussonin, Md.S. *J. Org. Chem.*, ***1993***, *58*, 6913

Ph HO CO_2Et

Et_3SiH , 3M $LiClO_4$
ether

Ph CO_2Et

89%

Wustrow, D.J.; Smith III, W.J.; Wise, L.D. *Tetrahedron Lett.*, ***1994***, *35*, 61

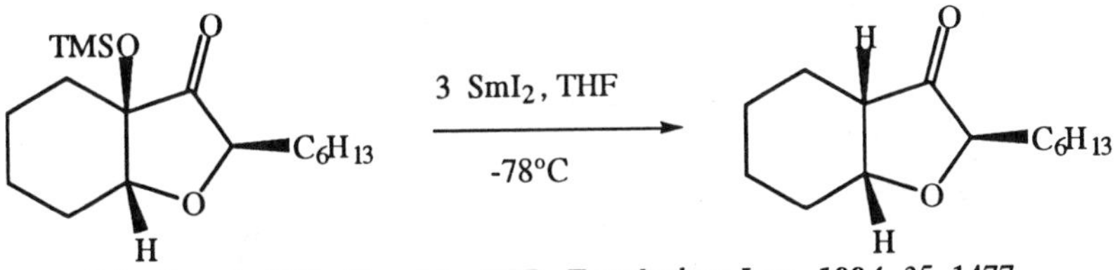
TMSO
O
C6H13
O
H
3 SmI2, THF
-78°C
H
O
C6H13
O
H

Linderman, R.J.; Cusack, K.P.; Kwochka, W.R. *Tetrahedron Lett.*, ***1994***, *35*, 1477

$$CH_3\text{—}(CH_2)_9\text{—}OH \xrightarrow[e^-]{PPh_3\text{ , }Et_4NBr\text{ , }MeCN} CH_3\text{—}(CH_2)_9\text{—}H \quad 94\%$$

Maeda, H.; Maki, T.; Eguchi, K.; Koide, T.; Ohmori, H. *Tetrahedron Lett.*, ***1994***, *35*, 4129

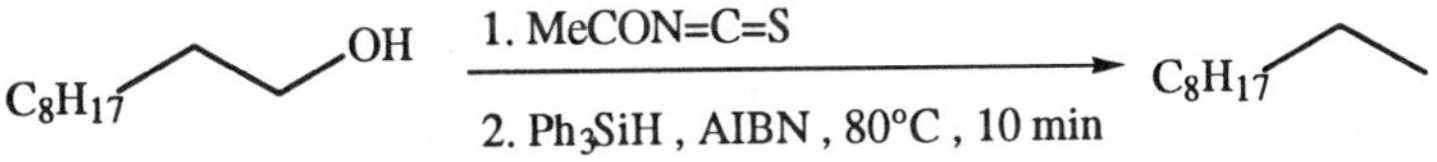
C8H17
OH
1. MeCON=C=S
2. Ph3SiH, AIBN, 80°C, 10 min
C8H17

48x83%

Oba, M.; Nishiyama, K. *Synthesis*, ***1994***, 624

Also via: Section 160 (Halides and Sulfonates).

SECTION 154: HYDRIDES FROM ALDEHYDES

For the conversion RCHO → R-Me, etc., see Section 64 (Alkyls from Aldehydes).

NO ADDITIONAL EXAMPLES

SECTION 155: HYDRIDES FROM ALKYLS, METHYLENES AND ARYLS

NO ADDITIONAL EXAMPLES

SECTION 156: HYDRIDES FROM AMIDES

NO ADDITIONAL EXAMPLES

SECTION 157: HYDRIDES FROM AMINES

This section lists examples of the conversion RNH_2 (or R_2NH) → R-H.

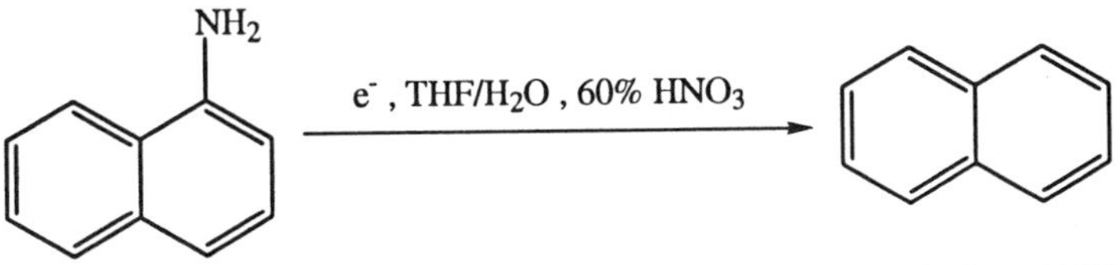
NH2
e-, THF/H2O, 60% HNO3

67%

Torii, S.; Okumoto, H.; Satoh, H.; Minoshima, T.; Kurozumi, S. *Synlett*, ***1995***, 439

SECTION 158: HYDRIDES FROM ESTERS

This section lists examples of the reactions $RCO_2R' \rightarrow R\text{-}H$ and $RCO_2R' \rightarrow R'H$.

OCHO

C_8H_{17}

$Pd(acac)_2$, PBu_3 , rt

PhH , 3h

92%

C_8H_{17}

(alkyne:allene = 99:1)

Mandai, T.; Matsumoto, T.; Kawada, M.; Tsuji, J. *Tetrahedron Lett.*, ***1993***, *34*, 2161

Me Me Me O SMe S

t-BuMe$_2$SiO

hν (tungsten lamp)

Bu_3SnH , 0°C

O N S

other initiators were also used

Me Me H

Barton, D.H.R.; Parekh, S.I.; Tse, C.-L. *Tetrahedron Lett.*, ***1993***, *34*, 2733

O SMe S

5 eq. $PhSiH_3$, AIBN

toluene , reflux , 1h

quant.

Barton, D.H.R.; Jang, D.O.; Jaszverenyi, J.Cs. *Tetrahedron*, ***1993***, *49*, 2793, 7193

O OPh S

H_2 Si Si H_2

deoxygenated cyclohexane

AIBN , 80°C

99%

Gimisis, T.; Ballestri, M.; Ferreri, C.; Chatgilaloglu, C.; Boukherroub, R.; Manuel, G. *Tetrahedron Lett.*, ***1995***, *36*, 3897

OAc; Ph; Ph; O — TBAF•x H_2O , NMO , THF , rt; 1,3-propanedithiol → Ph; Ph; O 99%

Uelo, M.; Okamura, A.; Yamaguchi, J. *Tetrahedron Lett.*, ***1995***, *36*, 7467

SECTION 159: HYDRIDES FROM ETHERS, EPOXIDES AND THIOETHERS

This section lists examples of the reaction R-O-R' → R-H.

S*t*-Bu; Ph; N; Me — CO_2H; SH; TFA , 72°C , 1.5h → Ph; N; Me 84%

Hamel, P.; Zajac, N.; Atkinson, J.G.; Girard, Y. *Tetrahedron Lett.*, ***1993***, *34*, 2059

$C_{11}H_{23}$-S-Ph —5 eq. Ni_2B , MeOH-THF→ $C_{11}H_{24}$ 90%

2 $NiBr_2$ + 6 $NaBH_4$ → Ni_2B

Back, T.G.; Baron, D.L.; Yang, K. *J. Org. Chem.*, ***1993***, *58*, 2407

O; O; SePh — DMN* , ascorbic acid , iPrOH; hν , 3h → O; O 85%

DMN* = photoactivated 1,5-dimethoxynaphthalene

Pandey, G.; Rao, K.S.S.P.; Sekhar, B.B.V.S. *J. Chem. Soc. Chem. Commun.*, ***1993***, 1636

SECTION 160: HYDRIDES FROM HALIDES AND SULFONATES

This section lists the reductions of halides and sulfonates, R-X → R-H.

O; C_5H_{11}; Br — 57% HI , 25°C , 15 min → O; C_5H_{11} 98%

Penso, M.; Mottadelli, S.; Albanese, D. *Synth. Commun.*, ***1993***, *23*, 1385

Br Br MeLi , THF , -105°C Br (98:2 , *E:Z*) 52%

Grandjean, D.; Pale, P. *Tetrahedron Lett.*, ***1993***, *34*, 1155

MeO $OCH_2CH_2CH_2Br$ Bu_3SnH , AIBN , PhH reflux , 0.03M , 8h MeO $CH_2CH_2CH_2OH$ + MeO C_3H_7

an aryl translocation 24% 61%

Lee, E.; Lee, C.; Tae, J.S.; Whang, H.S.; Li, K.S. *Tetrahedron Lett.*, ***1993***, *34*, 2343

OH Cl Cl Cl Cl Cl $NaBH_2(OCH_2CH_2OMe)_2$ $NiCl_2$, $MeOCH_2CH_2OH$ THF , 68°C , 2h OH

Tabaei, S.H.; Pittman Jr., C.U. *Tetrahedron Lett.*, ***1993***, *34*, 3263

Cl Cl Cl Cl HCO_2H , Pd/C , toluene H_2O , Na_2CO_3 , 90°C quant.

Barren, J.P.; Baghel, S.S.; McCloskey, P.J. *Synth. Commun.*, ***1993***, *23*, 1601

Br PVP-$PdCl_2$, H_2 , 65°C ambient pressure , EtOH , NaOH quant.

PVP-$PdCl_2$ = palladium anchored on poly(N-vinyl-2-pyrrolidinone)

Zhang, Y.; Liao, S.; Xu, Y. *Tetrahedron Lett.*, ***1994***, *35*, 4599

C_8H_{17}—Br → (borohydride exchange resin, rt, 3h; 0.1 $Ni(OAc)_2$, MeOH) → C_8H_{17}—H quant.

Yoon, N.M.; Lee, H.J.; Ahn, J.H.; Choi, J. *J. Org. Chem.*, ***1994***, *59*, 4687

3.1 eq. Bu_3SnH, AIBN

H_2O, 90°C, 1d

CO_2H Br CO_2H 99%

Maitra, U.; Sarma, K.D. *Tetrahedron Lett.*, ***1994***, *35*, 7861

SECTION 161: HYDRIDES FROM HYDRIDES

NO ADDITIONAL EXAMPLES

SECTION 162: HYDRIDES FROM KETONES

This section lists examples of the reaction $R_2C\text{-}(C{=}O)R \rightarrow R_2C\text{-}H$.

Et_3SiH, $BF_3{\bullet}OEt_2$, CH_2Cl_2

O Ph Ph → Ph Ph 92%

Smonou, I. *Synth. Commun.*, ***1994***, *24*, 1999

$NaBH_3CN$, $BF_3{\bullet}OEt_2$, THF

rt, 12h

HO O → HO 94%

Srikrishna, A.; Sattigeri, J.A.; Viswajanani, R.; Yelamaggad, C.V. *Synlett*, ***1995***, 93

1. $ClCO_2Me$
2, $NaBH_3$

OH OH O → O_2CO_2Me OH 94x76%

Mitchell, D.; Doecke, C.W.; Hay, L.A.; Koenig, T.M.; Wirth, D.D. *Tetrahedron Lett.*, ***1995***, *36*, 5335

1. Me_3SiCl, NEt_3
2. Et_3SiH, $TiCl_3$

NHCO$_2$Me, CO$_2$H, Ph — 95%

Yato, M.; Homma, K.; Ishida, A. *Heterocycles*, ***1995***, *41*, 17

$NaCNBH_3$, THF
BF_3•OEt_2, rt, 30 min

MeO — MeO + MeO

(3 : 2) 85%

Srikrishna, A.; Viswajanani, R.; Sattigeri, J.A.; Yelamaggad, C.V. *Tetrahedron Lett.*, ***1995***, *36*, 2347

SECTION 163: HYDRIDES FROM NITRILES

This section lists examples of the reaction, R-C≡N → R-H (includes reactions of isonitriles (R-N≡C).

SmI_2, THF-HMPA, 0°C

Ph, Ph, CN, CN — Ph, Ph, CN, H — 97%

Kang, H.-Y.; Hong, W.S.; Cho, Y.S.; Koh, H.Y. *Tetrahedron Lett.*, ***1995***, *36*, 7661

SECTION 164: HYDRIDES FROM ALKENES

NO ADDITIONAL EXAMPLES

SECTION 165: HYDRIDES FROM MISCELLANEOUS COMPOUNDS

Mg, EtOH, cat. $HgCl_2$
rt

OH, SO$_2$Ph, Ph — OH, Ph — 98%

Lee, G.H.; Choi, E.B.; Lee, E.; Pak, C.S. *Tetrahedron Lett.*, ***1993***, *34*, 4541

CHAPTER 12
PREPARATION OF KETONES

SECTION 166: KETONES FROM ALKYNES

1. thxBHBr•SMe_2 , 25°C , 12h
2. NaOH
3. pH 7 buffer
4. H_2O_2

O

98.5%

Cha, J.S.; Min, S.J.; Kim, J.M.; Kwon, O.O. *Tetrahedron Lett.*, ***1993***, *34*, 5113

$SiMe_3$

CO-H_2O , $[Rh(COD)Cl]_2$

NEt_3 , 160°C , PPh_3 , 4h

O

78%

Tekeuchi, R.; Yasue, H. *J. Org. Chem.*, ***1993***, *58*, 5386

MeCN , 760°C , 4h

$(CO)_5Cr$=C(OMe)Me

O

O

(11:1 , 33%)

Kim, O.K.; Wulff, W.D.; Jiang, W.; Ball, R.G. *J. Org. Chem.*, ***1993***, *58*, 5572

HO

Ph

1. 10% $Co_2(CO)_8$, DME

2. 20% $P(OPh)_3$, reflux

O

Ph

91%

Iwasawa, N.; Matsuo, T. *Chem. Lett.*, ***1993***, 997

Br—≡—Bu → (1. $Br_2BH \cdot SMe_2$; 2. propylene glycol; 3. $CH_2=CHCH_2MgCl$, THF, -78°C; 4. HMPT, THF, 50°C, HMPA; 5. oxidation) → Bu-CH2-CO-CH2-CH=CH2 95%

Brown, H.C; Soundararajan, R. *Tetrahedron Lett.*, ***1994***, *35*, 6963

Ph—≡—Ph → ($MeReO_3$, H_2O_2, tOH) → PhCO-COPh 80%

Zhu, Z.; Espenson, J.H. *J. Org. Chem.*, ***1995***, *60*, 7728

SECTION 167: KETONES FROM ACID DERIVATIVES

PhCOPh → (PhCOCl; $TiCl_4$-Zn, 3% HCl) → PhCOCH(Ph)Ph 69% + Ph2C=CPh2 24%

Shi, D.; Chen, J.; Chai, W.; Chen, W.; Kao, T. *Tetrahedron Lett.*, ***1993***, *34*, 2963

C_6H_{13}-B(tetramethylborolane) → (1. $(C_6H_{13})_2Zn$; 2. CuCN•2 LiCl; 3. PhCOCl) → PhCOC_6H_{13} 98%

Langer, F.; Waas, J.; Knochel, P. *Tetrahedron Lett.*, ***1993***, *34*, 5261

PhCH2COCl → (e^-, Bu_4NBF_4, rt; meCN) → PhCH2COCH2Ph 80%

Folest, J.-C.; Pereira-Martins, E.; Troupel, M.; Périchon, J. *Tetrahedron Lett.*, ***1993***, *34*, 7571

PhCH2CH2CO_2H → (1. $(Me_2N=CHCl)Cl$, Py, MeCN/THF; 2. 5 eq. NaN_3, 0°C; 3. 0°C → rt) → PhCH2CH2CON_3 90%

Affandi, H.; Bayguen, A.V.; Read, R.W. *Tetrahedron Lett.*, ***1994***, *35*, 2729

MeO

$C_5H_{11}COOH$, $(PhCO)_2O$, 1d
20% ($SiCl_4$ + 3 $AgClO_4$)

CH_2Cl_2, rt

O
C_5H_{11}
MeO
67%

Suzuki, K.; Kitagawa, H.; Mukaiyama, T. *Bull. Chem. Soc. Jpn.*, ***1993***, *66*, 3729

O
Ph Cl
Bu_3Sn $SnBu_3$
$Pd(PPh_3)_4$, dioxane
100°C, 30 h
O
Ph Ph
O
50%

Echavarren, A.M.; Pérez, M.; Castaño, A.M.; Cuerva, J.M. *J. Org. Chem.*, ***1994***, *59*, 4179

O O
HO_2C
1. $(PhO)_2PON_3$, NEt_3
dioxane, reflux
2. 2 M HCl
O
O H
H
O
61% (2.8:1 *cis:trans*)

Booker-Milburn, K.I.; Cowell, J.K.; Harris, L.J. *Tetrahedron Lett.*, ***1994***, *35*, 3883

O
t-BuCl/Li/ ultrasound
LiO
O
, THF
O
O
76%

Aurell, M.J.; Einhorn, C.; Einhorn, J.; Luche, J.L. *J. Org. Chem.*, ***1995***, *60*, 8

$PhCO_2Li$
iPrCl, Li, THF, ultrasound, rt
O
Ph
63%

Aurell, M.J.; Danhui, Y.; Einhorn, J.; Einhorn, C.; Luche, J.L. *Synlett*, ***1995***, 459

$NaIO_4$, MeOH , TEBA / DB-18-c6 — 92%

Kore, A.R.; Sagar, A.D.; Salunkhe, M.M. *Org. Prep. Proceed. Int.*, ***1995***, *27*, 373

$[Ph_3GaBu]Li$, THF-hexane , 0°C — 82%

Han, Y.; Fang, L.; Tao, W.-T.; Huang, Y.-Z. *Tetrahedron Lett.*, ***1995***, *36*, 1287

PhMgBr , CuBr , LiBr , THF , 0°C — 75%

Babudri, F.; Fiandanese, V.; Marchese, G.; Punzi, A. *Tetrahedron Lett.*, ***1995***, *36*, 7305

EtMgBr , $Ni(dppe)Cl_2$ / THF — 70%

Malanga, C.; Aronica, L.A.; Lardicci, L. *Tetrahedron Lett.*, ***1995***, *36*, 9185

SECTION 168: KETONES FROM ALCOHOLS AND THIOLS

0.2% *bis*-Ru catalyst , THF / 65°C — C_4H_9 — 99%

Bäckvall, J.-E.; Andreasson, U. *Tetrahedron Lett.*, ***1993***, *34*, 5459

1. *t*-BuOOH , H^+ / 2. DBU — 80x58%

Antonioletti, R.; Arista, L.; Bonadies, F.; Locati, L.; Scettri, A. *Tetrahedron Lett.*, ***1993***, *34*, 7089

1. $RuCl_3 \cdot n\ H_2O$, $Co(OAc)_2 \cdot 4\ H_2O$ / EtOAc , CH_3CHO , 20°C / 2. 1M Na_2SO_3 — 89%

Murahashi, S.-I.; Naota, T.; Hirai, N. *J. Org. Chem.*, ***1993***, *58*, 7318

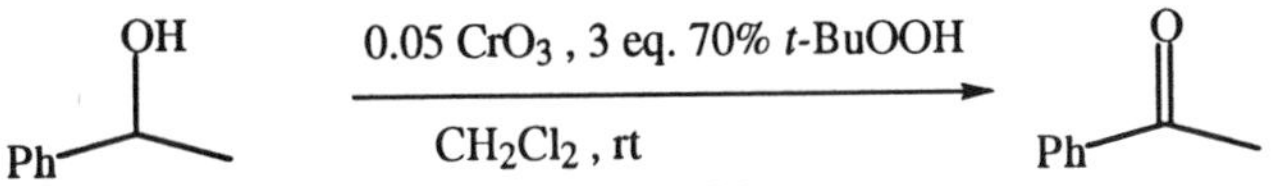
OH
Ph
0.05 CrO_3 , 3 eq. 70% *t*-BuOOH
CH_2Cl_2 , rt
O
Ph

96%

Muzart, J.l; Ajjou, A.N'A. *Synthesis,* ***1993,*** 785

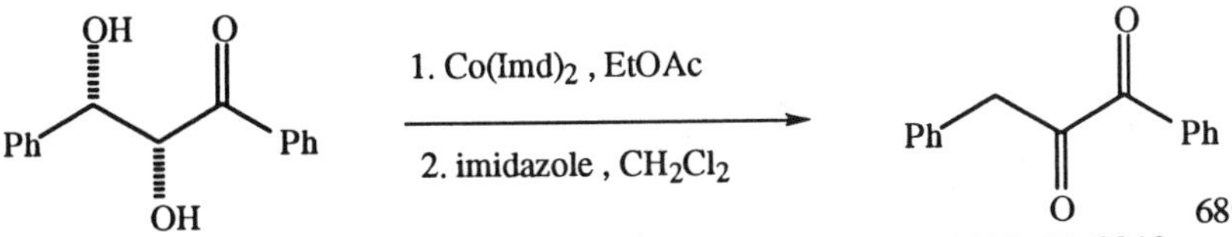
OH
O
Ph
Ph
OH
1. $Co(Imd)_2$, EtOAc
2. imidazole , CH_2Cl_2
O
Ph
Ph
O

68%

Kang, S.-K.; Park, D.-C.; Rho, H.-S.; Han, S.-M. *Synth. Commun.,* ***1993,*** *23,* 2219

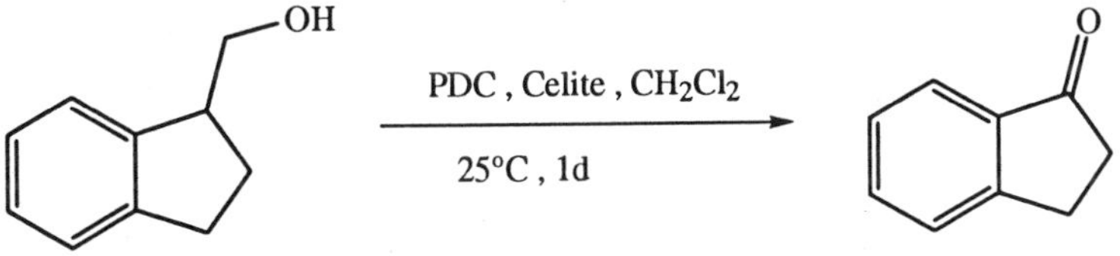
OH
PDC , Celite , CH_2Cl_2
25°C , 1d
O

55%

Bijoy, P.; Subba Rao, G.S.R. *Synth. Commun.,* ***1993,*** *23,* 2701

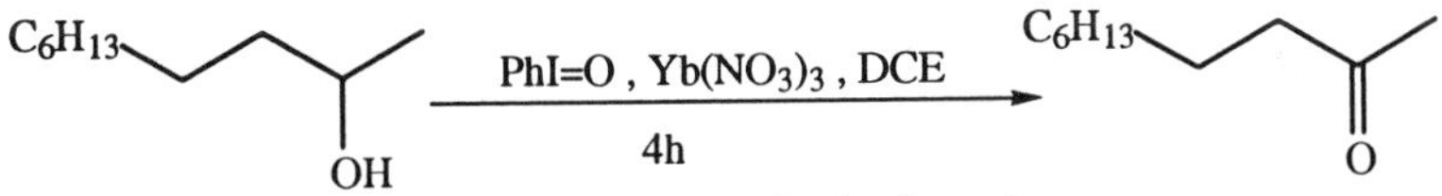
C_6H_{13}
OH
PhI=O , $Yb(NO_3)_3$, DCE
4h
C_6H_{13}
O

aldehydes can also be formed quant.

Yokoo, T.; Matsumoto, K.; Oshima, K.; Utimoto, K. *Chem. Lett.,* ***1993,*** 571

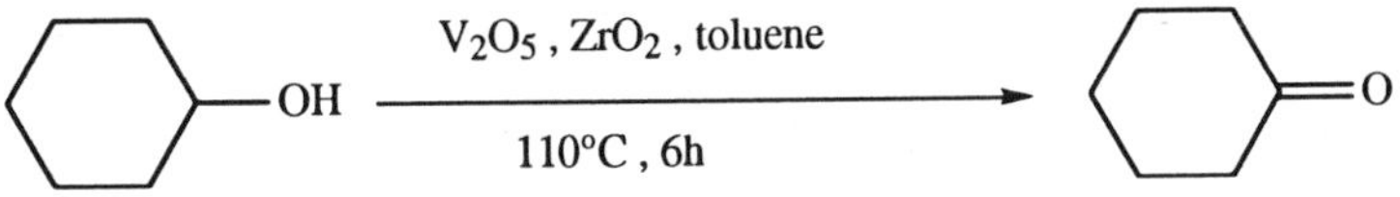
OH
V_2O_5 , ZrO_2 , toluene
110°C , 6h
O

aldehydes can also be formed 88%

Nakamura, H.; Matsuhashi, H.; Arata, K. *Chem. Lett.,* ***1993,*** 749

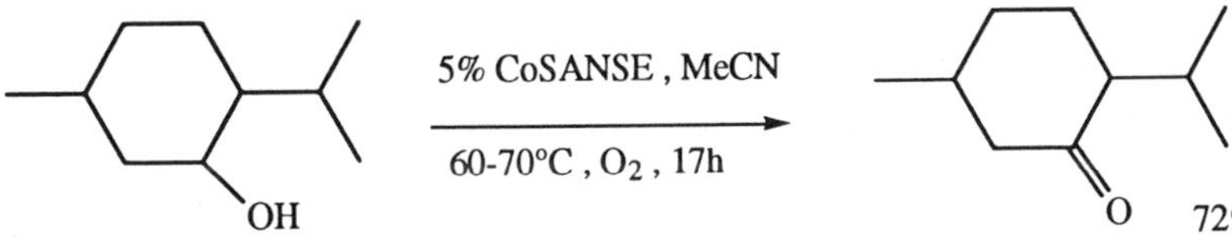
OH
5% CoSANSE , MeCN
60-70°C , O_2 , 17h
O

72%

CoSANSE = [bis(salicylidene-N-(methyl-3-hydroxypropionate))] Co

Punniyamurthy, T.; Iqbal, J. *Tetrahedron Lett.,* ***1994,*** *35,* 4007

PhI , 3% $Pd(OAc)_2$, DMF
3 eq. KOAc , 80°C , 3d
Ph
O
OH
40%

Larock, R.C.; Yum, E.K.; Yang, H. *Tetrahedron*, ***1994***, *50*, 305

, O_2
OH
CoSANSE
O
79%
CoSANSE = chiral cobalt catalyst

Kalra, S.J.S.; Punniyamurthy, T.; Iqbal, J. *Tetrahedron Lett.*, ***1994***, *35*, 4847

t-BuOOH , Et_4NOH , rt
[OsO_4] , aq. *t*-BuOH , 12h
HO H H OH
HO H H O
70%

Beck, C.; Seifert, K. *Tetrahedron Lett.*, ***1994***, *35*, 7221

0.5% Ru complex , 20% Co complex
0.5 O_2 , toluene , 100°C , 36h
2,5-di*t*-butyl-1,4-benzoquinone
OH
O
87%

Wang, G.-Z.; Andreasson,U.; Bäckvall, J.-E. *J. Chem. Soc. Chem. Commun.*, ***1994***, 1037

cat. $PdCl_2$, cat. Adogen 464
9.3h , Na_2CO_3 , DCE , reflux
OH
O
also for primary alcohol → aldehydes
73%

Aït-Mohand, S.; Hénin, F.; Muzart, J. *Tetrahedron Lett.*, ***1995***, *36*, 2473

$RuCl_3$•n H_2O , EtOAc
30% $MeCO_3H$
t-Bu O
t-Bu O
96%

Murahashi, S.-I.; Naota, T.; Oda, Y.; Hirai, N. *Synlett*, ***1995***, 733

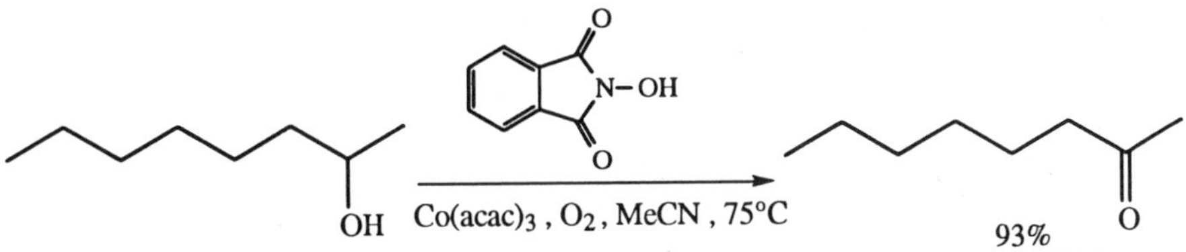

OH $Co(acac)_3$, O_2 , MeCN , 75°C 93% O

Iwahama, T.; Sakaguchi, S.; Nishiyama, Y.; Ishii, Y. *Tetrahedron Lett.*, ***1995***, *36*, 6923

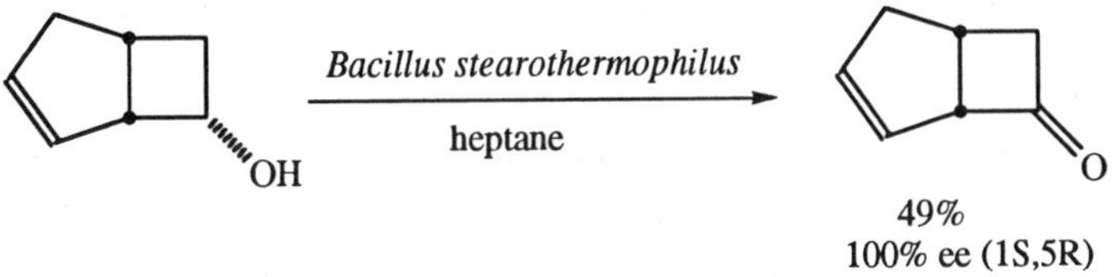

Bacillus stearothermophilus

heptane

OH O

49%
100% ee (1S,5R)

Fantin, G.; Fogagnolo, M.; Giovannini, P.P.; Medici, A.; Pedrini, P.; Poli, S. *Tetrahedron Lett.*, ***1995***, *36*, 441

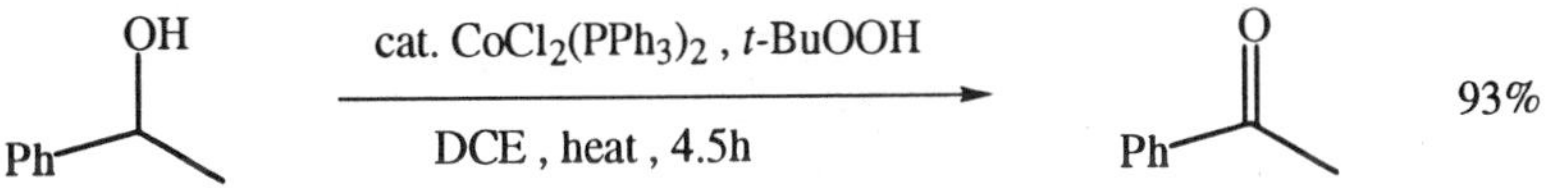

OH cat. $CoCl_2(PPh_3)_2$, *t*-BuOOH O

Ph DCE , heat , 4.5h Ph 93%

aldehydes can also be formed

Iyer, S.; Varghese, J.P. *Synth. Commun.*, ***1995***, *25*, 2261

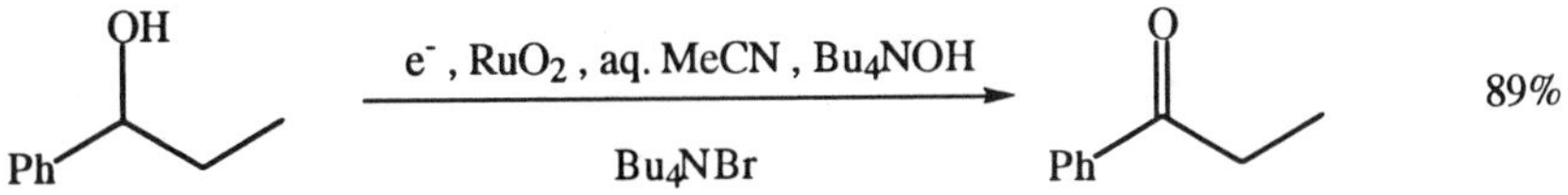

OH e^- , RuO_2 , aq. MeCN , Bu_4NOH O 89%

Ph Bu_4NBr Ph

Torii, S.; Yoshida, A. *Chem. Lett.*, ***1995***, 369

Related Methods: Section 48 (Aldehydes from Alcohols and Phenols).

SECTION 169: KETONES FROM ALDEHYDES

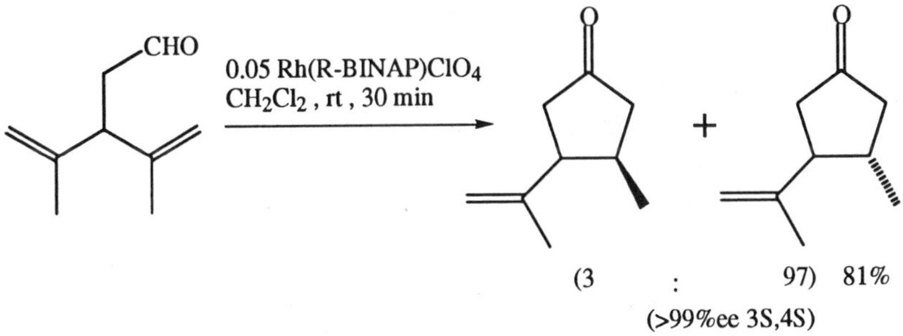

CHO 0.05 Rh(R-BINAP)ClO_4
CH_2Cl_2 , rt , 30 min

O O

(3 : 97) 81%
(>99%ee 3S,4S)

Wu, X.-M.; Funakoshi, K.; Sakai, K. *Tetrahedron Lett.*, ***1993***, *34*, 5927

F—C6H4—CHO → (butanal, $CoCl_2$ / O_2) → 4,4'-difluorobenzil 71%

PhCHO leads to the acid

Punniyamurthy, T.; Kalra, S.J.S.; Iqbal, J. *Tetrahedron Lett.*, ***1994***, *35*, 2959

PhCHO → (1. *t*-$BuMe_2SiCBr_2Li$, -78°C → rt; 2. *sec*-BuLi; 3. PhCHO; 4. HMPA) → PhC(O)CH=CHPh 59%

Shinokubo, H.; Oshima, K.; Utimoto, K. *Tetrahedron Lett.*, ***1994***, *35*, 3741

CHO → ([Rh(S-BINAP)] ClO_4, CH_2Cl_2 / 25°C) → 3-methylcyclopentanone 59% (78% ee, S)

Barnhart, R.W.; Wang, X.; Noheda, P.; Bergens, S.H.; Whelan, J.; Bosnich, B. *J. Am. Chem. Soc.*, ***1994***, *116*, 1821

PhCHO → (Me_3SiCBr_3, $CrBr_3/LiAlH_4$ / THF, 60°C) → PhC(O)CH3 80%

Hodgson, D.M.; Comina, P.J. *Synlett*, ***1994***, 663

PhCHO → (1. benzotriazole (N, N, N, H), EtOH / $CH(OEt)_3$, H^+, THF 70%; 2. BuLi, THF, -78°C; 3. EtBr 4. H^+ 94%) → PhC(O)CH2CH3

Katritzky, A.R.; Lang, H.; Wang, Z.; Zhang, Z.; Song, H. *J. Org. Chem.*, ***1995***, *60*, 7619

PhCHO → ($C_8H_{17}ZrCpCl$, 20% $ZnBr_2$ / THF, 3h, 25°C) → PhCHO → PhC(O)C_8H_{17} 83%

Zheng, B.; Srebnik, M. *J. Org. Chem.*, ***1995***, *60*, 3278

SECTION 170: KETONES FROM ALKYLS, METHYLENES AND ARYLS

This section lists examples of the reaction, $R\text{-}CH_2\text{-}R' \rightarrow R(C{=}O)\text{-}R'$.

Ph Ph — 5% CoSANSE, MeCN; cat. 2-oxocyclopentene carboxylate; 60°C, 15h, O_2 → Ph(C=O)Ph 69%

CoSANSE = [bis(salicylidene-N-(methyl-3-hydroxypropionate))] Co

Punniyamurthy, T.; Iqbal, J. *Tetrahedron Lett.*, ***1994***, *35*, 4003

OMe — Ac_2O, Me_2S/BF_3; CH_2Cl_2, -78°C → rt → OMe, O 76%

Kiselyov, A.S.; Harvery, R.G. *Tetrahedron Lett.*, ***1995***, *36*, 4005

SECTION 171: KETONES FROM AMIDES

Me N OMe O O OMe N Me — 1. 2 eq. PhLi; 2. H^+ → Ph(C=O)(C=O)Ph 75%

Sibi, M.P.; Marvin, M.; Sharma, R. *J. Org. Chem.*, ***1995***, *60*, 5016

N(C=O)–Ph — 1. 2.2 BuLi, -78°C; 2. AcOH; 3. K_2CO_3, 65°C → Bu(C=O)–Ph 82%

Brandänge, S.; Holmgren, E.; Leijonmarck, H.; Rodriguez, B. *Acta Chem. Scand. B.*, ***1995***, *49*, 922

N—CO_2Et — 1. 2 eq. PhLi; 2. H_3O^+ → Ph(C=O)Ph 95%

Prakash, G.K.S.; York, C.; Liao, Q.; Kotian, K.; Olah, G.A. *Heterocycles*, ***1995***, *40*, 79

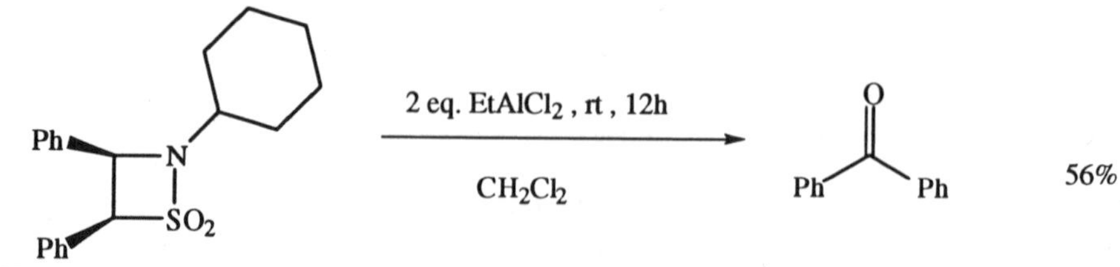

2 eq. $EtAlCl_2$, rt, 12h

CH_2Cl_2

56%

Kataoka, T.; Iwama, T. *Tetrahedron Lett.*, ***1995***, *36*, 245

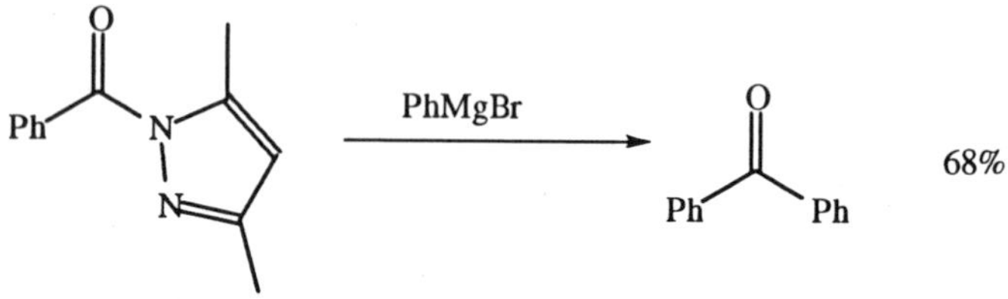

PhMgBr

68%

Kashima, C.; Kita, I.; Takahashi, K.; Hosomi, A. *J. Heterocyclic Chem.*, ***1995***, *32*, 25

REVIEW:

"Chemistry of N-Methoxy N-Methyl Amides. Applications in Synthesis. A Review," Sibi, M.P. *Org. Prep. Proceed. Int.*, ***1993***, *25*, 5

SECTION 172: KETONES FROM AMINES

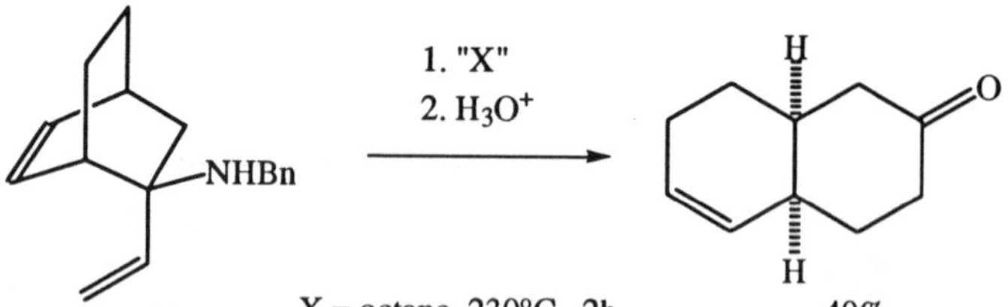

1. "X"
2. H_3O^+

X = octane, 230°C, 2h 40%
X = *t*-BuOK/BuLi/THF/-50°C, 2h 73%

Sprules, T.J.; Galpin, J.D.; Macdonald, D. *Tetrahedron Lett.*, ***1993***, *34*, 247

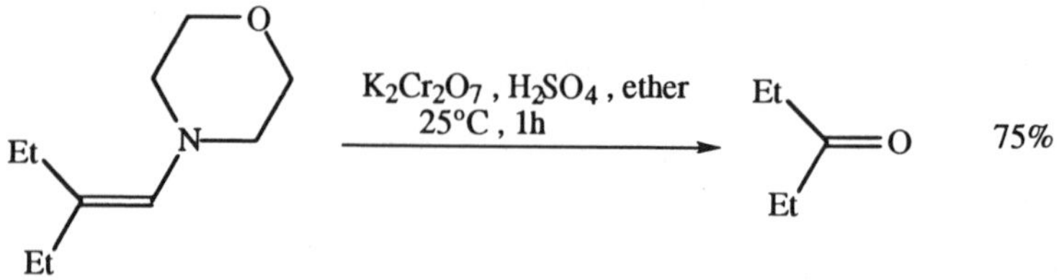

$K_2Cr_2O_7$, H_2SO_4, ether
25°C, 1h

75%

Harris, C.E.; Lee, L.Y.; Dorr, H.; Singaram, B. *Tetrahedron Lett.*, ***1995***, *36*, 2921

OH, N; TS-1 , H_2O_2 , acetone; reflux , 4h; O; 75%

TS-1 = zeolite titanium silicalite

Joseph, R.; Sudalai, A.; Ravindranathan, T. *Tetrahedron Lett.*, ***1994***, *35*, 5493

NH_2, NH_2; O, O; O; 86% O

Gagnon, J.L.; Zajac Jr., W.W. *Tetrahedron Lett.*, ***1995***, *36*, 1803

SECTION 173: KETONES FROM ESTERS

I, CO_2Et, Ph; 1. 2.1 eq. SmI_2; 2. H_3O^+; O, Ph; 74%

Molander, G.A.; McKie, J.A. *J. Org. Chem.*, ***1993***, *58*, 7216

O, O; 1. $Bu_3SnOSnBu_3$; 2. MeI , LiBr ($-CO_2$); 3. H^+; Me, O; 70%

Shibata, I.; Nishio, M.; Baba, A.; Matsuda, H. *Chem. Lett.*, ***1993***, 1953

SECTION 174: KETONES FROM ETHERS, EPOXIDES AND THIOETHERS

Ph, O, O; SiO_2/hexane/CH_2Cl_2; rt , 1d; Ph, O, Ph, O; 98%

Rao, T.B.; Rao, J.M. *Synth. Commun.*, ***1993***, *23*, 1527

$C_7H_{15}CH_2SePh$; OTBDMS, Ph, , aq. MeCN , hν; 1,4-dicyanonaphthalene; C_7H_{15}, O, Ph; 70%

Pandey, G.; Sochanchingwung, R. *J. Chem. Soc. Chem. Commun.*, ***1994***, 1945

OMe

e^-, MeCN - H_2O

$LiNO_3$

O

54%

Shono, T.; Yamamoto, Y.; Takigawa, K.; Maekawa, H.; Ishifune, M.; Kashimura, S. *Chem. Lett.*, ***1994***, 1045

O

5% $Pd(OAc)_2$, 15% PPh_3
PhH, reflux

can also generate aldehydes

O

92

Kulasegaram, S.; Kulawiec, R.J. *J. Org. Chem.*, ***1994***, *59*, 7195

Bu

O

Bu

O

+

Bu

CHO

MABR, CH_2Cl_2, -20°C (0 : 100) 73%
SbF_3, PhH, 25°C (85 : 15) 62%

Maruoka, K.; Murase, N.; Bureau, R.; Ooi, T.; Yamamoto, H. *Tetrahedron*, ***1994***, *50*, 3663

Br

O

Ph O

Bu_3SnH, AIBN, PhH
reflux, 18h

Ph O

+

Ph

O

50% 32%

Crich, D.; Yao, Q. *J. Chem. Soc. Chem. Commun.*, ***1993***, 1265

C_7H_{15}

O

Me_4FeLi_2, ether, -78°C

45 min

C_7H_{15}

O

81%

Kauffmann, T.; Neiteler, C.; Neiteler, G. *Chem. Ber.*, ***1994***, *127*, 659

OBz

O

BF_3•OEt_2, CH_2Cl_2, 0°C

O

OBz

84%

Fujioka, H.; Kitagaki, S.; Imai, R.; Kondo, M.; Okamoto, S.; Yoshida, Y.; Akai, S.; Kita, Y. *Tetrahedron Lett.*, ***1995***, *36*, 3219

SECTION 175: KETONES FROM HALIDES AND SULFONATES

HO / Br (3-bromophenol) → 1. t-BuLi, THF, -78°C; 2. MeO–N(Me)–C(=O)–cyclohexyl, -78°C → rt → HO–C$_6$H$_4$–C(=O)–cyclohexyl 60%

Selnick, H.G.; Bourgeois, M.L.; Butcher, J.W.; Radzilowski, E.M. *Tetrahedron Lett.*, ***1993***, *34*, 2043

I–C$_6$H$_4$–NO_2 → $PhB(OH)_2$, CO, 80°C, 5h; 3% $PdCl_2(MeCN)_2$; 3 eq. K_2CO_3, anisole → Ph–C(=O)–C$_6$H$_4$–NO_2 89%

Ishiyama, T.; Kizaki, H.; Miyaura, N.; Suzuki, A. *Tetrahedron Lett.*, ***1993***, *34*, 7595

NC–C$_6$H$_4$–I → 1. Zn/Ag-graphite, THF, 25°C; 2. benzoyl chloride → NC–C$_6$H$_4$–C(=O)–Ph 89%

Fürstner, A.; Singer, R.; Knochel, P. *Tetrahedron Lett.*, ***1994***, *35*, 1047

Br (2-bromobutane) → 1. Riecke zinc, THF, reflux, 2.5h; 2. PhCOCl, CuCN, LiBr → Ph–C(=O)–CH(CH$_3$)CH$_2$CH$_3$ 95%

Hanson, M.V.; Brown, J.D.; Rieke, R.D.; Niu, Q.J. *Tetrahedron Lett.*, ***1994***, *35*, 7205

$(OC)_5M$=C(OMe)Ph → 1. MeLi, LiBr, $ClCH_2I$, -78°C → 20°C; 2. H_2O, 20°C → Me–C(=O)–Ph

M = Mo 80%
= Cr 86%

Barluenga, J.; Bernad Jr., P.L.; Concellón, J.M. *Tetrahedron Lett.*, ***1994***, *35*, 9471

C_6H_{13}CH$_2$CH$_2$I → 1. Zn, THF, 40°C; 2. $CoBr_2$, CO, THF, NMP, 25°C, 8h → C_8H_{17}–C(=O)–C_8H_{17} 59%

Devasagayaraj, A.; Knochel, P. *Tetrahedron Lett.*, ***1995***, *36*, 8411

, 5 atm CO

$PdCl_2/PPh_3$, NEt_3, PhH
120°C, 20h

60%

Satoh, T.; Itaya, T.; Okuro, K.; Miura, M.; Nomura, M. *J. Org. Chem.*, ***1995***, *60*, 7267

OBn I OTf

1. OEt OTBDMS, BuLi
THF, -78°C
2. aq. HF, MeCN

OBn O

88x99%

Hosoya, T.; Hasegawa, T.; Kuriyama, Y.; Matsumoto, T.; Suzuki, K. *Synlett*, ***1995***, 177

Br Ph Ph Br

2 eq. DMSO, microwave

O Ph Ph O 75%

Villemin, D.; Hammadi, M. *Synth. Commun.*, ***1995***, *25*, 3145

Related Methods: Section 177 (Ketones from Ketones).
Section 55 (Aldehydes from Halides).

SECTION 176: KETONES FROM HYDRIDES

This section lists examples of the replacement of hydrogen by ketonic groups, R-H → R(C=O)-R'. For the oxidation of methylenes, R_2CH_2 → $R_2C=O$, see section 170 (Ketones from Alkyls).

1% $RuCl_2(PPh_3)_3$, PhH
t-BuOOH, 10% aq. Na_2SO_3
rt

O + OH

41% conversion

47% 7%

Murahashi, S.; Oda, Y.; Naota, T.; Kuwabara, T. *Tetrahedron Lett.*, ***1993***, *34*, 1299

Ph Ph

CrO_2Cl_2, CCl_4
ultrasound

O Ph Ph 99%

Luzzio, F.A.; Moore, W.J. *J. Org. Chem.*, ***1993***, *58*, 512

$BaRu(O)_2(OH)_3$, LiCl , CH_2Cl_2

AcOH , 23°C , 1h

69%

Lau, T.-C.; Mak, C.-K. *J. Chem. Soc. Chem. Commun.*, ***1993***, 766

N—CrO_3H , CH_2Cl_2

-20°C , 3h

O 98%

Banwell, M.G.; Haddad, N.; Huglin, J.A.; MacKay, M.F.; Reum, M.E.; Ryan, J.H.; Turner, K.A. *J. Chem. Soc. Chem. Commun.*, ***1993***, 954

OMe

0.2 $Yb(OTf)_3$, $MeNO_2$

2 Ac_2O , 18h

O OMe

Kawada, A.; Mitamura, S.; Kobayashi, S. *J. Chem. Soc. Chem. Commun.*, ***1993***, 1157

OMe

Ac_2O , 0.2 $Sc(OTf)_3$

$MeNO_2$, 50°C , 4h

O OMe

89%

Kawada, A.; Mitamura, S.; Kobayashi, S. *Synlett*, ***1994***, 545

Ph Ph

10% N-hydroxyphthalimide

PhCN , O_2 , 100°C , 20h

O Ph Ph

73%

Ishii, Y.; Nakayama, K.; Takeno, M.; Sakaguchi, S.; Iwahama, T.; Nishiyama, Y. *J. Org. Chem.*, ***1995***, *60*, 3934

MeO

$Hf(OTf)_4$, Ac_2O , $LiClO_4$—$MeNO_2$

MeO O

95%

Hachiya, I.; Moriwaki, M.; Kobayashi, S. *Tetrahedron Lett.*, ***1995***, *36*, 409

PhCOCl , toluene , reflux

0.1 [ReBr(CO)$_5$]

91% (11:4:85 *o:m:p*)

Kusama, H.; Narasaka, K. *Bull. Chem. Soc. Jpn.*, ***1995***, *68*, 2379

SECTION 177: KETONES FROM KETONES

This section contains alkylations of ketones and protected ketones, ketone transpositions and annulations, ring expansions and ring openings and dimerizations. Conjugate reductions and Michael alkylations of enone are listed in Section 74 (Alkyls from Alkenes).

For the preparation of enamines or imines from ketones, see Section 356 (Amine-Alkene).

1. $CF_3SO_3SiMe_3$, MeCN
Me_3Si-S-$SiMe_3$, rt
2. aq. $NaHCO_3$

76%

Degl'Innocenti, A.; Capperucci, A.; Mordini, A.; Reginato, G.; Ricci, A.; Cerreta, F. *Tetrahedron Lett.*, ***1993***, *34*, 873

1. PhCN , $TiCl_4$-Zn
2. 3% HCl

86% 7%

Gao, J.; Hu, M-Y.; Chen, J-x.; Yuan, S.; Chen, W-x *Tetrahedron Lett.*, ***1993***, *34*, 1617

$SnCl_4$, $MeNO_2$

86%

Patra, P.K.; Patro, B.; Ila, H.; Junjappa, H. *Tetrahedron Lett.*, ***1993***, *34*, 3951

$Si(iPr)_3$

$TiCl_4$, CH_2Cl_2
-78°C → -20°C

$Si(iPr)_3$ 92%

(5:1 *anti:syn*)

Knölker, H.-J.; Graf, R. *Tetrahedron Lett.*, ***1993***, *34*, 4765

1. LDA , THF , TBDMSCl
2. hν , DCN , aq. MeCN
Pyrex filter

MeO O MeO O

98x72%

Pandey, G.; Krishna, A.; Girija, K.; Karthikeyan, M. *Tetrahedron Lett.*, ***1993***, *34*, 6631

1. $LiN(TMS)_2$
2. $MnBr_2$
3. MeI , 22°C , 4h

95%

Reetz, M.T.; Haning, H. *Tetrahedron Lett.*, ***1993***, *34*, 7395

hν , $LiClO_4$
3h , NEt_3

Me 70%

Cossy, J.; Furet, N. *Tetrahedron Lett.*, ***1993***, *34*, 8107

1. KH , ether , 3h
2. BuBr , ether , rt , 18h

Bu 42%

\+ 56% cyclohexanone

Bates, R.B.; Taylor, S.R. *J. Org. Chem.*, ***1993***, *58*, 4469

C_3H_7 C_3H_7

1. PhMeN-MnMe , 20°C
THF/DMSO
2. MeI

C_3H_7 C_3H_7

93% Me

Cahiez, G.; Figadère, B.; Cléry, P. *Tetrahedron Lett.*, ***1994***, *35*, 3065

OTMS

CF_3CO_2Ag , MeI

CH_2Cl_2 , -78°C → rt

O O

+

(6.1 : 1) 78%

Angers, P.; Canonne, P. *Tetrahedron Lett.*, ***1994***, *35*, 367

PhSe O

Bu_3SnH , AIBN

O 53% (1:1)

Chen, L.; Gill, G.B.; Pattenden, G. *Tetrahedron Lett.*, ***1994***, *35*, 2593

$OSiMe_3$

1. MeLi
2. $MnCl_2$
3. $PhCH_2Br$

O

Ph

89%

Cahiez, G.; Chau, K.; Cléry, P. *Tetrahedron Lett.*, ***1994***, *35*, 3069

O

Bu Bu

$MeCHN_2$, Me_3Al ,0°C

CH_2Cl_2 , 1h

O

Bu Bu

Me

55%

Maruoka, K.; Concepcion, A.B.; Yamamoto, H. *J. Org. Chem.*, ***1994***, *59*, 4725

I

Cl

O

2.5 eq. Bu_3SnH , AIBN

8h slow addition

57% O

Dowd, P.; Zhang, W.; Mahmood, K. *Tetrahedron Lett.*, ***1994***, *35*, 5563

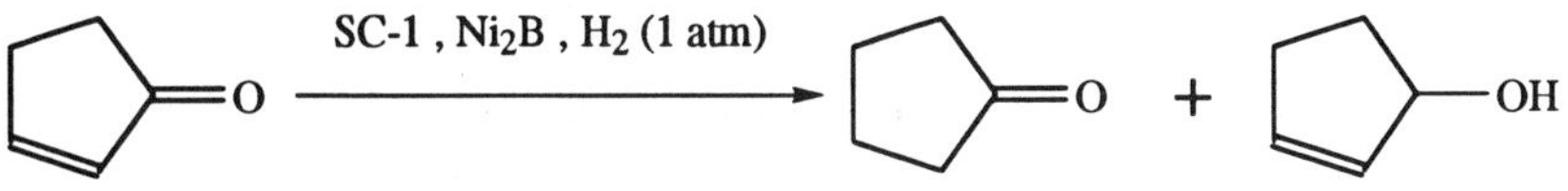
SC-1 , Ni_2B , H_2 (1 atm)
O
O
+
OH

(>99 : 1) >99%

Belisle, C.M.; Young, Y.M.; Singaram, B. *Tetrahedron Lett., **1994**, 35*, 5595

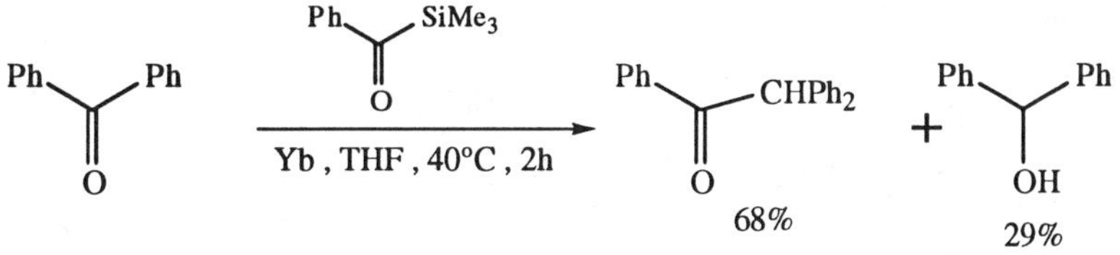
Ph
SiMe3
O
Ph
Ph
O
Yb , THF , 40°C , 2h
Ph
CHPh2
O
+
Ph
Ph
OH
68%
29%

Taniguchi, Y.; Nagafuji, A.; Makioka, Y.; Takaki, K.; Fujiwara, Y. *Tetrahedron Lett., **1994**, 35*, 6897

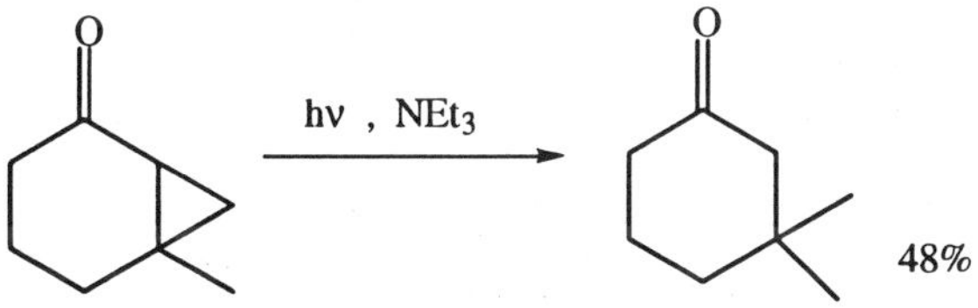
O
hν , NEt3
O
48%

Kirschberg, T.; Mattay, J. *Tetrahedron Lett., **1994**, 35*, 7217

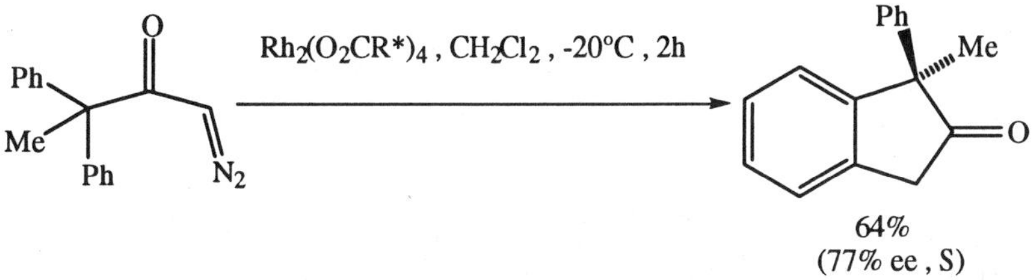
O
Ph
Me
Ph
N2
Rh2(O2CR*)4 , CH2Cl2 , -20°C , 2h
Ph
Me
O
64%
(77% ee , S)

Watanabe, N.; Ohtake, Y.; Hashimoto, S.; Shiro, M.; Ikegami, S. *Tetrahedron Lett., **1995**, 36*, 1491

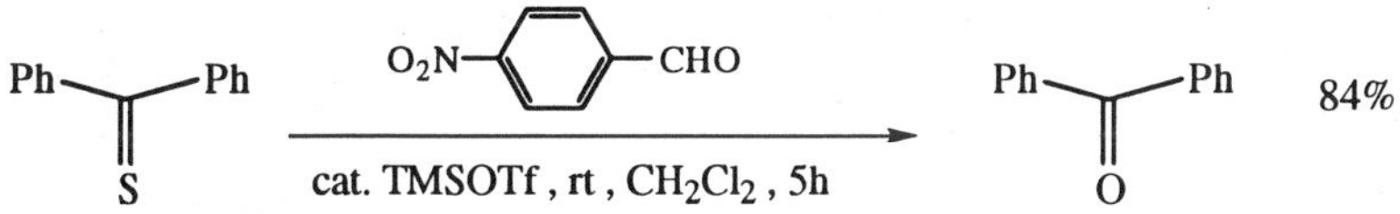
Ph
Ph
S
O2N
CHO
cat. TMSOTf , rt , CH2Cl2 , 5h
Ph
Ph
O
84%

Ravindranathan, T.; Chavan, S.P.; Awachat, M.M.; Kelkar, S.V. *Tetrahedron Lett., **1995**, 36*, 2277

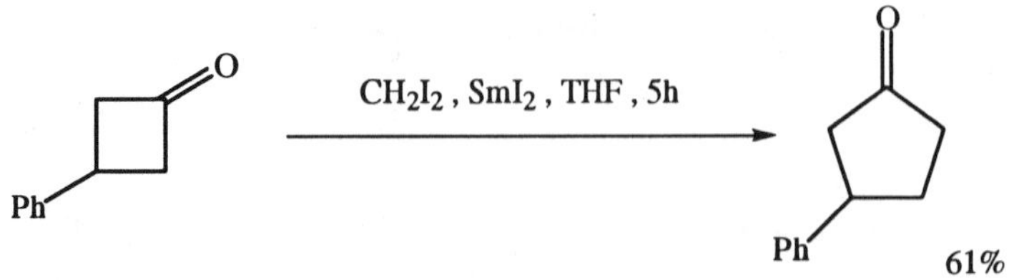
O
CH_2I_2 , SmI_2 , THF , 5h
Ph
O
Ph

61%

Fukuzawa, S.; Tsuchimoto, T. *Tetrahedron Lett.*, ***1995***, *36*, 5937

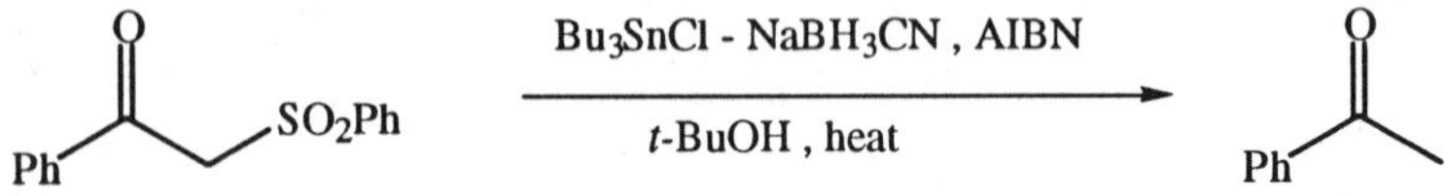
O
SO_2Ph
Ph
Bu_3SnCl - $NaBH_3CN$, AIBN
t-BuOH , heat
O
Ph

86%

Giovannini, R.; Petrini, M. *Synlett*, ***1995***, 973

REVIEW:

"Organotin Enolates in Organic Synthesis. A Review," Shibata, I.; Baba, A. *Org. Prep. Proceed. Int.*, ***1994***, *26*, 123

Related Methods: Section 49 (Aldehydes from Aldehydes).

SECTION 178: KETONES FROM NITRILES

NO ADDITIONAL EXAMPLES

SECTION 179: KETONES FROM ALKENES

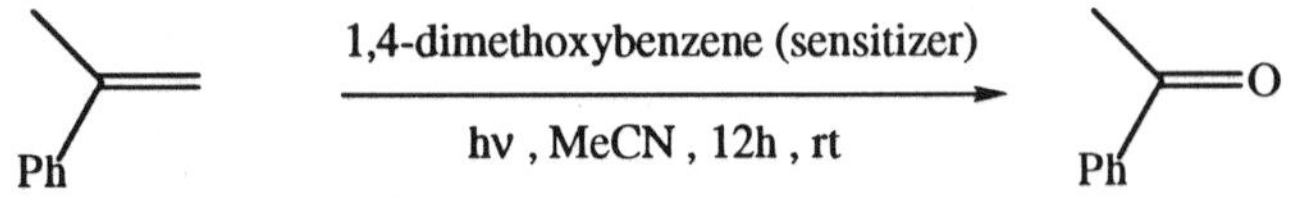
Ph
1,4-dimethoxybenzene (sensitizer)
hν , MeCN , 12h , rt
O
Ph

90% conversion (quant.)

Bhalerao, U.T.; Sridhar, M. *Tetrahedron Lett.*, ***1993***, *34*, 4341

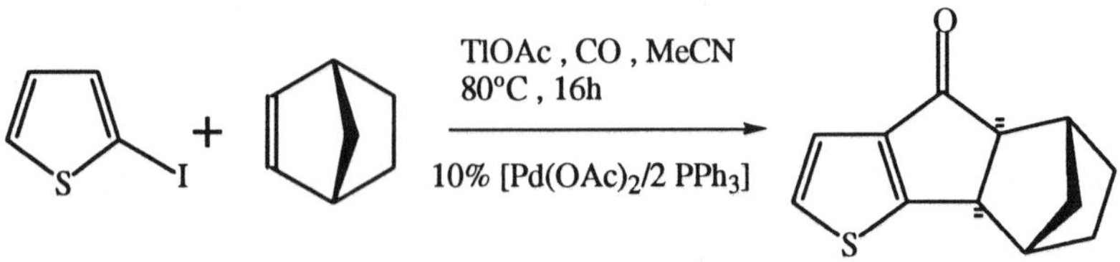
S
I
+
TlOAc , CO , MeCN
80°C , 16h
10% [$Pd(OAc)_2$/2 PPh_3]
O
S

75%

Grigg, R.; Khalil, H.; Levett, P.; Virica, J.; Sridharan, V. *Tetrahedron Lett.*, ***1994***, *35*, 3197

10% $Pd(OAc)_2$, 20% PPh_3
Et_4NCl , CO

Bu_3Sn–(2-furyl)

88%

Grigg, R.; Redpath, J.; Sridharan, V.; Wilson, D. *Tetrahedron Lett.*, ***1994***, *35*, 4429

1. $Et_2BH{\cdot}SMe_2$, ether , 25°C
2. Et_2Zn , neat , 0°C
3. $CuCN{\cdot}2\ LiCl$, THF , 0°C
4. PhCOCl

Ph

86%

Langer, F.; Devasagayaraj, A.; Chavant, P.-Y.; Knochel, P. *Synlett*, ***1994***, 410

Me_3SiN_3 , CH_2Cl_2 , CrO_3

N_3

59%

Reddy, M.V.R.; Kumareswaran, R.; Vankar, Y.D. *Tetrahedron Lett.*, ***1995***, *36*, 6751

1. [$CoCl_2{\cdot}THF/PhEt_2N{:}BH_3$, PhH , CO]
2. CO , 6h
3. 25°C , 36h
4. H_2O_2 , NaOH , NaO*t*-Bu

$C_{10}H_{21}$ $C_{10}H_{21}$ $C_{10}H_{21}$

70%

Rao, M.L.N.; Periasamy, M. *Tetrahedron Lett.*, ***1995***, *36*, 9069

, $Ni(cod)_2$

2 PPh_3 , DCE

+

(>20 : 1) 99%

Lautens, M.; Edwards, L.G.; Tam, W.; Lough, A.J. *J. Am. Chem. Soc.*, ***1995***, *117*, 10276

1. $Cl_2C{=}C{=}O$
2. Bu_3SnH , AIBN

85% (84:16)

Dowd, P.; Zhang, W.; Geib, S.J. *Tetrahedron*, ***1995***, *51*, 3435

See also: Section 134 (Ethers from Alkenes).
Section 174 (Ketones from Ethers).

SECTION 180: KETONES FROM MISCELLANEOUS COMPOUNDS

Conjugate reductions and reductive alkylations of enones are listed in Section 74 (Alkyls from Alkenes).

Ph, =N-NHPh → (dimethyldioxirane), TFP, CH_2Cl_2, 20°C, 30 min → Ph, =O

92%

TFP = 1,1,1-trifluoropropane

Altamura, A.; Curci, R.; Edwards, J.O. *J. Org. Chem.*, ***1993***, *58*, 7289

=NOH → acetone, (dimethyldioxirane), 0°C, 10 min → =O

80%

Olah, G.A.; Liao, Q.; Lee, C.-S.; Prakash, G.K.S. *Synlett*, ***1993***, 427

Ph, Ph, =N-OH → O_3 , CH_2Cl_2, -2°C, 2h → Ph, Ph, =O 95%

Yang, Y.; Li, T.; Li, Y. *Synth. Commun.*, ***1993***, *23*, 1121

N—NHTs, Ph, Me → $PhI(OAc)_2$, MeCN , rt , 10 min → O, Ph, Me

87%

Zeng, H.; Chen, Z.-C. *Synth. Commun.*, ***1993***, *23*, 2497

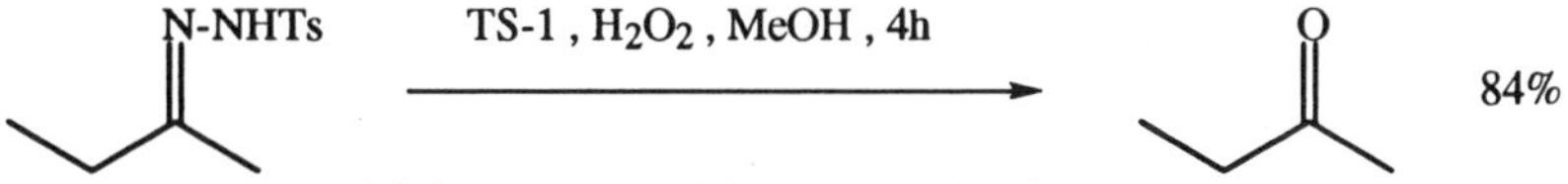

N-NHTs → $TS\text{-}1, H_2O_2, MeOH, 4h$ → O 84%

TS-1 = titanium silicate molecular sieves

Kumar, P.; Hegde, V.R.; Pandey, B.; Ravindranathan, T. *J. Chem. Soc. Chem. Commun.*, ***1993***, 1553

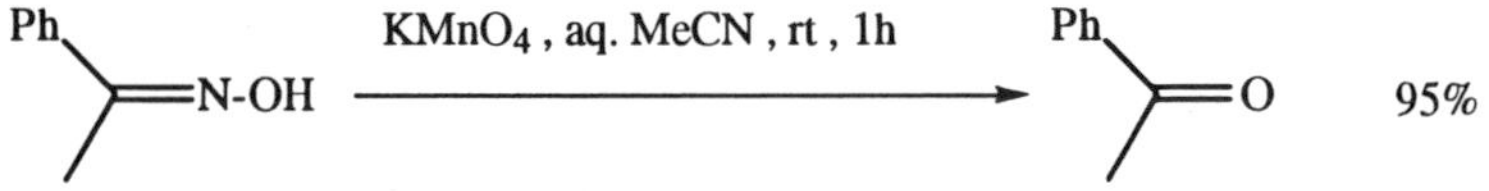

Ph =N-OH → $KMnO_4$, aq. MeCN, rt, 1h → Ph =O 95%

Wali, A.; Ganeshpure, P.A.; Satish, S. *Bull. Chem. Soc. Jpn.*, ***1993***, *66*, 1847

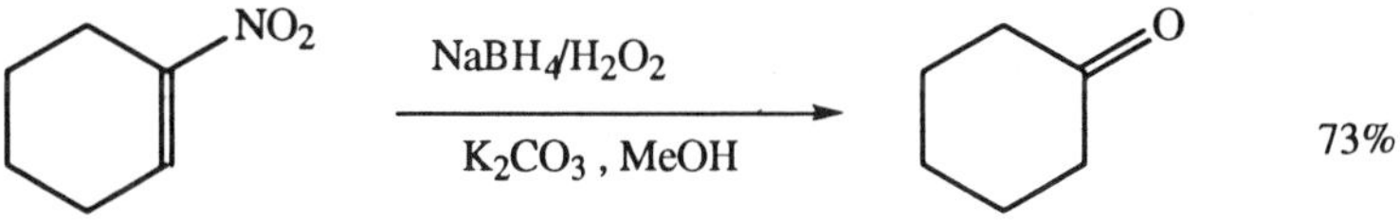

NO_2 → $NaBH_4/H_2O_2$, K_2CO_3, MeOH → O 73%

Ballini, R.; Bosica, G. *Synthesis*, ***1994***, 723

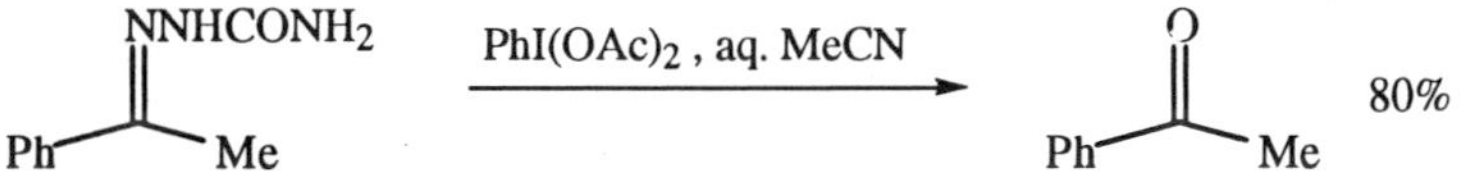

$NNHCONH_2$ Ph Me → $PhI(OAc)_2$, aq. MeCN → O Ph Me 80%

Chen, D.W.; Chen, Z.C. *Synthesis*, ***1994***, 777

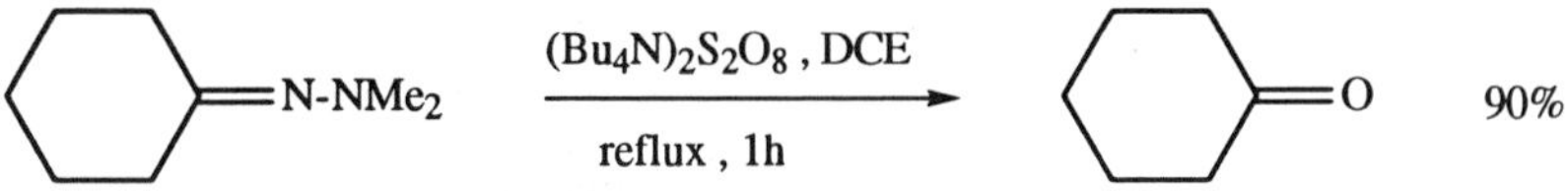

=N-NMe_2 → $(Bu_4N)_2S_2O_8$, DCE, reflux, 1h → =O 90%

Choi, H.C.; Kim, Y.H. *Synth. Commun.*, ***1994***, *24*, 2307

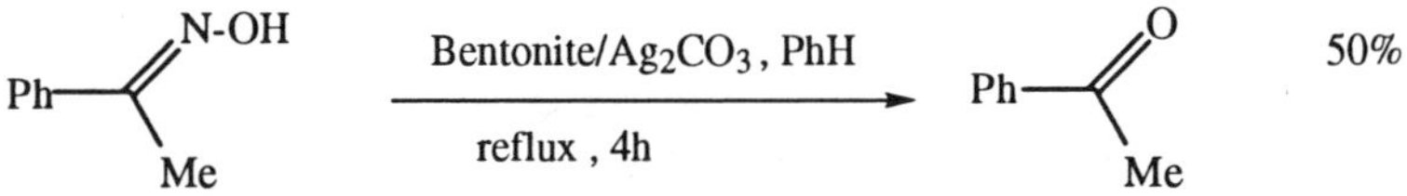

N-OH Ph Me → Bentonite/Ag_2CO_3, PhH, reflux, 4h → O Ph Me 50%

Sanabria, R.; Miranda, R.; Lara, V.; Delgado, F. *Synth. Commun.*, ***1994***, *24*, 2805

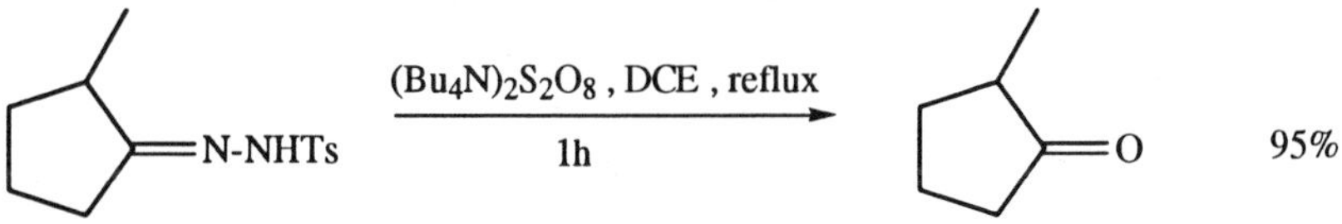

=N-NHTs → $(Bu_4N)_2S_2O_8$, DCE, reflux, 1h → =O 95%

Chen, F.; Yang, J.; Zhang, H.; Guan, C.; Wan, J. *Synth. Commun.*, ***1995***, *25*, 3163

REVIEW:

"Macrocycle Synthesis: Cyclic Ketones, Ketoalkenes, Diketones and Dienes of Ring Size C_{21} to C_{26}," Forbes, M.D.E.; Dang, Y. *Org. Prep. Proceed. Int.*, ***1993***, *25*, 309

SECTION 180A: PROTECTION OF KETONES

[Rh(MeCN)$_3$(triphos)] (OTf)$_2$

acetone, rt, 4h

90%

Ma, S.; Venanzi, L.M. *Tetrahedron Lett.*, ***1993***, *34*, 8071

HS SH , Sm/I_2

rt

92%

Zhang, Y.; Yu, Y.; Lin, R. *Org. Prep. Proceed. Int.*, ***1993***, *25*, 365

MeO OH OH

80%

p-TsOH, PhH, reflux, cat. TMSI

OMe

DDQ, H_2O, CH_2Cl_2

95%

McDonald, C.E.; Nice, L.E.; Kennedy, K.E. *Tetrahedron Lett.*, ***1994***, *35*, 57

Me_3SiO $SiMe_3$ $OSiMe_3$

$CF_3SO_3SiMe_3$, CH_2Cl_2 -78°C

94%

$SiMe_3$

$LiBF_4$, MeCN, heat 96%

Lillie, B.M.; Avery, M.A. *Tetrahedron Lett.*, ***1994***, *35*, 969

OEt OEt

AcCl, 2% $SmCl_3$, rt

15 min

95%

Wu, S.-H.; Ding, Z.B. *Synth. Commun.*, ***1994***, *24*, 2173

OSiMe$_3$ → cat. $Pd(OAc)_2$, 2 NaOAc , O_2 / DMSO → O 58%

Larock, R.C.; Hightower, T.R.; Kraus, G.A.; Hahn, P.; Zheng, D. *Tetrahedron Lett.*, ***1995***, *36*, 2423

S S Ph → SeO_2 , AcOH , rt , 25 min / aldehydes can also be used → O Ph 98%

Haroutounian, S.A. *Synthesis*, ***1995***, 39

O O → $CuSO_4/SiO_2$, $CHCl_3$ / 20°C , 2d → O 70%

Caballero, G.M.; Gros, E.G. *Synth. Commun.*, ***1995***, *25*, 395

t-Bu O → HO OH , CH_2Cl_2 / cat. TMSOTf , iPrOTMS -20°C , 3h → *t*-Bu O O 99%

Kurihara, M.; Miyata, N. *Chem. Lett.*, ***1995***, 263

See Section 362 (Ester-Alkene) for the formation of enol esters and Section 367 (Ether-Alkenes) for the formation of enol ethers. Many of the methods in Section 60A (Protection of Aldehydes) are also applicable to ketones.

CHAPTER 13
PREPARATION OF NITRILES

SECTION 181: NITRILES FROM ALKYNES

NO ADDITIONAL EXAMPLES

SECTION 182: NITRILES FROM ACID DERIVATIVES

NO ADDITIONAL EXAMPLES

SECTION 183: NITRILES FROM ALCOHOLS AND THIOLS

$C_6H_{13}CH(OH)Me$ —[$Me_2C(CN)OH$, MeCN , DEAD]→ $C_6H_{13}CH(CN)Me$ 58%

Wilk, B.K. *Synth. Commun.*, ***1993***, *23*, 2481

SECTION 184: NITRILES FROM ALDEHYDES

CHO —[1. Me_2NNH_2 , MeOH , rt; 2. "crude" , MeOH , 0°C, 2.4 MMPP•6 H_2O]→ CN 88%

MMPP = magnesium monoperoxyphthalate

Fernández, R.; Gasch, C.; Lassalwtta, J-M.; Llera, J-M.; Vázquez, J. *Tetrahedron Lett.*, ***1993***, *34*, 141

SECTION 185: NITRILES FROM ALKYLS, METHYLENES AND ARYLS

NO ADDITIONAL EXAMPLES

SECTION 186: NITRILES FROM AMIDES

$\xrightarrow[3\ NEt_3]{2\ PPh_3,\ 2\ CCl_4}$ 97%

other reagents are also given that lead to this conversion

Walters, M.A.; Hoem, A.B.; Arcand, H.R.; Hegeman, A.D.; McDonough, C.S. *Tetrahedron Lett.*, ***1993***, *34*, 1453

$\xrightarrow[\text{(dark), MS 4Å, 12h}]{Ag_2O\text{, EtI, PhH, 25°C}}$ Ph—CN 82%

Sznaidman, M.L.; Crasto, C.; Hecht, S.M. *Tetrahedron Lett.*, ***1993***, *34*, 1581

$\xrightarrow[TBAHSO_4,\ Na_3PO_4 \cdot 12\ H_2O]{NaOCl,\ NaBr,\ PhH/H_2O}$ 55%

Correia, J. *Synthesis*, ***1994***, 1127

SECTION 187: NITRILES FROM AMINES

NO ADDITIONAL EXAMPLES

SECTION 188: NITRILES FROM ESTERS

$\xrightarrow[\text{THF, reflux, 16h}]{Me_3SiCN,\ Pd(PPh_3)_4}$

98% (*E*:*Z*, >9:1)

Tsuji, Y.; Yamada, N.; Tanaka, S. *J. Org. Chem.*, ***1993***, *58*, 16

SECTION 189: NITRILES FROM ETHERS, EPOXIDES AND THIOETHERS

NO ADDITIONAL EXAMPLES

SECTION 190: NITRILES FROM HALIDES AND SULFONATES

1. BuLi, THF, -100°C
2. ZnI_2, -100°C → 0°C
3. TosCN, -78°C → rt, 3h

84%

Klement, I.; Lennick, K.; Tucker, C.E.; Knochel, P. *Tetrahedron Lett.*, ***1993***, *34*, 4623

Br (1-bromobutane) —[$Me_2C(CN)OH$, MeCN, TMG]→ CN (pentanenitrile) 71%

TMG = 1,1,3,3-tetramethylguanidine

Dowd, P.; Wilk, B.K.; Wlostowski, M. *Synth. Commun.*, ***1993***, *23*, 2323

Ph_2CHCl —[$TiCl_4$, TMSCN, CH_2Cl_2, 0°C]→ Ph_2CH—CN 93%

Zieger, H.E.; Wo, S. *J. Org. Chem.*, ***1994***, *59*, 3838

SECTION 191: NITRILES FROM HYDRIDES

NO ADDITIONAL EXAMPLES

SECTION 192: NITRILES FROM KETONES

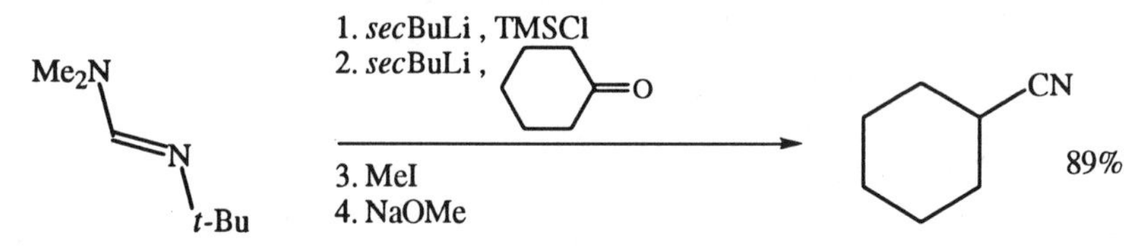

also works with aldehyde substrates

Santiago, B.; Meyers, A.I. *Tetrahedron Lett.*, ***1993***, *34*, 5839

PhC(O)CF_3 —[1. $Me(Cl)AlNH_2$; 2. *t*-BuOK]→ Ph—CN 85%

Kende, A.S.; Liu, K. *Tetrahedron Lett.*, ***1995***, *36*, 4035

SECTION 193: NITRILES FROM NITRILES

Conjugate reductions and Michael alkylations of alkene nitriles are found in Section 74D (Alkyls from Alkenes).

NO ADDITIONAL EXAMPLES

SECTION 194: NITRILES FROM ALKENES

NO ADDITIONAL EXAMPLES

SECTION 195: NITRILES FROM MISCELLANEOUS COMPOUNDS

Cl–C6H4–CH=N-OH —[Envirocat EPZG, 100°C; neat]→ Cl–C6H4–CN 83%

Dandgar, B.P.; Jagtap, S.R.; Ghodeshwar, S.B.; Wadgaonkar, P.P. *Synth. Commun.*, ***1995***, *25*, 2993

CHAPTER 14
PREPARATION OF ALKENES

SECTION 196: ALKENES FROM ALKYNES

I

Ph-C≡C-Ph , DMF , 120°C

10% $Pd(OAc)_2$, 20% PPh_3

2 TlOAc , 36h

Ph Ph

45%

Grigg, R.; Kennewell, P.; Teasdale, A.; Sridharan, V. *Tetrahedron Lett.*, ***1993***, *34*, 153

OTs

$(PhMe_2Si)_2CuCNLi_2$

THF , -78°C → rt

$SiMe_2Ph$

75%

Fleming, I.; de Marigorta, E.M. *Tetrahedron Lett.*, ***1993***, *34*, 1201

C_3H_7—≡—C_3H_7

1. Br , rt

$Cp_2Zr(CH_2{=}CH_2)$. 1d

2. H^+

C_3H_7 C_3H_7

68%

Takahashi, T.; Kondakov, D.Y.; Suzuki, N. *Tetrahedron Lett.*, ***1993***, *34*, 6571

N-OBn

$(Me_3Si)_3SiH$, hν (sunlamp)

2h , hexane

BnO-N

70%

Pattenden, G.; Schulz, D.J. *Tetrahedron Lett.*, ***1993***, *34*, 6787

10% $Pd(OAc)_2$
20% PPh_3, DMF
Tl(OAc), 130°C
18h

O_2N NO_2 I EtO_2C CO_2Et O

60%

Brown, S.; Clarkson, S.; Grigg, R.; Sridharan, V. *Tetrahedron Lett.*, ***1993***, *34*, 157

1. $Cp_2Zr(H)Cl$, CH_2Cl_2
20°C, 20 min
2. 5% $AgCl_4$
10 min

TBDMSO HO

92%

Wipf, P.; Xu, W. *J. Org. Chem.*, ***1993***, *58*, 825

1. *t*-BuLi, pentane/ether
2. -78°C → rt → -78°C
3. D_2O

Ph I Ph D

88%

Bailey, W.F.; Ovaska, T.V. *J. Am. Chem. Soc.*, ***1993***, *115*, 3080

1. PhSCu, ether, BuLi, -35°C
2. add alkyne via syringe pump

t-Bu OAc *t*-Bu

86%

Nantz, M.H.; Bender, D.M.; Janaki, S. *Synthesis*, ***1993***, 577

1. *t*-BuLi, pentane/ether, -78°C
2. +20°C, 1h
3. MeOH

I $(CH_2)_3$

85%

Bailey, W.F.; Ovaska, T.V. *Chem. Lett.*, ***1993***, 819

Et—≡—Et → 1. CH_2=CH_2 , Cp_2ZrBu_2 rt , 1h; 2. H^+ → 91%

Takahashi, T.; Xi, Z.; Rousset, C.J.; Suzuki, N. *Chem. Lett.*, ***1993***, 1001

Ph—≡—Et → HCO_2H , NEt_3 , 3h; cat. $Pd_2(dba)_3$-PBu_3 → 89% *cis* + 5% *trans*

Tani, K.; Ono, N.; Okamoto, S.; Sato, F. *J. Chem. Soc. Chem. Commun.*, ***1993***, 386

Me_3Si—≡ → CO , EtOH , $PdCl_2(PP_3)_2$, PhH; 5 $SnCl_2$, 90°C , 1.5h → $SiMe_3$ EtO_2C 77%

Takeuchi, R.; Sugiura, M. *J. Chem. Soc., Perkin Trans. 1.*, ***1993***, 1031

Bu—≡ → 1. 3 eq $AlMe_3$, 0.2 Cp_2ZrCl_2 1.5 eq. H_2O , CH_2Cl_2 , -70°C; 2. 3M HCl → Bu + Bu (97 : 3) quant.

Wipf, P.; Lim, S. *Angew. Chem. Int. Ed. Engl.*, ***1993***, *32*, 1068

≡—C_8H_{17} → Ph_3SiH , BEt_3 , PhH → Ph_3Si C_8H_{17} 42% (12:1 *Z:E*)

Miura, K.; Oshima, K.; Utimoto, K. *Bull. Chem. Soc. Jpn.*, ***1993***, *66*, 2356

C_3H_7—≡—C_3H_7 → SiO_2—O—Si—H; $Pd(PPh_3)_4$, AcOH , 16h → C_3H_7 C_3H_7 97%

Kini, A.D.; Nadkarni, D.V.; Fry, J.L. *Tetrahedron Lett.*, ***1994***, *35*, 1507

I O → $CrCl_2$, cat. $NiCl_2$; DMF , 25°C , 18h → O 52%

Hodgson, D.M.; Wells, C. *Tetrahedron Lett.*, ***1994***, *35*, 1601

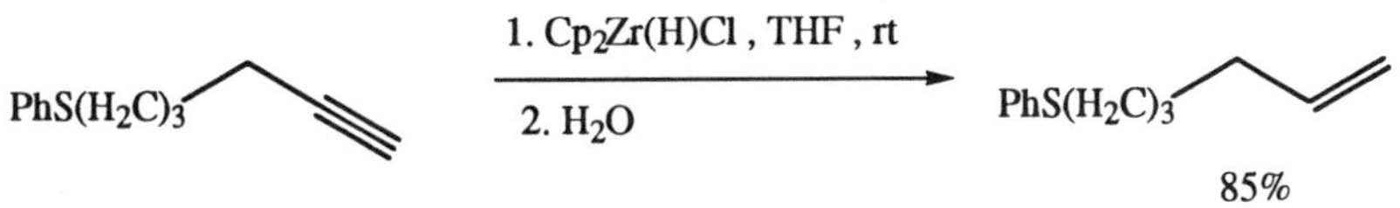

$PhS(H_2C)_3$ (alkyne) → 1. $Cp_2Zr(H)Cl$, THF, rt; 2. H_2O → $PhS(H_2C)_3$ (alkene) 85%

Lipshutz, B.H.; Lindsley, C.; Bhandari, A. *Tetrahedron Lett.*, ***1994***, *35*, 4669

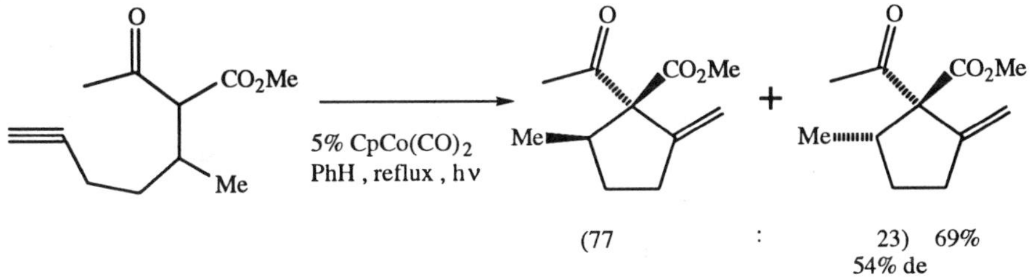

5% $CpCo(CO)_2$, PhH, reflux, hν

(77 : 23) 69%
54% de

Cruciani, P.; Aubert, C.; Malacria, M. *Tetrahedron Lett.*, ***1994***, *35*, 6677

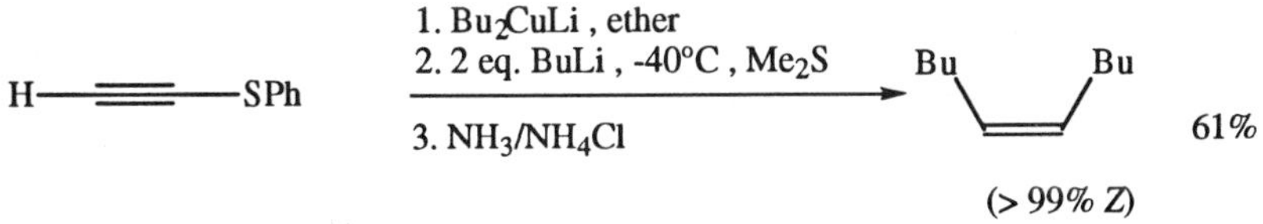

H—≡—SPh → 1. Bu_2CuLi, ether; 2. 2 eq. BuLi, -40°C, Me_2S; 3. NH_3/NH_4Cl → Bu, Bu 61%

(> 99% *Z*)

Creton, I.; Marek, I.; Brasseur, D.; Jestin, J.-L.; Normant, J.-F. *Tetrahedron Lett.*, ***1994***, *35*, 6877

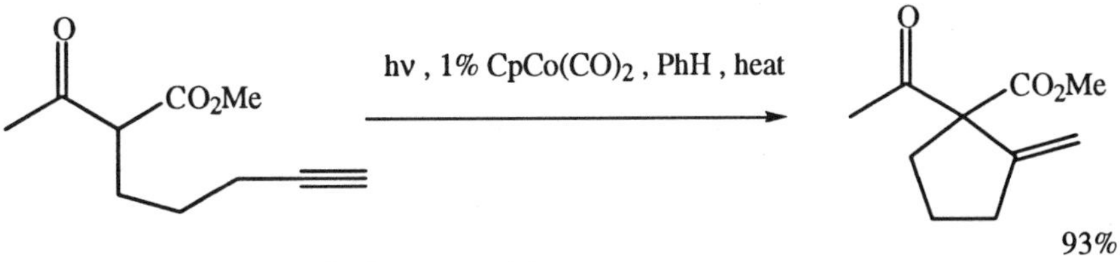

hν, 1% $CpCo(CO)_2$, PhH, heat

93%

Stammler, R.; Malacria, M. *Synlett*, ***1994***, 92

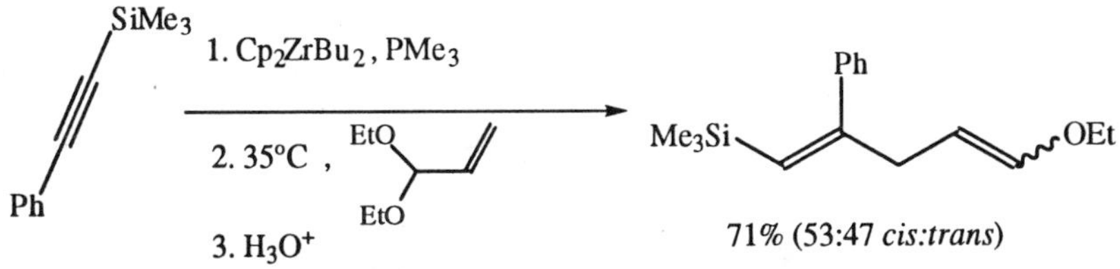

$SiMe_3$, Ph → 1. Cp_2ZrBu_2, PMe_3; 2. 35°C, EtO, EtO; 3. H_3O^+ → Me_3Si, Ph, OEt

71% (53:47 *cis:trans*)

Takahashi, T.; Kondakov, D.Y.; Suzuki, N. *Chem. Lett.*, ***1994***, 259

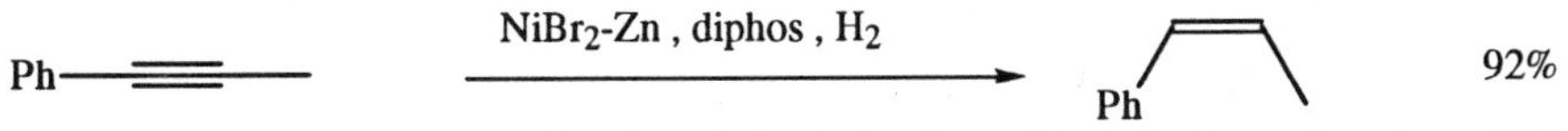

Ph—≡— → $NiBr_2$-Zn, diphos, H_2 → Ph 92%

Sakai, M.; Takai, Y.; Mochizuki, H.; Sasaki, K.; Sakakibara, Y. *Bull. Chem. Soc. Jpn.*, ***1994***, *67*, 1984

OSiEt3

3% $(PCy_3)_2Cl_2Ru$=CHCH=CPh_2

CH_2Cl_2 , (0.6 M) , 25•C , 8h

Me OSiEt3 Me 95%

Kim, S.-H.; Bowden, N.; Grubbs, R.H. *J. Am. Chem. Soc.*, ***1994***, *116*, 10801

Ph

Me_3SiI , Me_2Zn , dioxane

5% $Pd(PPh_3)_4$, 25°C, 8 h

Ph SiMe3 61%

Chatani, N.; Amishiro, N.; Morii, T.; Yamashita, T.; Murai, S. *J. Org. Chem.*, ***1995***, *60*, 1834

OHC Me

20% $Cp_2Ti(PMe_3)_2$, $HSi(OEt)_3$

OSi(OEt)3 Me 73%

Crowe, W.E.; Rachita, M.J. *J. Am. Chem. Soc.*, ***1995***, *117*, 6787

Br Ph

SmI_2 , THF , DMPU

Ph 83%

Zhou, Z.; Larouche, D.; Bennett, S.M. *Tetrahedron*, ***1995***, *51*, 11623

SECTION 197: ALKENES FROM ACID DERIVATIVES

C_6H_{13} CO_2H

0.01 $PdCl_2(PPh_3)_2$, PPh_3

250°C , N_2

C_6H_{13} 98%

Miller, J.A.; Nelson, J.A.; Byrne, M.P. *J. Org. Chem.*, ***1993***, *58*, 18

O Ph Cl

, Me_3Si-$SiMe_3$, toluene

5% $Pd(dba)_2$, 80°C , 4h

Ph SiMe3 86%

Obora, Y.; Tsuji, Y.; Kawamura, T. *J. Am. Chem. Soc.*, ***1995***, *117*, 9814

SECTION 198: ALKENES FROM ALCOHOLS AND THIOLS

H-montmorillonite, dioxane (air free), reflux, 6h

(81 : 19) 91%

Kantam, M.L.; Santhi, P.L.; Siddiqui, M.F. *Tetrahedron Lett.*, ***1993***, *34*, 1185

1. $SOCl_2$, NEt_3, 0°C
2. DBU, 0°C → rt

NMe(OMe)

(96-98% ee) 92%

Bennani, Y.L.; Sharpless, K.B. *Tetrahedron Lett.*, ***1993***, *34*, 2083

$Mo(acac)_3$, dioxane, reflux

(1 : 1) 92%

Kantam, M.L.; Prasad, A.D.; Santhi, P.L. *Synth. Commun.*, ***1993***, *23*, 45

Ph_3BiBr_2/I_2, 45°C, 3h

73% + 8%

Dorta, R.L.; Suárez, E.; Betancor, C. *Tetrahedron Lett.*, ***1994***, *35*, 5035

SECTION 199: ALKENES FROM ALDEHYDES

1. 2 LDA
2. PhCHO

78% (>98:2, *E:Z*)

Coutrot, Ph.; Grison, C.; Gérardin-Charbonnier, C.; Lecouvery, M. *Tetrahedron Lett.*, ***1993***, *34*, 2767

CHO, Ph, $SiMe_2t$-Bu → ($Me_2CH_2PPh_3^+ Br^-$; BuLi, THF, -78°C) → Ph, $SiMe_2t$-Bu 92%

[94:6 *E:Z* (>98%ee, R)]

Bhushan, V.; Lohray, B.B.; Enders, D. *Tetrahedron Lett.*, ***1993***, *34*, 5067

C_7H_{15}, CHO → (1. H_2S, THF, -30°C 2. NH_2NH_2 3. -30°C, 4h 4. $MgSO_4$, -30°C 5. $CaCO_3$, $Pb(OAc)_4$ 6. THF, reflux 7. PPh_3, THF, reflux) → C_7H_{15}, C_7H_{15} 80x93%

Collazo, L.R.; Guziec Jr., F.S. *J. Org. Chem.*, ***1993***, *58*, 43

OAc, Br → ($C_7H_{15}CHO$, Zn, $CrCl_3$, DMF/THF) → C_7H_{15} 56% (91:9 *E:Z*)

Knecht, M.; Boland, W. *Synlett*, ***1993***, 837

PhCHO → ($CH_2(CN)_2$, CdI_2, neat; heaat, 5 min) → Ph, CN, CN 95%

Prajapati, D.; Sandhu, J.S. *J. Chem. Soc., Perkin Trans. 1.*, ***1993***, 739

MeO_2C, CHO → ($Bu_3SnCH_2Br_2$, 25°C; Li, $CrCl_2$, DMF/THF) → MeO_2C, $SnBu_3$ 61%

Hodgson, D.M.; Boulton, L.T.; Maw, G.N. *Tetrahedron Lett.*, ***1994***, *35*, 2231

CHO → ([PPh_3=CHMe/I_2/ 2 $NaNTMS_2$]) → I 42%

(10:1 *Z:E*)

Chen, J.; Wang, T.; Zhao, K. *Tetrahedron Lett.*, ***1994***, *35*, 2827

PhCHO → ($(Me_3Si)_2CBr_2$, $CrCl_2$; DMF, 25°C) → Ph, $SiMe_3$, $SiMe_3$ 84%

Hodgson, D.M.; Comina, P.J. *Tetrahedron Lett.*, ***1994***, *35*, 9469

$C_6H_{13}CHO$ → (EtO_2C–CH=CH–CH$_2$–PPh_3^+ Br^- , dioxane; Ph_2S_2 , hν , K_2CO_3 , H_2O , 90°C , 2h) → C_6H_{13}–CH=CH–CH$_2$–CH$_2$–CO_2Et

70% (78:22 *E:Z*)
(without hν 12:18 *E:Z*)

Matikainen, J.K.; Kaltia, S.; Hase, T. *Synlett*, *1994*, 817

EtO_2C–CH$_2$–$SiMe_3$ → (PhCHO , DMSO; cat. CsF , rt → 100°C) → EtO_2C–CH=CH–Ph

93%

Bellassoued, M.; Ozanne, N. *J. Org. Chem.*, *1995*, *60*, 6582

PhCHO → (Cp_2Ti titanacyclobutene with Me_3Si, $SiMe_3$, $SiMe_3$; 25°C) → Ph–CH=CH–$SiMe_3$ 80%

Petasis, N.A.; Staszewski, J.P.; Fuk, D.-K. *Tetrahedron Lett.*, *1995*, *36*, 3619

$(PhO)_2P(O)CH_2CO_2Et$ → (1. KHMDS , THF , 1h, 5 eq. 18-crown-5, -78°C; 2. PhCHO) → Ph–CH=CH–CO_2Et + Ph–CH=CH–CO_2Et

(99 : 1) 98%

Ando, K. *Tetrahedron Lett.*, *1995*, *36*, 4107

Ph–CH$_2$–CHO → (Me_3SiCH=C=O , BF_3•OEt_2) → Ph–CH$_2$–CH=CH–CO_2H

90% (1:1 *E:Z*)

Black, T.H.; Zhang, Y.; Huang, J.; Smith, D.C.; Yates, B.E. *Synth. Commun.*, *1995*, *25*, 15

Related Methods: Section 207 (Alkenes from Ketones).

SECTION 200: ALKENES FROM ALKYLS, METHYLENES AND ARYLS

This section contains dehydrogenations to form alkenes and unsaturated ketones, esters and amides. It also includes the conversion of aromatic rings to alkenes. Reduction of aryls to dienes is found in Section 377 (Alkene-Alkene). Hydrogenation of aryls to alkanes and dehydrogenations to form aryls are included in Section 74 (Alkyls from Alkenes).

C_8K, THF

42%

Weitz, I.S.; Rabinovitz, M. *J. Chem. Soc., Perkin Trans. 1.*, ***1993***, 117

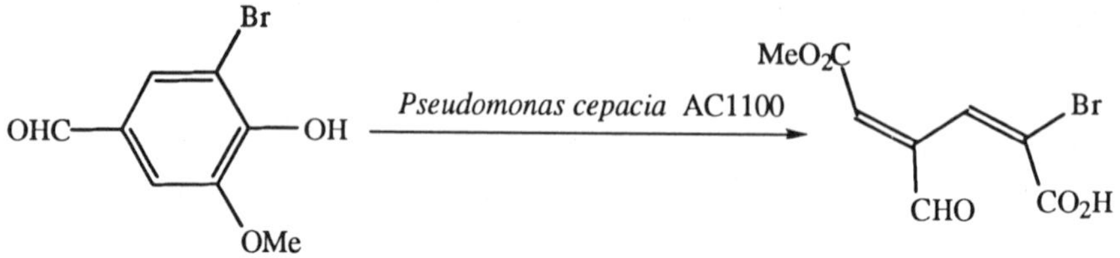

Br, OHC, OH, OMe

Pseudomonas cepacia AC1100

MeO_2C, Br, CHO, CO_2H

Artaud, I.; Tomasi, I.; Martin, G.; Petre, D.; Mansuy, D. *Tetrahedron Lett.*, ***1995***, *36*, 869

SECTION 201: ALKENES FROM AMIDES

Related Methods: Section 65 (Alkyls from Alkyls).
Section 74 (Alkyls from Alkenes).

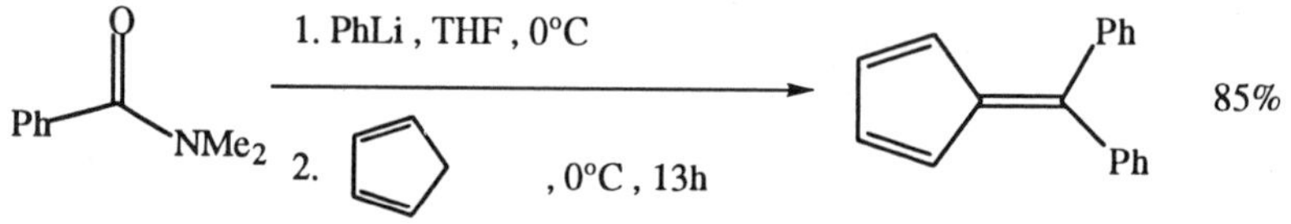

O, Ph, NMe_2

1. PhLi, THF, 0°C

2. [cyclopentadiene], 0°C, 13h

Ph, Ph

85%

Jurata, H.; Ekinaka, T.; Kawase, T.; Oda, M. *Tetrahedron Lett.*, ***1993***, *34*, 3445

SECTION 202: ALKENES FROM AMINES

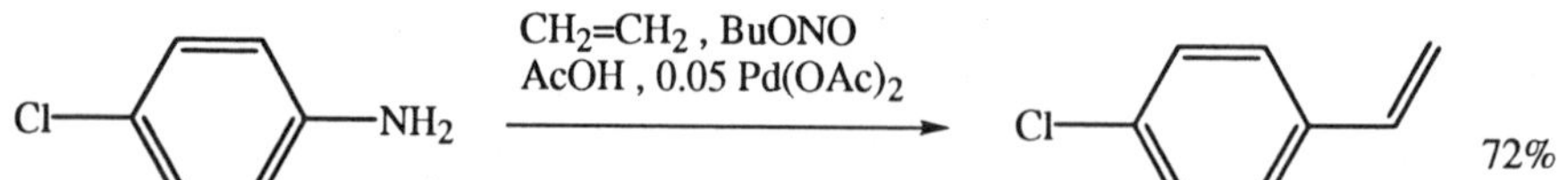

Cl, NH_2

CH_2=CH_2, BuONO
AcOH, 0.05 $Pd(OAc)_2$

Cl

72%

Beller, M.; Fischer, H.; Kühlein, K. *Tetrahedron Lett.*, ***1994***, *35*, 8773

SECTION 203: ALKENES FROM ESTERS

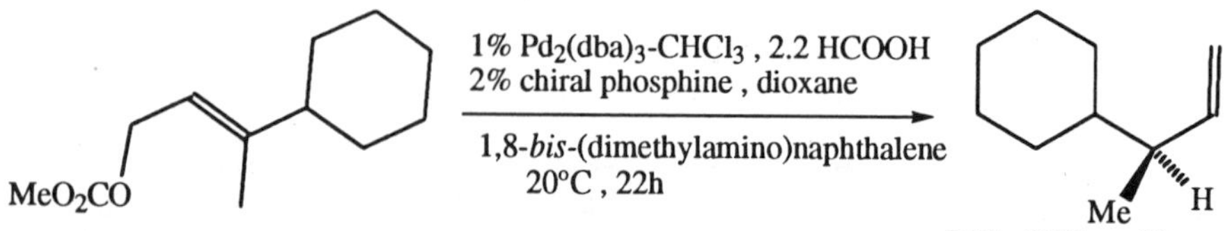

MeO_2CO

1% $Pd_2(dba)_3$-$CHCl_3$, 2.2 HCOOH
2% chiral phosphine, dioxane

1,8-*bis*-(dimethylamino)naphthalene
20°C, 22h

Me, H

96% (85% ee, R)

Hayashi, T.; Iwamura, H.; Naito, M.; Matsumoto, Y.; Uozumi, Y.; Miki, M.; Yanagi, K. *J. Am. Chem. Soc.*, ***1994***, *116*, 775

SECTION 204: ALKENES FROM ETHERS, EPOXIDES AND THIOETHERS

Te , THF

$LiBHEt_3$

OAc OH Ph Ph 65%

Dittmer, D.C.; Zhang, Y.; Discordia, R.P. *J. Org. Chem.*, ***1994***, *59*, 1004

Bu_3SnH , cat. BEt_3

toluene , 0°C

OH Ph S OH Ph

90%

Uenishi, J.; Kubo, Y. *Tetrahedron Lett.*, ***1994***, *35*, 6697

sec-BuLi , -78°C → rt , 90 min

O Bu Bu Bu Bu 76%

Doris, E.; Deschoux, L.; Mioskowski, C. *Tetrahedron Lett.*, ***1994***, *35*, 7943

SECTION 205: ALKENES FROM HALIDES AND SULFONATES

Bu_3SnH , AIBN , Tol , reflux

Br Ph Ph 71%

Destabel, C.; Kilburn, J.D.; Knight, J. *Tetrahedron Lett.*, ***1993***, *34*, 3151

1. Li , THF , rt , 1h
2. $ZnCl_2$, THF , 0°C → rt
3. , THF , 66°C

Br I I 85%

5% "$Pd(PPh_3)_2$" , 10h

Katz, T.J.; Gilbert, A.M.; Huttenloch, M.E.; Min-Min, G.; Brintzinger, H.H. *Tetrahedron Lett.*, ***1993***, *34*, 3551

Me_3Si / Br / Bu / Br —MeLi, 20°C→ Bu / $SiMe_3$ 82%

Baird, M.S.; Dale, C.M.; Al Dulayym, J.B. *J. Chem. Soc., Perkin Trans. 1.*, ***1993***, 1373

$C_{14}H_{16}$ / Br —2 eq. $Me_2S{=}CH_2$, THF; -10°C → rt→ $C_{14}H_{16}$ 92%

Alcaraz, L.; Harnett, J.J.; Mioskowski, C.; Martel, J.P.; Le Gall, T.; Shin, D.-S.; Falck, J.R. *Tetrahedron Lett.*, ***1994***, *35*, 5453

$C_8H_{17}ZnI$ —1. CuCN•2 LiCl, DMPU; 2. *E*-hexenyl iodide, 60°C, 12h→ C_8H_{17} / Bu quant.

Marquais, S.; Cahiez, G.; Knochel, P. *Synlett*, ***1994***, 849

Br / Ph / Ph / Br —SmI_2, THF, 5 min→ Ph / Ph 95%

Yanada, R.; Bessho, K.; Yanada, K. *Chem. Lett.*, ***1994***, 1279

I / MeO —Bu_3Sn / Ph; 10% Pd/C, 10% CuI, 20% $AsPh_3$; NMP, 80°C→ Ph / MeO 82%

Roth, G.P.; Farina, V.; Liebeskind, L.S.; Peña-Cabrera, E. *Tetrahedron Lett.*, ***1995***, *36*, 2191

Br / Br —EtMgBr, THF, 0°C; $Ni(dppe)Cl_2$→ quant.

Malanga, C.; Aronica, L.A.; Lardicci, L. *Tetrahedron Lett.*, ***1995***, *36*, 9189

SECTION 206: ALKENES FROM HYDRIDES

For conversions of methylenes to alkenes ($RCH_2R' \rightarrow RR'C{=}CH_2$), see Section 200 (Alkenes from Alkyls).

NO ADDITIONAL EXAMPLES

SECTION 207: ALKENES FROM KETONES

$C_{12}H_{25}C(O)Me$ → (Cp_2Ti(cyclopropyl)$_2$; PhMe, 50°C) → $C_{12}H_{25}(Me)C{=}$cyclopropylidene 65%

Petasis, N.A.; Bzowej, E.I. *Tetrahedron Lett.*, ***1993***, *34*, 943

Cyclohexanone → ($(MeO)_2POCH_2CO_2Me$; LiOH, THF, MS, reflux, 4h) → cyclohexylidene-$CH{-}CO_2Me$ 85%

Bonadies, F.; Cardilli, A.; Lattanzi, A.; Orelli, L.R.; Screttri, A. *Tetrahedron Lett.*, ***1994***, *35*, 3383

4-*t*-Bu-cyclohexanone → (Sm, SmI_2, CH_3CHBr_2; cat. $CrCl_3$, THF, rt) → 4-*t*-Bu-cyclohexylidene=CH(Me) 71%

Matsubara, S.; Horiuchi, M.; Takai, K.; Utimoto, K. *Chem. Lett.*, ***1995***, 259

$Ph_2P(O)CH(CH_2Ph)C(O)CH_3$ → (1. EtLi, $CeCl_3$, THF, -78°C; 2. KH, DMF, 50°C) → $PhCH_2CH{=}C(CH_3)CH_2CH_3$ 95% (92:8 *E:Z*)

Bartoli, G.; Marcantoni, E.; Sambri, L.; Tamburini, M. *Angew. Chem. Int. Ed. Engl.*, ***1995***, *34*, 2046

Related Methods: Section 199 (Alkenes from Aldehydes).

SECTION 208: ALKENES FROM NITRILES

NO ADDITIONAL EXAMPLES

SECTION 209: ALKENES FROM ALKENES

Allyl 1-phenylallyl ether → (2% $Cl_2(PCy_3)_2Ru{=}CH{-}CH{=}CHPh$; PhH, 20°C) → 2-phenyl-2,5-dihydrofuran 84%

Fu, G.C.; Nguyen, S.T.; Grubbs, R.H. *J. Am. Chem. Soc.*, ***1993***, *115*, 9856

, 10% $Sc(OTf)_3$

CH_2Cl_2 , 0°C , 13h

new Diels-Alder catalyst

Kobayashi, S.; Hachiya, I.; Araki, M.; Ishitani, H. *Tetrahedron Lett.*, ***1993***, *34*, 3755

OTBS ArN=Mo(OR)=CHR catalyst

PhH , 20°C , 30 min

OTBS

91%

Fu, G.C.; Grubbs, R.H. *J. Am. Chem. Soc.*, ***1993***, *115*, 3800

2 Ph + C_6H_{13}

1% $Mo[OC(CF_3)_2Me]_2(NAr)CHCMe_2Ph$

Ph C_6H_{13}

89%

Crowe, W.E.; Zhang, Z.J. *J. Am. Chem. Soc.*, ***1993***, *115*, 10998

B $SiMe_3$

1. PhBr , THF , $Pd(PPh_3)_4$
NaOH , reflux , 12h
2. NaOH , H_2O_2

Ph $SiMe_3$

83%

Soderquist, J.A.; Colbert, J.C. *Tetrahedron Lett.*, ***1994***, *35*, 27

I I Ph C_8H_{17}

SmI_2 PhH/HMPA , rt , 10 min

C_8H_{17} Ph

76%

Kunishima, M.; Hioki, K.; Tani, S.; Kato, A. *Tetrahedron Lett.*, ***1994***, *35*, 7253

O O

8% $(PCy_3)_2Cl_2Ru$=CHCH=CPh_2

PhH , 55°C , 3h

O O

75%

Miller, S.J.; Kim, S.-H.; Chen, Z.-R.; Grubbs, R.H. *J. Am. Chem. Soc.*, ***1995***, *117*, 2108

Ts

1. 5% $Pd(PPh_3)_4$, 4 eq. $ZnEt_2$
ether
2. aq. NH_4Cl

AcO ene reaction

(96 : 4) 44%

Oppolzer, W.; Schröder, F. *Tetrahedron Lett.*, ***1994***, *35*, 7935

Ni(dppe)Cl_2 , iPrMgBr , THF
Me_3SiCl , rt , (seconds)

quant.

Malanga, C.; Urso, A.; Lardicci, L. *Tetrahedron Lett.*, ***1995***, *36*, 1133

2% $(OAr)_2Cl_2W{=}O$, 90°C , 1h

1,2,4-trichlorobenzene

Ar = 2,4-dibromophenyl

68% (97% ee)

Nugent, W.A.; Feldman, J.; Calabrese, J.C. *J. Am. Chem. Soc.*, ***1995***, *117*, 8992

REVIEW:

"Reagent-Controlled Asymmetric Diels-Alder Reactions," Oh, T.; Reilly, M. *Org. Prep. Proceed. Int.*, ***1994***, *26*, 129

SECTION 210: ALKENES FROM MISCELLANEOUS COMPOUNDS

1. BuLi , THF , -78°C
2. $C_{11}H_{23}CO_2Me$
3. KO*t*-Bu
4. EtI
5. $CeCl_3$, $NaBH_4$
6. THF , 105°C , 20h

t-Bu

$C_{11}H_{23}$

63x96x93x92% (102:1 *E:Z*)

Denmark, S.E.; Amburgey, J. *J. Am. Chem. Soc.*, ***1993***, *115*, 10386

t-BuO$_2$S–CH$_2$CH$_2$–C$_{10}$H$_{21}$ → (2% Pd(acac)$_2$, BuLi, THF, reflux) → CH$_2$=CH–C$_{10}$H$_{21}$

70%

Gai, Y.; Jin, L.; Julia, M.; Verpeaux, J.-N. *J. Chem. Soc. Chem. Commun.*, ***1993***, 1625

t-BuO$_2$S–CH$_2$CH$_2$–C$_3$H$_7$ → (BuLi, 2% Fe(acac)$_3$; 0°C → reflux) → C$_3$H$_7$CH$_2$CH=CHCH$_2$C$_3$H$_7$

79% (76:24 *E:Z*)

Jin, L.; Julia, M.; Verpeaux, J.N. *Synlett*, ***1994***, 215

PhCH$_2$–SO$_2$–CH$_2$Ph → (KOH, Al$_2$O$_3$, CBr$_2$F$_2$-t-BuOH) → PhCH=CHPh

96%

Chan, T.-L.; Fong, S.F.; Li, Y.; Man, T.-O.; Poon, C.-D. *J. Chem. Soc. Chem. Commun.*, ***1994***, 1771

PhCH$_2$CH$_2$C(SO$_2$Ph)=CHPh → (8 eq. SmI$_2$, THF, DMPU; 35 min) → PhCH$_2$CH$_2$CH=CHPh

95% (*E:Z* 9:1)

Keck, G.E.; Savin, K.A.; Weglarz, M.A. *J. Org. Chem.*, ***1995***, *60*, 3194

(C$_9$H$_{19}$)$_2$CH–O–P(O)(OEt)$_2$ → (Lawesson's reagent, xylene; reflux, 22h) → C$_9$H$_{19}$CH=CHC$_8$H$_{17}$

quant.

Shimagaki, M.; Fujieda, Y.; Kimura, T.; Makata, T. *Tetrahedron Lett.*, ***1995***, *36*, 719

PhCH(OAc)CH$_2$SO$_2$Ph → (Mg°, cat. HgCl$_2$, EtOH; rt, 2h) → PhCH=CH$_2$

98%

Lee, G.H.; Lee, H.K.; Choi, E.B.; Kim, B.T.; Pak, C.S. *Tetrahedron Lett.*, ***1995***, *36*, 5607

REVIEW:

"Rare Earth Metal Trifluoromethanesulfonates as Water-Tolerated Lewis Acid Catlaysts in Organic Synthesis," Kobayashi, S. *Synlett*, ***1994***, 679

CHAPTER 15

PREPARATION OF OXIDES

This chapter contains reactions which prepare the oxides of nitrogen, sulfur and selenium. Included are *N*-oxides, nitroso and nitro compounds, nitrile oxides, sulfoxides, selenoxides and sulfones. Oximes are considered to be amines and appear in those sections. Preparation of sulfonic acid derivatives are found in Chapter Two and the preparation of sulfonates in Chapter Ten.

SECTION 211: OXIDES FROM ALKYNES

OH OTBS

1. $ClPPh_2$, NEt_3, THF, -78°C

2. (1,4-cyclohexadiene), PhCl, 75°C

CO_2Me

OTBS O P Ph Ph

MeO_2C

99x?%

Grissom, J.W.; Slattery, B.J. *Tetrahedron Lett.*, ***1994***, *35*, 5137

SECTION 212: OXIDES FROM ACID DERIVATIVES

SO_2Cl

$SnMe_3$ Me

PhCl, 130°C

2d

S O_2 Me

95%

Neumann, W.P.; Wicemec, C. *Chem. Ber.*, ***1993***, *126*, 763

SECTION 213: OXIDES FROM ALCOHOLS AND THIOLS

SH

1. (dimethyldioxirane), -40°C

CH_2Cl_2

2. air

SO_2H 74%

Gu, D.; Harpp, D.N. *Tetrahedron Lett.*, ***1993***, *34*, 67

SECTION 214: OXIDES FROM ALDEHYDES

NO ADDITIONAL EXAMPLES

SECTION 215: OXIDES FROM ALKYLS, METHYLENES AND ARYLS

NO ADDITIONAL EXAMPLES

SECTION 216: OXIDES FROM AMIDES

NO ADDITIONAL EXAMPLES

SECTION 217: OXIDES FROM AMINES

Ph N(OH) Ph — 0,05 $Pr_4N\ RuO_4$, 1.5 NMO / MeCN , rt , 2d → Ph $N^{\oplus}$($O^{\ominus}$) Ph

75% (>50:1 *Z*:*E*)

Goti, A.; De Sarlo, F.; Romani, M. *Tetrahedron Lett.*, ***1994***, *35*, 6571

H_2N–C₆H₄–CO_2H — Oxone® / aq. acetone → O_2N–C₆H₄–CO_2H 73%

Webb, K.S.; Seneviratne, V. *Tetrahedron Lett.*, ***1995***, *36*, 2377

SECTION 218: OXIDES FROM ESTERS

OAc — Na_2SO_2Tol , THF / 0.01 $Pd(PPh_3)_4$, rt / chiral phosphine , 72h → SO_2Tol + OS(=O)Tol

(10 : 1) 83%

Eichelmann, H.; Gais, H.-J. *Tetrahedron Asymmetry*, ***1995***, *6*, 643

SECTION 219: OXIDES FROM ETHERS, EPOXIDES AND THIOETHERS

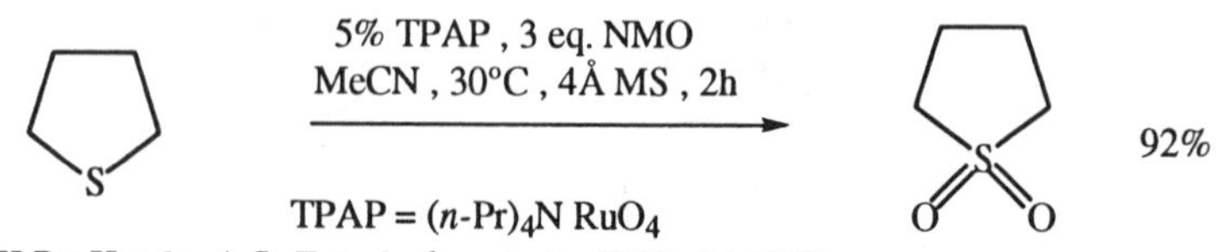

Guertin, K.R.; Kende, A.S. *Tetrahedron Lett.*, ***1993***, *34*, 5369

Ph–S–Me → [$C(NO_2)_4$, hν (Pyrex); CH_2Cl_2] → Ph–S(=O)–Me 92%

Ramkumar, D.; Sankararaman, S. *Synthesis*, ***1993***, 1057

Ph–S–Me → [H_2O_2, MeCN, K_2CO_3; MeOH, 0°C, 2h] → Ph–S(=O)–Me 82%

Page, P.C.B.; Graham, A.E.; Bethell, D.; Park, B.K. *Synth. Commun.*, ***1993***, *23*, 1507

Ph–S–Me → [MnO_2, Me_3SiCl, MeOH] → Ph–S(=O)–Me 99%

Bellesia, F.; Ghelfi, F.; Pagnoni, U.M.; Pinetti, A. *Synth. Commun.*, ***1993***, *23*, 1759

Bu–S–Bu → [*t*-BuOOH, H_2O, 70°C, 30 min] → Bu–S(=O)–Bu quant.

Fringuelli, F.; Pellegrino, R.; Pizzo, F. *Synth. Commun.*, ***1993***, *23*, 3157

Ph–S–CH=CH$_2$ → [cyclohexanone monooxygenase] → Ph–S(=O)–CH=CH$_2$

73% (99% ee, R)

Secundo, F.; Carrea, G.; Dallavalle, S.; Franzosi, G. *Tetrahedron Asymmetry*, ***1993***, *4*, 1981

HO–CH$_2$–C$_6$H$_4$–S–Me → [1.35 eq. oxone; rt, 1.2h] → HO–CH$_2$–C$_6$H$_4$–S(=O)$_2$–Me

with 0.65 eq. Oxone - obtain 47% sulfoxide 83%

Webb, K.S. *Tetrahedron Lett.*, ***1994***, *35*, 3457

PhCH$_2$–S–CH$_3$ → [1.1 oxaziridine (N–C_4F_9, C_3F_3, F); $CHCl_3$, $CFCl_3$, -40°C] → PhCH$_2$–S(=O)–CH$_3$ 95%

with 2.7 eq., obtain the sulfone (91%)

DesMarteau, D.D.; Petrov, V.A.; Montanaari, V.; Pregnolato, M.; Resnati, G. *J. Org. Chem.*, ***1994***, *59*, 2762

Ph–S–Ph → ($NaIO_4$, cat. $RuCl_3 \cdot H_2O$; 1h) → $Ph–SO_2–Ph$ quant.

Su, W. *Tetrahedron Lett.*, ***1994***, *35*, 4955

mcpba

(8 : 3)

Glass, R.S.; Singh, W.P.; Hay, B.A. *Tetrahedron Lett.*, ***1994***, *35*, 5809

1. [H_2O_2, ether, DBU, -28°C (-)-camphorsulfonylimine]

2. aq. Na sulfite

quant (35% ee, R)

Page, P.C.B.; Heer, J.P.; Bethell, D.; Collington, E.W.; Andrews, D.M. *Tetrahedron Lett.*, ***1994***, *35*, 9629

NaOCl, TEMPO, KBr

Bu_4NCl, CH_2Cl_2, satd. $NaHCO_3$

87%

Siedlecka, R.; Skarzewski, J. *Synthesis*, ***1994***, 401

PhIO, cat. TsOH, MeCN, 25°C

82%

Cavicchioni, G. *Synth. Commun.*, ***1994***, *24*, 2223

1. MnO_2-35% aq. HCl, MeOH 0°C → 10°C, 0.75h

2. NaOH

99%

Fabretti, A.; Ghelfi, F.; Grandi, R.; Pagnoni, U.M. *Synth. Commun.*, ***1994***, *24*, 2393

$NaBrO_2$, rt, wet zeolite F-9

CH_2Cl_2, 1h

82%

Hirano, M.; Kudo, H.; Morimoto, T. *Bull. Chem. Soc. Jpn.*, ***1994***, *67*, 1492

Ph–S–Me → (PhIO, MeCN, 1h; 10% Mn salen catalyst) → Ph–S(=O)–Me

57% (62% ee, R)

Noda, K.; Hosoya, N.; Irie, R.; Yamashita, Y.; Katsuki, T. *Tetrahedron*, ***1994***, *50*, 9609

p-Tol–S–Me → (chiral oxaziridine, CH_2Cl_2; 0°C) → *p*-Tol–S(=O)–Me

70% ee

Jennings, W.B.; Kochanewyczm, M.J.; Lovely, C.J.; Boyd, D.R. *J. Chem. Soc. Chem. Commun.*, ***1994***, 2569

Ph–S–Me → ($PhMe_2COOH$, $Ti(OiPr)_4$, H_2O; (RR)-DET, CH_2Cl_2, -20°C) → Ph–S(=O)–Me 77% (99% ee, R)

Brunel, J.-M.; Diter, P.; Duetsch, M.; Kagan, H.B. *J. Org. Chem.*, ***1995***, *60*, 8086

Bu–S–Bu → (1.5 M HNO_3, 1% $FeBr_3$, 30 min) → Bu–S(=O)–Bu 84%

Suárez, A.R.; Rossi, L.I.; Martín, S.E. *Tetrahedron Lett.*, ***1995***, *36*, 1201

Bu–S–Bu → (SiO_2, CH_2Cl_2, *t*-BuOOH; 30 min) → Bu–S(=O)–Bu 86%

with 2 eq. *t*-BuOOH, obtain 83% of sulfone

Breton, G.W.; Fields, J.D.; Kropp, P.J. *Tetrahedron Lett.*, ***1995***, *36*, 3825

$PhCH_2$–S–Me → (cyclohexanone monooxygenase) → $PhCH_2$–S(=O)–Me

97% (54% ee, R)

Pasta, P.; Carrea, G.; Holland, H.L.; Dallavalle, S. *Tetrahedron Asymmetry*, ***1995***, *6*, 933

Ph–S–Me → (O_2, *m*-xylene, Mn (III) salen complex; *t*-BuCHO, rt) → Ph–S(=O)–Me 66% (51% ee)

Nagata, T.; Imagawa, K.; Yamada, T.; Mukaiyama, T. *Bull. Chem. Soc. Jpn.*, ***1995***, *68*, 3241

SECTION 220: OXIDES FROM HALIDES AND SULFONATES

PhI , CuI , DMF , heat

$PhSO_2Na$ ⟶ $PhSO_2Ph$ 60%

Suzuki, H.; Abe, H. *Tetrahedron Lett.*, ***1995***, *36*, 6239

SECTION 221: OXIDES FROM HYDRIDES

NO_2-O_3 , CH_2Cl_2

-10°C , 3h

Me ⟶ Me, NO_2

51% (57:2:41 *o:m:p*)

Suzuki, H.; Murashima, T.; Kozai, I.; Mori, T. *J. Chem. Soc., Perkin Trans. 1.*, ***1993***, 1591
Suzuki, H.; Mori, T. *J. Chem. Soc., Perkin Trans. 1.*, ***1995***, 291
From aryl esters: meta-nitro is the major product
Suzuki, H.; Tomaru, J.-i.; Murashima, T. *J. Chem. Soc., Perkin Trans. 1.*, ***1994***, 2413
From aryl acetates: ortho:para predominates (60:40)
Suzuki, H.; Tatsumi, A.; Ishibashi, T.; Mori, T. *J. Chem. Soc., Perkin Trans. 1.*, ***1995***, 339

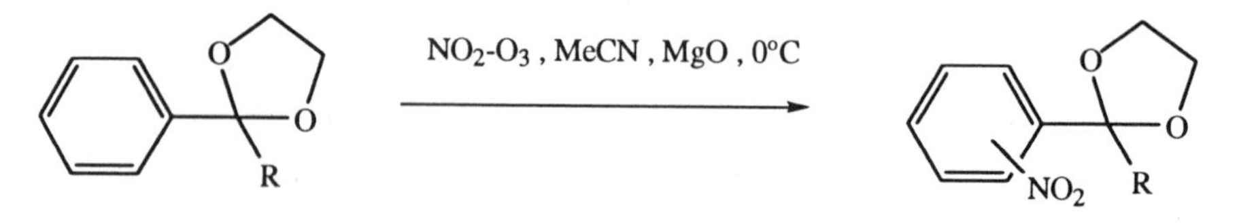

R = Et 95% (14:31:55 *o:m:p*)
R = H 0%

Suzuki, H.; Yonezawa, S.; Mori, T. *Bull. Chem. Soc. Jpn.*, ***1995***, *68*, 1535

REVIEW:

"Ozone-Mediated Nitration of Aromatic Compounds with Lower Oxides of Nitrogen," Mori, T.; Suzuki, H. *Synlett*, ***1995***, 383

SECTION 222: OXIDES FROM KETONES

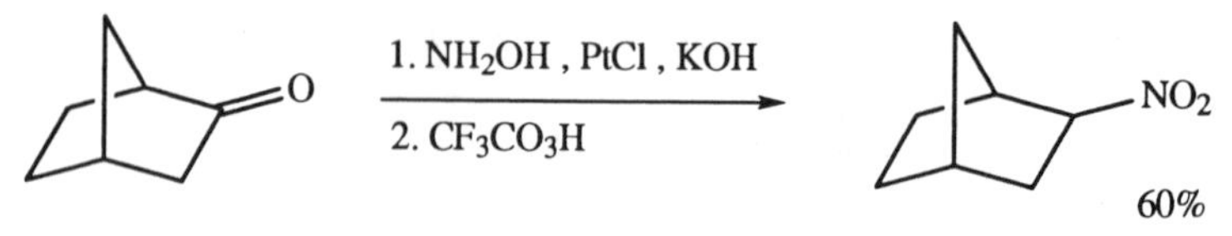

60%

(1:13 *exo:endo*)

Olah, G.A.; Ramaiah, P.; Prakash, C.K.S. *J. Org. Chem.*, ***1993***, *58*, 763

SECTION 223: OXIDES FROM NITRILES

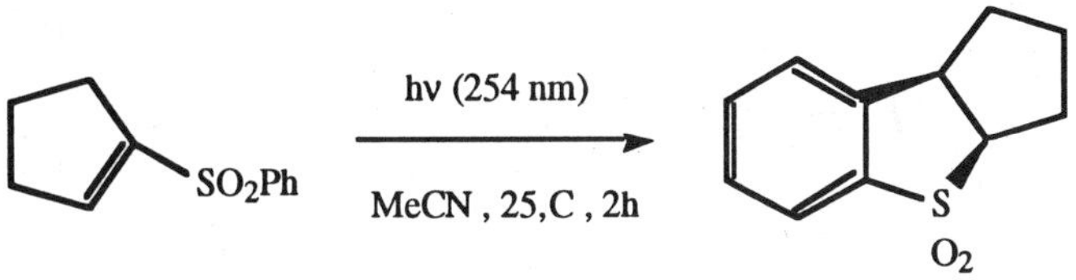

hv (254 nm)

MeCN , 25,C , 2h

SO_2Ph

S

O_2

55%

Li, C.; Fuchs, P.L. *Tetrahedron Lett.*, ***1993***, *34*, 1855

SECTION 224: OXIDES FROM ALKENES

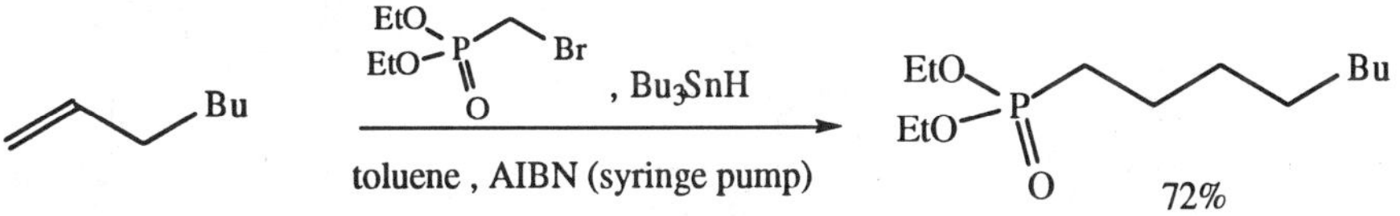

Bu

EtO, EtO, P, O, Br

, Bu_3SnH

toluene , AIBN (syringe pump)

EtO, EtO, P, O, Bu

72%

Bałczewski, P.; Mikołajczyk, M. *Synthesis*, ***1995***, 392

SECTION 225: OXIDES FROM MISCELLANEOUS COMPOUNDS

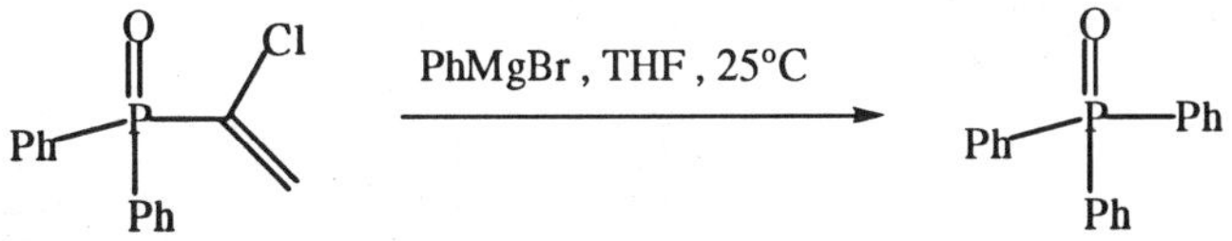

O, Cl, Ph, P, Ph

PhMgBr , THF , 25°C

O, Ph, P, Ph, Ph

71%

Cardellicchio, C.; Fiandanese, V.; Naso, F.; Pietrusiewicz, K.M.; Wiśiewski, W. *Tetrahedron Lett.*, ***1993***, *34*, 3135

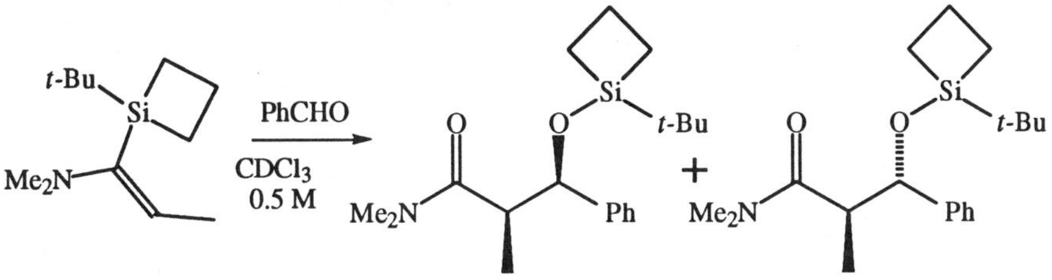

t-Bu, Si, Me_2N

PhCHO

$CDCl_3$ 0.5 M

O, O, Si, *t*-Bu, Me_2N, Ph

+

O, O, Si, *t*-Bu, Me_2N, Ph

(9 : 91) 84%

Denmark, S.E.; Griedel, B.D.; Coe, D.M. *J. Org. Chem.*, ***1993***, *58*, 988

PhO$_2$S SO$_2$Ph

1. LiNAPH

2. Br

PhO$_2$S

Ph

Ph

79%

Yu, J.; Cho, H.-S.; Chandrasekhar, S.; Falck, J.R.; Mioskowski, C. *Tetrahedron Lett.*, ***1994***, *35*, 5437

O

Me P

Ph Cl

MgBr

THF, rt

O

P Me

Ph

61%

(>98% ee, S)

Cardellicchio, C.; Fiandanese, V.; Naso, F.; Pacifico, S.; Koprowski, M.; Pietrusiewicz, K.M. *Tetrahedron Lett.*, ***1994***, *35*, 6343

CHAPTER 16

PREPARATION OF DIFUNCTIONAL COMPOUNDS

SECTION 300: ALKYNE - ALKYNE

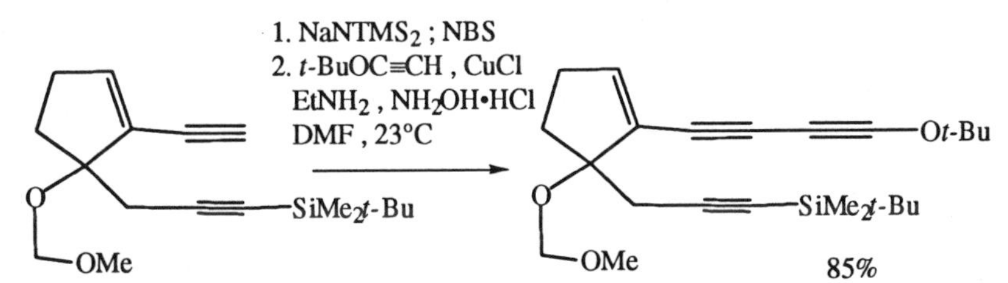

Magriotis, P.A.; Vourloumis, D.; Scott, M.E.; Tarli, A. *Tetrahedron Lett.*, ***1993***, *34*, 2071

SECTION 301: ALKYNE - ACID DERIVATIVES

NO ADDITIONAL EXAMPLES

SECTION 302: ALKYNE - ALCOHOL, THIOL

CHO → 1. (B-allenyl-9-BBN), ether, 25°C, 15 min; 2. H_2O_2, NaOH → OH 82%

Brown, H.C.; Khire, U.R.; Racherla, U.S. *Tetrahedron Lett.*, ***1993***, *34*, 15

MsO → 1. Bu_3ZnI, -85°C → 0°C; 2. 3 eq. $ZnCl_2$; 3. iPrCHO → Bu, OH 89% (98:2 *anti:syn*)

Katsuhira, T.; Harada, T.; Maejima, K.; Osada, A.; Oku, A. *J. Org. Chem.*, ***1993***, *58*, 6166

$SbBu_4$ / $SiMe_3$ — PhCHO , -78°C; BuLi/THF/hexane → Ph, OH, $SiMe_3$ + OH, Ph, Me_3Si

(94 : 6) 80%

Zhang, L-T.; Mo, X-S.; Huang, J-L.; Huang, Y.-Z. *Tetrahedron Lett.*, ***1993***, *34*, 1621

O, S, *p*-Tol, $SiMe_3$ — 1. 2 eq. LDA , THF , -100°C; 2. 2 eq. PhCHO , THF; 3. NaH ,THF , 0°C → OH, Ph

83% (>99%ee, S)

Kusuda, S.; Kawamura, K.; Ueno, Y.; Toru, T. *Tetrahedron Lett.*, ***1993***, *34*, 6587

O, Ph—S, Cl, Ph — 1. 3 eq. *t*-BuLi; 2. EtCHO , -78°C → -30°C → HO, Et, Ph

67%

Satoh, T.; Hayashi, Y.; Yamakawa, K. *Bull. Chem. Soc. Jpn.*, ***1993***, *66*, 1866

CHO, $SiMe_2Ph$, H — 1. *o*-DCB , reflux; 2. TBAF → OH

49%

Jin, J.; Smith, D.T.; Weinreb, S.M. *J. Org. Chem.*, ***1995***, *60*, 5366

PHCHO — $(PhC{\equiv}C)_2AlEt_2^-\ Na^+$; Tol , 0°C , 1h → Ph, OH, Ph 95%

Ahn, J.H.; Joung, M.J.; Yoon, N.M. *J. Org. Chem.*, ***1995***, *60*, 6173

C_8H_{17}—≡—I — Bu(C=O)Bu; SmI_2 , HMPA/PhH → C_8H_{17}, OH, Bu, Bu

78%

Kunishima, M.; Tanaka, S.; Kono, K.; Hioki, K.; Tani, S. *Tetrahedron Lett.*, ***1995***, *36*, 3707

Ph—≡—H → (t-BuCHO, GaI_3, NEt_3; THF, rt) → Ph—≡—CH(OH)(t-Bu) 80%

Han, Y.; Huanag, Y.-Z. *Tetrahedron Lett.*, ***1995***, *36*, 7277

PhCHO → (Ph—≡—$SnBu_3$; $InCl_3$, MeCN, 25°C, 30 min) → Ph(HO)CH—≡—Ph 92%

Yasuda, M.; Miyai, T.; Shibata, I.; Baba, A.; Nomura, R.; Matsuda, H. *Tetrahedron Lett.*, ***1995***, *36*, 9497

≡—CH_2Br → (PhCHO, Zn, THF; satd. NH_4Cl, rt) → PhCH(OH)CH_2C≡CH + PhCH(OH)CH=C=CH_2

(89 : 11) 68%

Yavari, I.; Riazi-Kermani, F. *Synth. Commun.*, ***1995***, *25*, 2923

SECTION 303: ALKYNE - ALDEHYDE

Ph—≡—CH_2SPh → (1. SO_2Cl_2, CCl_4, 0°C; 2. $CuCl_2$, H_2O) → Ph—≡—CHO 47%

Fortes, C.C.; Garrote, C.F.D. *Synth. Commun.*, ***1993***, *23*, 2869

SECTION 304: ALKYNE - AMIDE

NO ADDITIONAL EXAMPLES

SECTION 305: ALKYNE - AMINE

HC≡C—CH(C_5H_{11})—O—P(O)(OEt)$_2$ → (2 Et_2NH_2, 1% CuCl; THF, 50°C, 2h) → HC≡C—CH(C_5H_{11})NEt_2 91%

Imada, Y.; Yuasa, M.; Nakamura, I.; Murahashi, S.-I. *J. Org. Chem.*, ***1994***, *59*, 2282

3-benzyloxazolidine → (F_2B—≡—$SiMe_3$) → $PhCH_2N(CH_2CH_2OH)CH_2$—≡—$SiMe_3$ 50%

Wu, M.-J.; Yan, D.-S.; Tsai, H.-W.; Chen, S.-H. *Tetrahedron Lett.*, ***1994***, *35*, 5003

Br $BuNH_2$, 3% CuBr , MeCN → NHBu 90%

Geri, R.; Polizzi, C.; Lardicci, L.; Caporusso, A.M. *Gazz. Chim. Ital.*, ***1994***, *124*, 241

OTf, N; TMS—≡ , DMF; $Pd_2(dba)_3 \cdot CHCl_3$, P(*o*-Tol)$_3$ iPr$_2$NEt → TMS, N 93%

Okita, T.; Isobe, M. *Tetrahedron*, ***1995***, *51*, 3737

SECTION 306: ALKYNE - ESTER

NO ADDITIONAL EXAMPLES

SECTION 307: ALKYNE - ETHER, EPOXIDE, THIOETHER

Ph—C≡C—Br $\frac{1}{2}$ PhSeSePh, CuI , HMPA → Ph—C≡C—SePh 75%

Braga, A.L.; Reckziegel, A.; Menezes, P.H.; Stefani, H.A. *Tetrahedron Lett.*, ***1993***, *34*, 393

C_5H_{11}—≡—H PhSeBr , CuI , DMF → C_5H_{11}—≡—SPh 82%

Braga, A.L.; Silveira, C.C.; Reckziegel, A.; Menezes, P.H. *Tetrahedron Lett.*, ***1993***, *34*, 8041

Ph—≡ PhI(OAc)$_2$, PhSeSePh , CH_2Cl_2, 40°C → Ph—≡—SePh 81%

Tingoli, M.; Tiecco, M.; Testaferri, L.; Balducci, R. *Synlett*, ***1993***, 211

HO; ≡—C(Me)$_2$—OCOCF$_3$, MeCN; DBU , $CuCl_2 \cdot 2\ H_2O$, 0°C → O 83%

Godfrey Jr., J.D.; Mueller, R.H.; Sedergran, T.C.; Soundararajan, N.; Colandrea, V.J. *Tetrahedron Lett.*, ***1994***, *35*, 6405

SECTION 308: ALKYNE - HALIDE

DMSO , 15°C

DBU

80%

Ratovelomanana, V.; Rollin, Y.; Gébéhenne, C.; Gosmini, C.; Périchon, J. *Tetrahedron Lett.*, ***1994***, *35*, 4777

Ph—≡—H $\xrightarrow[CH_2Cl_2]{(collidine)_2I^{\oplus}\ PF_6^{\ominus}}$ Ph—≡—I 75%

Brunel, Y.; Rousseau, G. *Tetrahedron Lett.*, ***1995***, *36*, 2619

SECTION 309: ALKYNE - KETONE

1. PhCOCOCl , NEt_3

2. FVP (500*C)

72% x 82%

Aitken, R.A.; Hérion, H.; Janosi, A.; Raut, S.V.; Seth, S.; Shannon, I.J.; Smith, F.C. *Tetrahedron Lett.*, ***1993***, *34*, 5621

SECTION 310: ALKYNE - NITRILE

Ph—≡—H → Ph—≡—CN 72%

1. CuCN , TMSCl , H_2O
DMSO , MeCN , 72h

2. cat. NaI , 50°C

Luo, F.-T.; Wang, R.-T. *Tetrahedron Lett.*, ***1993***, *34*, 5911

SECTION 311: ALKYNE - ALKENE

CO_2Me ; C_6H_{13}

10% $Pd(OAc)_2$, 30% PPh_3
THF , reflux , 1h

C_6H_{13} 96%

Mandai, T.; Tsujiguchi, Y.; Matsuoka, S. *Tetrahedron Lett.*, ***1993***, *34*, 7615

Bu_3Sn SiMe$_3$ CO$_2$Et OTf 1. Pd(dba)$_2$, Ph$_3$As NMepyrrolidone , 40°C 2. aq. HCl CO$_2$Et SiMe$_3$ 90%

Lipshutz, B.H.; Alami, M. *Tetrahedron Lett.*, ***1993***, *34*, 1433

Bu Cl C_5H_{11} H 0.05 Pd(PPh)$_4$, 0.1 CuI 2 eq. NEt$_3$, PhH , 1h Bu C C_5H_{11} 91%

Gueugnot, S.; Linstrumelle, G. *Tetrahedron Lett.*, ***1993***, *34*, 3853

C_5H_{11} I OH N H , Pd(PPh$_3$)$_4$, 15 min C_5H_{11} OH 93%

Alami, M.; Ferri, F.; Linstrumelle, G. *Tetrahedron Lett.*, ***1993***, *34*, 6403

PhI 1. norbornadiene , Bu_3Sn—≡—Ph 1% PdCl$_2$(PPh$_3$)$_2$, 80-100°C , 20h 2. heat Ph Ph 54%

Kosugi, M.; Kimura, T.; Oda, H.; Migita, T. *Bull. Chem. Soc. Jpn.*, ***1993***, *66*, 3522

Cl Cl 1. C_5H_{11} , Pd(PPh$_3$)$_4$, CuI BuNH$_2$, rt 2. PdCl$_2$(PhCN)$_2$, piperidine rt ≡—SiMe$_3$ C_5H_{11} C_5H_{11} 72%

Alami, M.; Crousse, B.; Linstrumelle, G. *Tetrahedron Lett.*, ***1994***, *35*, 3543

Ph$_3$P O Ph FVP (500°C) Ph 59%

Aitken, R.A.; Boeters, C.; Morrison, J.J. *J. Chem. Soc., Perkin Trans. 1.*, ***1994***, 2473

1. sec-BuLi, THF, -45°C
2. $ZnBr_2$ -45°C → 20°C
3. H_3O^+

$SiMe_3$ $SiMe_3$ Me_3Si $SiMe_3$

85%

Meyer, C.; Marek, I.; Normant, J.-F.; Platzer, N. *Tetrahedron Lett.*, ***1994***, *35*, 5645

Bu—≡—H + Ph

5% $HRh(CO)(PPh_3)_3$
35% PEt_3
THF, reflux, 9h

Ph Ph

68% (>20:1 endo:exo)
(>20:1 E:Z)

Yamaguchi, M.; Omata, K.; Hirama, M. *Tetrahedron Lett.*, ***1994***, *35*, 5689

Ph—≡—$SnBu_3$

Cl, Dibal, THF
$Ni(acac)_2$, $P(OEt)_3$, reflux

Ph

77%

Cui, D.-M.; Hashimoto, N.; Ikeda, S.; Sato, Y. *J. Org. Chem.*, ***1995***, *60*, 5752

SECTION 312: CARBOXYLIC ACID - CARBOXYLIC ACID

NO ADDITIONAL EXAMPLES

SECTION 313: CARBOXYLIC ACID - ALCOHOL, THIOL

1. 1:3 $TiCl_4$:$LiAlH_4$, 1h
$CFCl_3$, THF, -5°C
2. aq. HCl

Me_2HC O Me_2HC OH CO_2H

62%

García, M.; del Campo, C.; Sinisterra, J.V.; Llama, E.F. *Tetrahedron Lett.*, ***1993***, *34*, 7973

SECTION 314: CARBOXYLIC ACID - ALDEHYDE

NO ADDITIONAL EXAMPLES

SECTION 315: CARBOXYLIC ACID - AMIDE

Me O O N H F_3C O

1. 2.2 LDA , $ZnCl_2$

2. H_3O^+

Me H N F_3C O CO_2H

86% (88% de)

Kazmaier, U.; Maier, S. *J. Chem. Soc. Chem. Commun.*, ***1995***, 1991

TFAHN O O

5 eq. LDA , 2.5 eq. quinone
$Al(OiPr)_3$

TFAHN CO_2H

70% (98% ds/86% ee)

Kazmaier, U.; Krebs, A. *Angew. Chem. Int. Ed. Engl.*, ***1995***, *34*, 2012

SECTION 316: CARBOXYLIC ACID - AMINE

O N CO_2Et Boc

1. PhMgBr , THF , -40°C
2. TFA , CH_2Cl_2

3. DDQ , CH_2Cl_2
4. 10% NaOH , reflux

Ph N CO_2H H

83x72x93%

Ezquerra, J.; Pedregal, C.; Rubio, A.; Valenciano, J.; Navio, J.L.G.; Alvarez-Builla, J.; Vaquero, J.J. *Tetrahedron Lett.*, ***1993***, *34*, 6317

Ph NPh

1. $OSiMe_3$ $OSiMe_3$, $ZnBr_2$, CH_2Cl_2

2. H_3O^+

NHPh Ph CO_2H

84% (35:65 *syn:anti*)

Mladenova, M.; Bellassoued, M. *Synth. Commun.*, ***1993***, *23*, 725

1. PHCHO
0.1 $[Au(c\text{-}hexNC)_2]BF_4$
bisphosphinoferrocene derivative, 25°C, 40h
2. 6N HCl

NC–CH_2–CO_2Me → Ph, CO_2H, HO, NH_2 + Ph, CO_2H, HO, NH_2

(94 : 6)

Soloshonok, V.A.; Hayashi, T. *Tetrahedron Lett.*, ***1994***, *35*, 2713

1. oxalyl chloride, ether
2. satd. $NaHCO_3$, reflux
3. $TsNHNH_2$, reflux
4. $NaBH_4$, THF

98x39%

Guan, X.; Borchardt, R.T. *Tetrahedron Lett.*, ***1994***, *35*, 3013

REVIEWS:

"Recent Developments in the Stereoselective Synthesis of α-Amino Acids," Duthaler, R.O. *Tetrahedron*, ***1994***, *50*, 1539

"Recent Stereoselective Synthetic Approaches to β-Amino Acids," Cole, D.C. *Tetrahedron*, ***1994***, *50*, 9517

Related Methods:

Section 315 (Carboxylic Acid - Amide).
Section 344 (Amide - Ester).
Section 351 (Amine - Ester).

SECTION 317: CARBOXYLIC ACID - ESTER

NO ADDITIONAL EXAMPLES

SECTION 318: CARBOXYLIC ACID - ETHER, EPOXIDE, THIOETHER

$TiCl_4$, Et_3SiH
CH_2Cl_2, -78°C

CO_2H + CO_2H

(85 : 15) 82%

Mead, K.T.; Pillai, S.K. *Tetrahedron Lett.*, ***1993***, *34*, 6997

SECTION 319: CARBOXYLIC ACID - HALIDE, SULFONATE

CO_2H 1. 2.2 eq. *sec*-BuLi/TMEDA THF, -90°C 2. 4 eq. Cl_3CCl_3, THF, -78°C 3. 4N HCl CO_2H Cl F F 71%

Moyroud, J.; Guesnet, J.-L.; Bennetau, B.; Mortier, J. *Tetrahedron Lett.*, ***1995***, *36*, 881

SECTION 320: CARBOXYLIC ACID - KETONE

Ph O H_2N NH_2 Na, *n*-C_3H_7OH, reflux Ph H_2N OH 98% (>98% ee)

Moody, H.M.; Kaptein, B.; Broxterman, Q.B.; Boesten, W.H.J.; Kamphuis, J. *Tetrahedron Lett.*, ***1994***, *35*, 1777

O OH 0.5 $Cu(ClO_4)_2$•6 H_2O O_2, MeCN, rt, 10-15 h O O OH O 96%

Cossy, J.; Belotti, D.; Bellosta, V.; Brocca, D. *Tetrahedron Lett.*, ***1994***, *35*, 6089

Also via: Section 360 (Ketone - Ester).

SECTION 321: CARBOXYLIC ACID - NITRILE

NO ADDITIONAL EXAMPLES

Also via: Section 361 (Nitrile - Ester).

SECTION 322: CARBOXYLIC ACID - ALKENE

CHO 1. CF_2=CHLi 2. H_3O^+ 3. H_3O^+ CO_2H 60% (*E*)

Tellier, F.; Sauvêtre, R. *Tetrahedron Lett.*, ***1993***, *34*, 5433

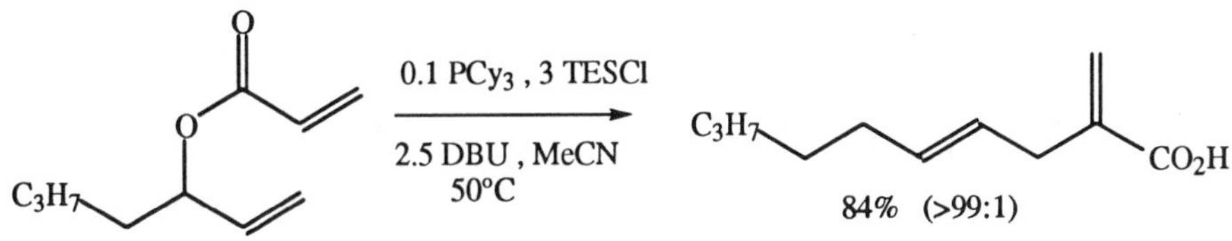

0.1 PCy3 , 3 TESCl
2.5 DBU , MeCN
50°C
C3H7
CO2H
84% (>99:1)

Hanamoto, T.; Baba, Y.; Inanaga, J. *J. Org. Chem.*, ***1993***, *58*, 299

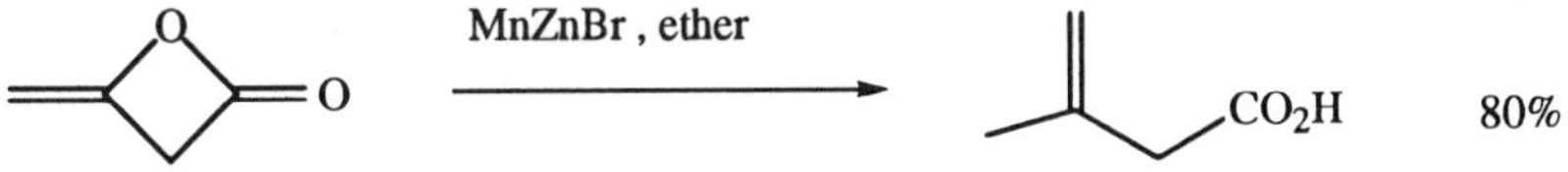

MnZnBr , ether
CO2H
80%

Duchêne, A.; Abarbi, M.; Parrain, J.-L.; Kitamura, M.; Noyori, R. *Synlett*, ***1994***, 524

Also via: Section 313 (Alcohol - Carboxylic Acids).
Section 349 (Amide - Alkene).
Section 362 (Ester - Alkene).
Section 376 (Nitrile - Alkene).

SECTION 323: ALCOHOL, THIOL - ALCOHOL, THIOL

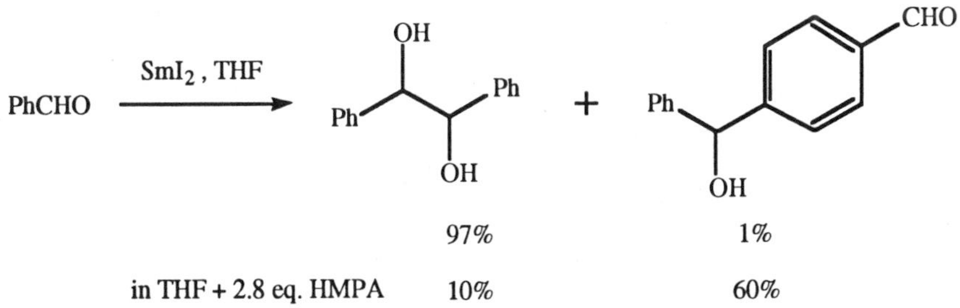

PhCHO
SmI2 , THF
OH
Ph
Ph
OH
+
CHO
Ph
OH
97%
1%
in THF + 2.8 eq. HMPA
10%
60%

Shiue, J-S.; Lin, C-C.; Fang, J-M. *Tetrahedron Lett.*, ***1993***, *34*, 335

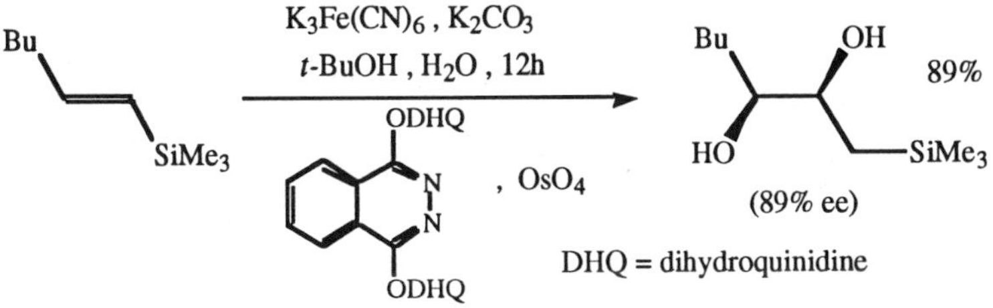

Bu
SiMe3
K3Fe(CN)6 , K2CO3
t-BuOH , H2O , 12h
ODHQ
N
N
ODHQ
, OsO4
Bu
OH
HO
SiMe3
89%
(89% ee)
DHQ = dihydroquinidine

Soderquist, J.A.; Rane, A.M.; López, C.J. *Tetrahedron Lett.*, ***1993***, *34*, 1893

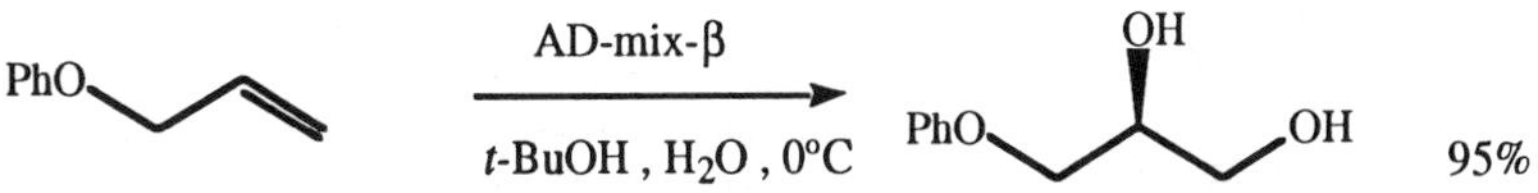

PhO
AD-mix-β
t-BuOH , H2O , 0°C
OH
PhO
OH
95%

Wang, Z-M.; Zhang, X-L.; Sharpless, K.B. *Tetrahedron Lett.*, ***1993***, *34*, 2267

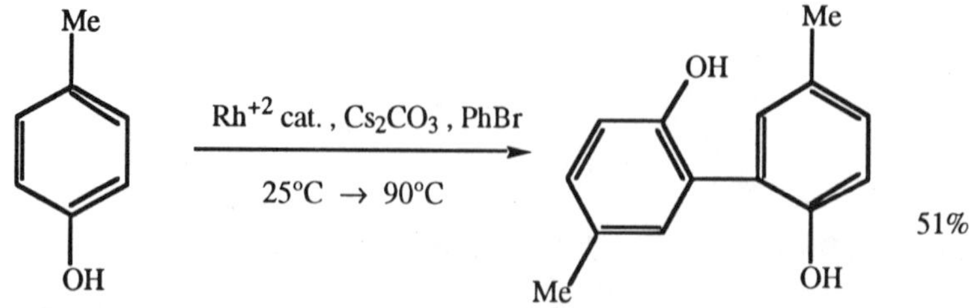

Me / OH (p-cresol) → Rh^{+2} cat., Cs_2CO_3, PhBr; 25°C → 90°C → biaryl diol (OH, Me, Me, OH) 51%

Barrett, A.G.M.; Itoh, T.; Wallace, E.M. *Tetrahedron Lett.*, ***1993***, *34*, 2233

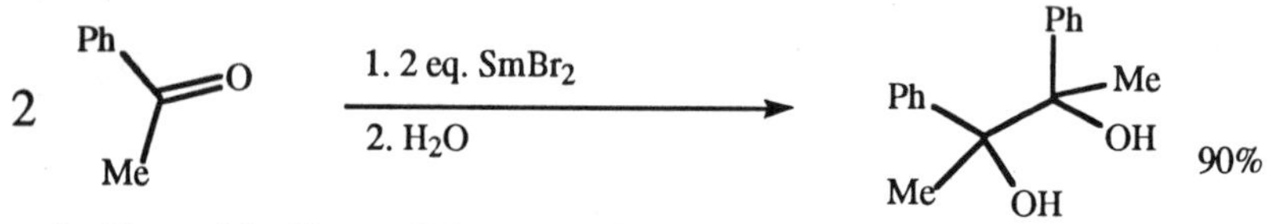

2 Ph(Me)C=O → 1. 2 eq. $SmBr_2$ 2. H_2O → Ph, Me, OH / Ph, Me, OH pinacol 90%

Lebrun, A.; Namy, J-L.; Kagan, H.B. *Tetrahedron Lett.*, ***1993***, *34*, 2311

Me_3Si ... C_5H_{11} → DHQ-CLB → Me_3Si ... (OH)(OH) C_5H_{11}

93% (71% ee, R)

Okamoto, S.; Tani, K.; Sato, F.; Sharpless, K.B.; Zargarian, D. *Tetrahedron Lett.*, ***1993***, *34*, 2509

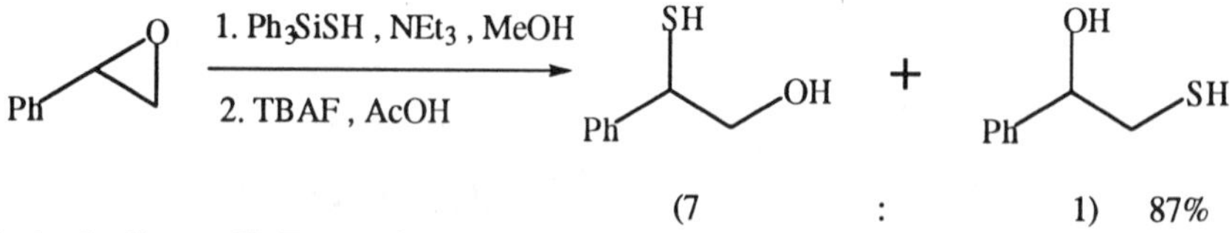

Ph epoxide → 1. Ph_3SiSH, NEt_3, MeOH 2. TBAF, AcOH → Ph(SH)CH–CH₂OH + Ph(OH)CH–CH₂SH

(7 : 1) 87%

Brittain, J.; Gareau, Y. *Tetrahedron Lett.*, ***1993***, *34*, 3363

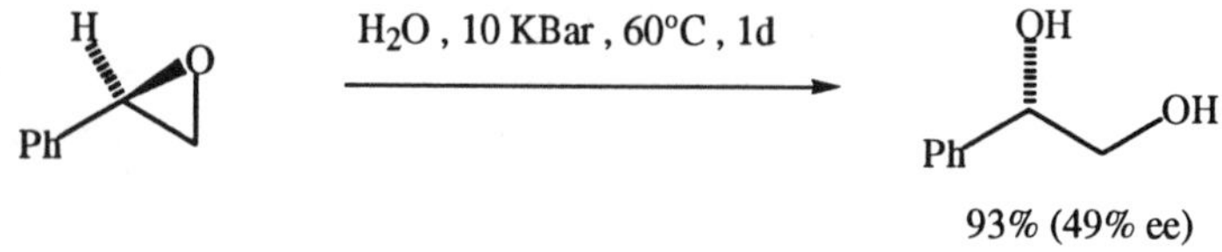

H, Ph epoxide → H_2O, 10 KBar, 60°C, 1d → Ph(OH)CH–CH₂OH

93% (49% ee)

Kotsuki, H.; Kataoka, M.; Nishizawa, H. *Tetrahedron Lett.*, ***1993***, *34*, 4037

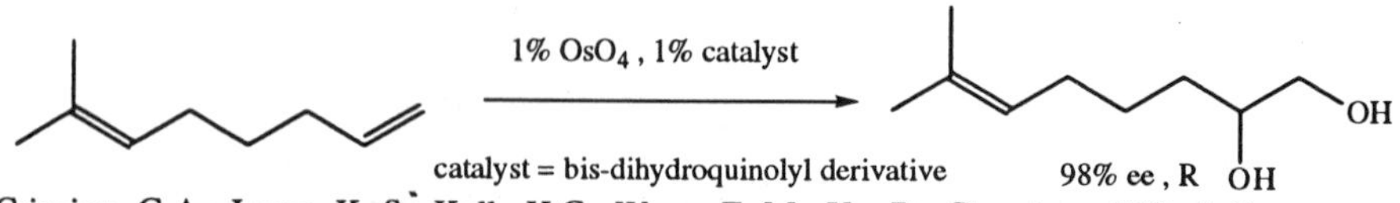

diene → 1% OsO_4, 1% catalyst → diol (OH, OH)

catalyst = bis-dihydroquinolyl derivative

98% ee, R

Crispino, G.A.; Jeong, K.-S.; Kolb, H.C.; Wang, Z.-M.; Xu, D.; Sharpless, K.B. *J. Org. Chem.*, ***1993***, *58*, 3785

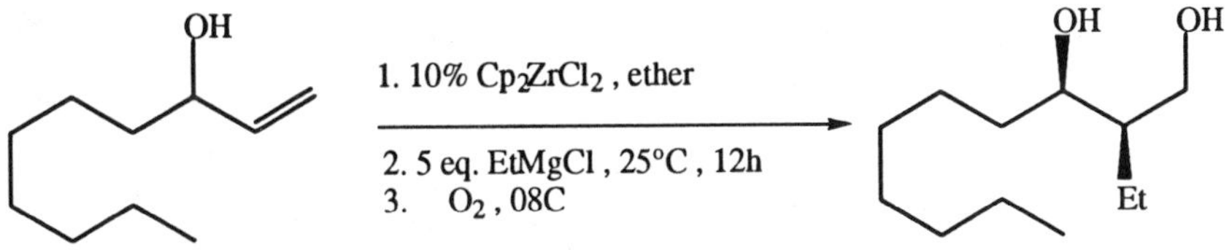

OH

1. 10% Cp_2ZrCl_2, ether

2. 5 eq. EtMgCl, 25°C, 12h

3. O_2, 08C

OH OH

Et

70% (98:2 *syn:anti*)

Hoveyda, A.H.; Morken, J.P. *J. Org. Chem.*, ***1993***, *58*, 4237

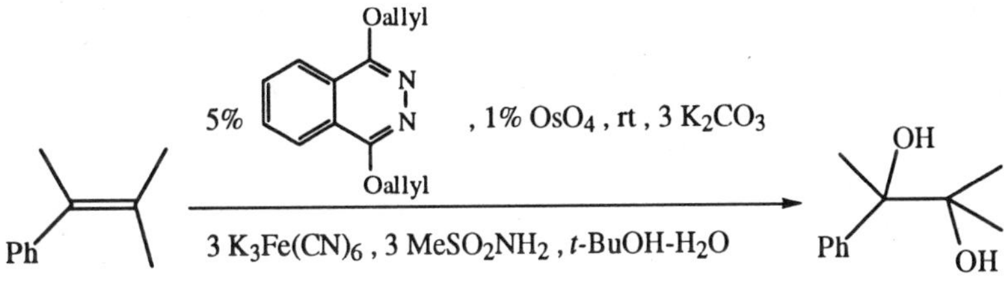

Oallyl

5% N N Oallyl, 1% OsO_4, rt, 3 K_2CO_3

Ph

3 $K_3Fe(CN)_6$, 3 $MeSO_2NH_2$, *t*-BuOH-H_2O

OH

Ph OH

82% (39% ee, R)

Morikawa, K.; Park, J.; Andersson, P.G.; Hashiyama, T.; Sharpless, K.B. *J. Am. Chem. Soc.*, ***1993***, *115*, 8463

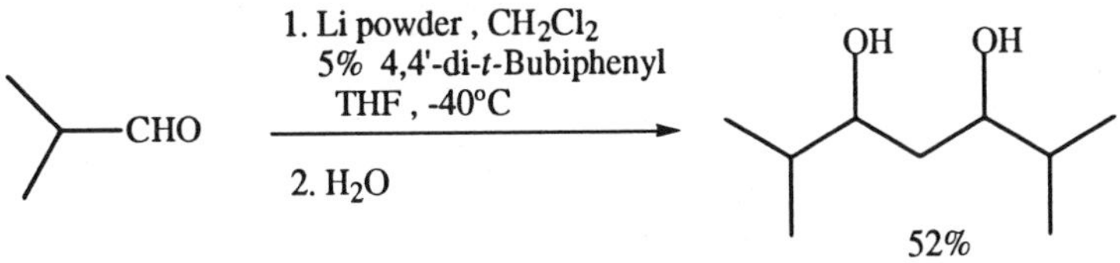

CHO

1. Li powder, CH_2Cl_2
5% 4,4'-di-*t*-Bubiphenyl
THF, -40°C

2. H_2O

OH OH

52%

Guijarro, A.; Yus, M. *Tetrahedron Lett.*, ***1994***, *35*, 253

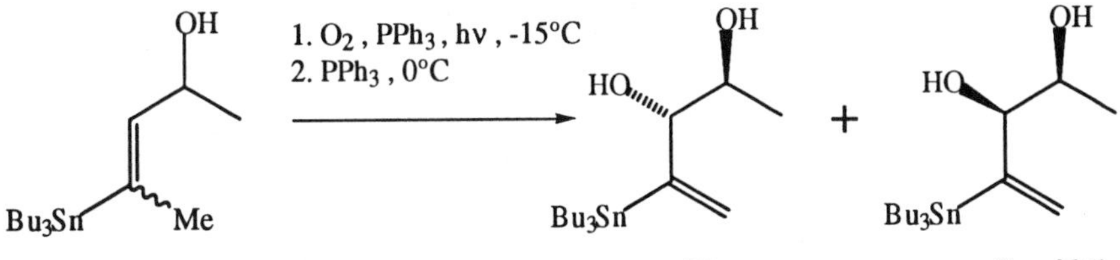

OH

Bu_3Sn Me

1. O_2, PPh_3, hν, -15°C
2. PPh_3, 0°C

HO OH

Bu_3Sn

\+

HO OH

Bu_3Sn

(95 : 5) 93%

Adam, W.; Gevert, O.; Klug, P. *Tetrahedron Lett.*, ***1994***, *35*, 1681

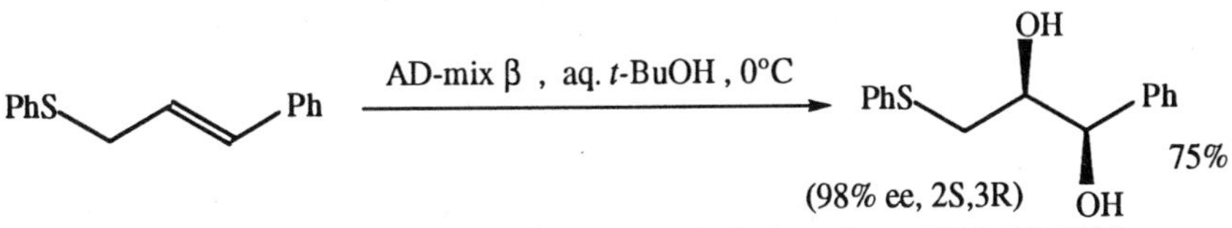

PhS Ph

AD-mix β, aq. *t*-BuOH, 0°C

OH

PhS Ph

OH

75%

(98% ee, 2S,3R)

Walsh, P.J.; Ho, P.T.; King, S.B.; Sharpless, K.B. *Tetrahedron Lett.*, ***1994***, *35*, 5129

OSiMe$_3$, Ph; iPrCHO, -78°C; iPr, O, TolO$_2$S—N, O, B, H

OH OH Ph + OH O Ph

60% 95:5 syn:anti (94% ee, syn)

21% 50% ee, S

Kaneko, Y.; Matsuo, T.; Kiyooka, S. *Tetrahedron Lett.*, ***1994***, *35*, 4107

O; Yb-TMS-Br; THF-HMPA; OTMS, TMSO; 62%

Taniguchi, Y.; Nakahashi, M.; Kuno, T.; Tsuno, M.; Makioka, Y.; Takaki, K.; Fujiwara, Y. *Tetrahedron Lett.*, ***1994***, *35*, 4111

S, S, O, N, O, O, Me, C_6H_{13}; *t*-BuSH, hν; O, O, O, C_6H_{13}, Me

57% (4:1 *syn:anti*)

Newcomb, M.; Dhanabalasingam, B. *Tetrahedron Lett.*, ***1994***, *35*, 5193

Ph, Ph; 0.1 K_2OsO_4, 3 $K_3Fe(CN)_6$, 3 K_2CO_3, $MeSO_2NH_2$ 0°C, 12h; 1% RO—(N=N)—OR; OH, OH, Ph, Ph

R = cinchonidine derivative; >99% (99% ee)

Corey, E.J.; Noe, M.C.; Grogan, M.J. *Tetrahedron Lett.*, ***1994***, *35*, 6405

2 *t*-BuCHO; $NbCl_3$, THF, rt; *t*-Bu, OH, HO, *t*-Bu

71% (>95:5 *d,l:meso*)

Szymoniak, J.; Besançon, J.; Moïse, C. *Tetrahedron*, ***1994***, *50*, 2841

PhCHO —(Al , KOH , MeOH , rt , 5 min)→ 87% (1:1 *d,l:meso*)

Khurana, J.M.; Sehgal, A. *J. Chem. Soc. Chem. Commun.*, ***1994***, 571

1. $^{d}Ipc_2BH$, ether , -78°C
2. PhCHO , -78°C
3. NaOH , H_2O_2

76% (>95% ee)

Brown, H.C.; Narla, G. *J. Org. Chem.*, ***1995***, *60*, 4686

BH_3 , THF

(16:84 meso:RR+SS
(99% ee SS)

Quallich, G.J.; Keavey, K.N.; Woodall, T.M. *Tetrahedron Lett.*, ***1995***, *36*, 4729

Bu_3SnH , PhH , 20°C
hν

46% (98:1 *cis:trans*)

Hays, D.S.; Fu, G.C. *J. Am. Chem. Soc.*, ***1995***, *117*, 7283

e^- , *t*-BuOH , H_2O , $K_3Fe(CN)_6$
K_2CO_3 , $(DHQD)_2PHAL$, $K_2OsO_2)OH)_4$

95% (97% ee , R)

Torii, S.; Liu, P.; Tanaka, H. *Chem. Lett.*, ***1995***, 319

REVIEWS:

"Synthesis of α,ω-Alkenediols. A Review," Patwardhan, S.A. *Org. Prep. Proceed. Int.*, ***1994***, *26*, 645

"The Oxygenation of Vinyl Cyclopropanes as an Entry Into Stereoselective 1,3-Diol Synthesis," Feldman, K.S. *Synlett*, ***1995***, 217

"Catalytic Asymmetric Dihydroxylation," Kolb, H.C.; Van Nieuwenhze, M.S.; Sharpless, K.B. *Chem. Rev.*, ***1994***, *94*,, 2483

Also via: Section 327 (Alcohol - Ester). Section 357 (Ester - Ester).

SECTION 324: ALCOHOL, THIOL - ALDEHYDE

1. PhCHO , NEt_3
2. H_3O^+

HO, Ph, CHO + CO_2H

51%

Barco, A.; Benetti, S.; De Risi, C.; Pollini, G.P.; Spalluto, G.; Zanirato, V. *Tetrahedron Lett.*, ***1993***, *34*, 3907

O

$[RhCl(CO)_2]_2$, 1-methylpyrazole

$HSiMe_2Ph$, CO (50 atm)

$OSiMe_2Ph$, CHO

82%

Fukumoto, Y.; Chatani, N.; Murai, S. *J. Org. Chem.*, ***1993***, *58*, 4187

O, C_6H_{13}, CHO

Ph_3P=CHOMe , $CHCl_3$,25•C

OH, C_6H_{13}, CHO

77%

Yu, L.; Wang, Z. *J. Chem. Soc. Chem. Commun.*, ***1993***, 232

Related Methods: Section 330 (Alcohol - Ketone).

SECTION 325: ALCOHOL, THIOL - AMIDE

O, $(Pri)_2N$, Cl

1. EtCHO , Li(powder)
3% $C_{10}H_8$, THF
2. H_2O

O, $(Pri)_2N$, OH

71%

Ramón, D.J.; Yus, M. *Tetrahedron Lett.*, ***1993***, *34*, 7115

NHTs, O

aq. NaOH , 100°C

Ts, N, OH

95%

Moulines, J.; Bats, J-P.; Hautefaye, P.; Nuhrich, A.; Lamidey, A-M. *Tetrahedron Lett.*, ***1993***, *34*, 2315

MeSO$_2$NH$_2$, t-BuOH , H$_2$O
rt , AD-mix-β^{TM}

Ph NEt$_2$ O → Ph OH O OH NEt$_2$ 96% (96% ee)

AD-mix-β^{TM} = 5% DHQD$_2$-PHAL (chichona alkaloid-phthalzine reagent), 2% K$_2$OsO$_2$(OH)$_4$
Bennani, Y.L.; Sharpless, K.B. *Tetrahedron Lett.*, ***1993***, *34*, 2079

CO$_2$Me NHCbz
1. Dibal , -78°C
2. OTMS OMe
3. TiCl$_2$(OiPr)$_2$

OH CO$_2$Me NHCbz + OH CO$_2$Me NHCbz
66% 10%

Kiyooka, S.; Suzuki, K.; Shirouchi, M.; Kaneko, Y.; Tanimori, S. *Tetrahedron Lett.*, ***1993***, *34*, 5729

CHO BocHN Me
Me$_2$CuLi , -55°C

BocHN OH Me Me + BocHN OH Me Me
(96 : 4) 84%

Reetz, M.T.; Rölfing, K.; Griebenow, N. *Tetrahedron Lett.*, ***1994***, *35*, 1969

NHBz Cl
1. BuLi , -78°C
2. LiNaphth , -78°C
3. PhCHO

NHBz OH Ph + NHBz OH Ph
(1 : 1) 72%

Foubelo, F.; Yus, M. *Tetrahedron Lett.*, ***1994***, *35*, 4831

NMe$_2$ O
1. LDA , THF , -78°C
2. CeCl$_3$, THF , rt
3. PhCHO , THF , -78°C

Ph NMe$_2$ OH O 68%

Shang, X.; Liu, H.-J. *Synth. Commun.*, ***1994***, *24*, 2485

O NMe$_2$ O H Me
MeAlCl$_2$

Me OH O H NMe$_2$ Me + HO Me O H NMe$_2$ Me
(97 : 3) 66%

Taniguchi, M.; Fujii, H.; Oshima, K.; Utimoto, K. *Bull. Chem. Soc. Jpn.*, ***1994***, *67*, 2514

NHC$_6$H$_{13}$ hν HO N O C$_6$H$_{13}$ 86%

Dittami, J.P.; Xu, F.; Qi, H.; Martin, M.W.; Bordner, J.; Decosta, D.L.; Kiplinger, J.; Reiche, P.; Ware, R. *Tetrahedron Lett.*, ***1995***, *36*, 4197

N O H (CH$_2$O)$_n$, acetone, K$_2$CO$_3$ / H$_2$O, ultrasound → N O OH 96%

Jouglet, B.; Oumoch, S.; Rosseau, G. *Synth. Commun.*, ***1995***, *25*, 3869

SECTION 326: ALCOHOL, THIOL - AMINE

CO$_2$Et N Ph 1. LDA, TiCl$_2$(OiPr)$_2$ THF / 2. *t*-BuCHO, THF / 3. H$_3$O$^+$ → NH$_2$ EtO$_2$C Ph OH + NH$_2$ EtO$_2$C Ph OH (>99 : 1) 78%

Kanemasa, S.; Mori, T.; Wada, E.; Tatsukawa, A. *Tetrahedron Lett.*, ***1993***, *34*, 677

OH Ph Ph OH 2 TsNCO, THF, 1d / 2% Pd[P(OiPr)$_3$]$_4$ / reflux → O Ph Ph O O 81%

Xu, D.; Sharpless, K.B. *Tetrahedron Lett.*, ***1993***, *34*, 951

t-BuPh$_2$O EtO$_2$CBnN OMe 1. SiMe$_3$, SnCl$_4$ / CH$_2$Cl$_2$, 0°C, 30 min / 2. TBAF, THF → HO EtO$_2$CBnN + HO EtO$_2$CBnN (1 : 1.1) 74%

Hioki, H.; Okauda, M.; Miyagi, W.; Itô, S. *Tetrahedron Lett.*, ***1993***, *34*, 6131

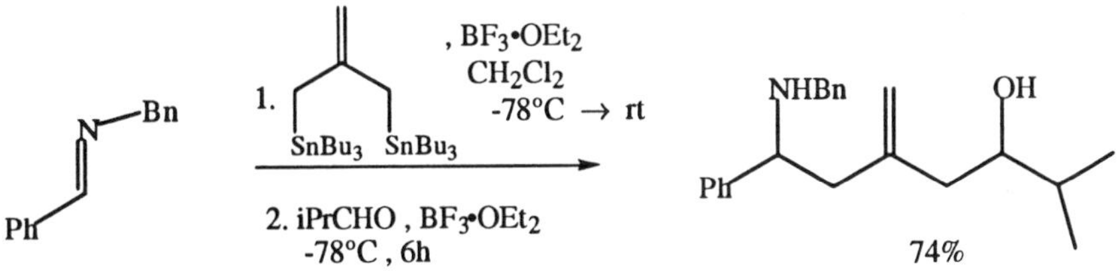

1. (2-methylene-1,3-propanediyl)bis(tributylstannane) [$SnBu_3$ $SnBu_3$], $BF_3{\bullet}OEt_2$, CH_2Cl_2, -78°C → rt

2. iPrCHO, $BF_3{\bullet}OEt_2$, -78°C, 6h

Ph, N-Bn → NHBn, OH, Ph — 74%

Keck, G.E.; Palani, A. *Tetrahedron Lett.*, ***1993***, *34*, 3223

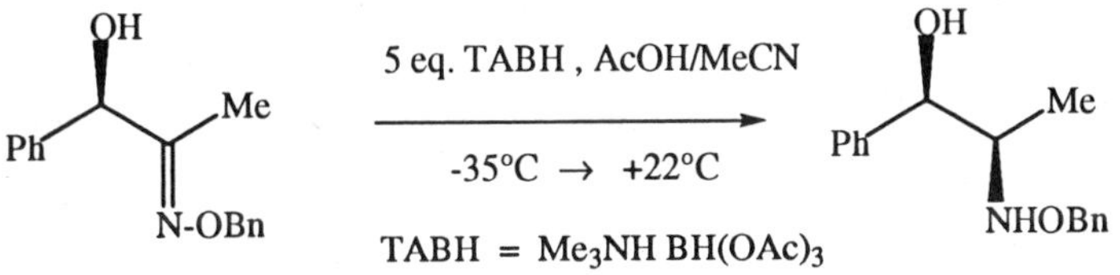

OH, Ph, Me, N-OBn — 5 eq. TABH, AcOH/MeCN, -35°C → +22°C → OH, Ph, Me, NHOBn

TABH = $Me_3NH\ BH(OAc)_3$

Williams, D.R.; Osterhout, M.H.; Reddy, J.P. *Tetrahedron Lett.*, ***1993***, *34*, 3271

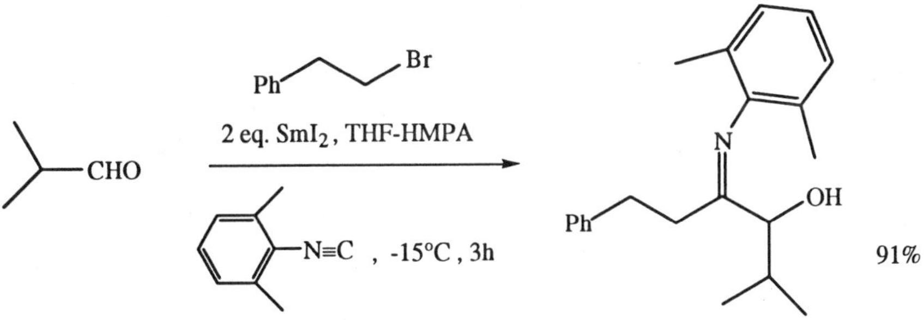

CHO — Ph–CH₂CH₂–Br, 2 eq. SmI_2, THF-HMPA; 2,6-dimethylphenyl–N≡C, -15°C, 3h → Ph, N, OH — 91%

Murakami, M.; Kawano, T.; Ito, H.; Ito, Y. *J. Org. Chem.*, ***1993***, *58*, 1458

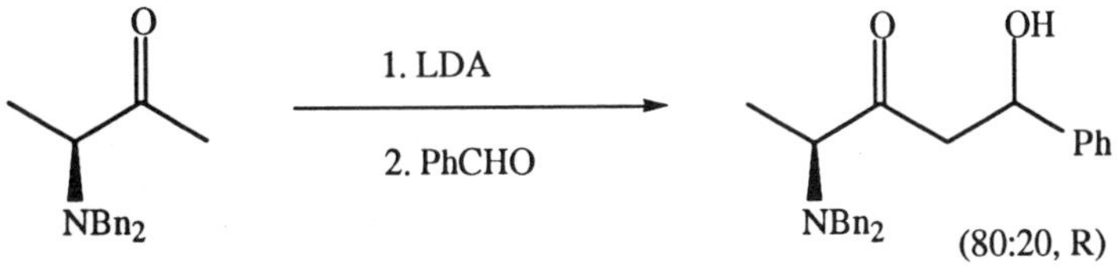

O, NBn_2 — 1. LDA, 2. PhCHO → O, OH, Ph, NBn_2 (80:20, R)

Lagu, B.R.; Crane, H.M.; Liotta, D.C. *J. Org. Chem.*, ***1993***, *58*, 4191

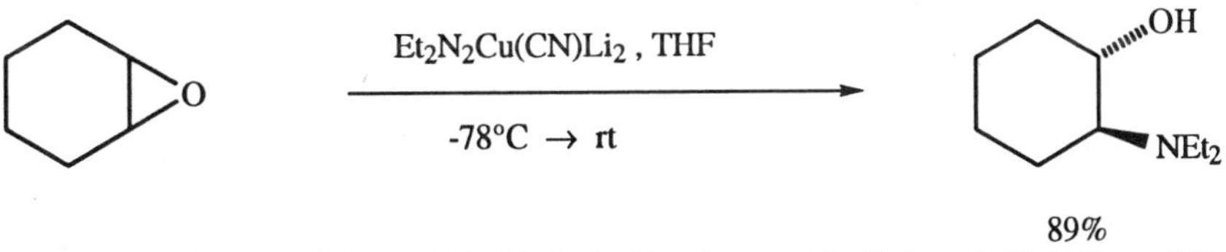

O — $Et_2N_2Cu(CN)Li_2$, THF, -78°C → rt → OH, NEt_2

89%

Yamamoto, Y.; Asao, N.; Meguro, M.; Tsukuda, N.; Nemoto, H.; Sadayori, N.; Wilson, J.G.; Nakamura, H. *J. Chem. Soc. Chem. Commun.*, ***1993***, 1201

$Yb(OTf)_3$, $NHEt_2$; CH_2Cl_2, 20 min — 97%

> 99% α-attack

Chini, M.; Crotti, P.; Favero, L.; Macchia, F.; Pineschi, M. *Tetrahedron Lett.*, ***1994***, *35*, 433

Et_2NH, $LiBF_4$, MeCN; 80°C, 18h — 88%

Chini, M.; Crotti, P.; Favero, L.; Macchia, F. *Tetrahedron Lett.*, ***1994***, *35*, 761

Bu_3SnH, AIBN; PhH, 9h — (21 : 79) 52%

Naito, T.; Tajiri, K.; Harimoto, T.; Ninomiya, I.; Kiguchi, T. *Tetrahedron Lett.*, ***1994***, *35*, 2205

CH_3NO_2, 2 eq. $AlCl_3$; 80°C, 5h — 70%

Sartori, G.; Bigi, F.; Maggi, R.; Tomasini, F. *Tetrahedron Lett.*, ***1994***, *35*, 2393

1. Li, cat. $C_{10}H_8$ 2. PhCHO 3. H_2O — PhHN~~Ph 68%

various electrophiles were used

Almena, J.; Foubelo, F.; Yus, M. *J. Org. Chem.*, ***1994***, *59*, 3210

MeCH=N-OMe, iPrOH; e^- — 93% (1:1)

Shono, T.; Kise, N.; Fujimoto, T.; Yamanami, A.; Nomura, R. *J. Org. Chem.*, ***1994***, *59*, 1730

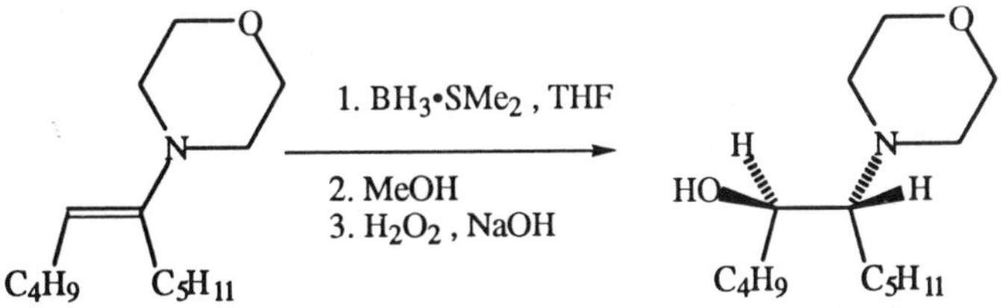

1. $BH_3 \cdot SMe_2$, THF
2. MeOH
3. H_2O_2, NaOH

C_4H_9 C_5H_{11} → C_4H_9 C_5H_{11} 45%

Goralski, C.T.; Hasha, D.L.; Nicholson, L.W.; Zakett, D.; Fisher, G.B.; Singaram, B. *Tetrahedron Lett.*, ***1994***, *35*, 3251

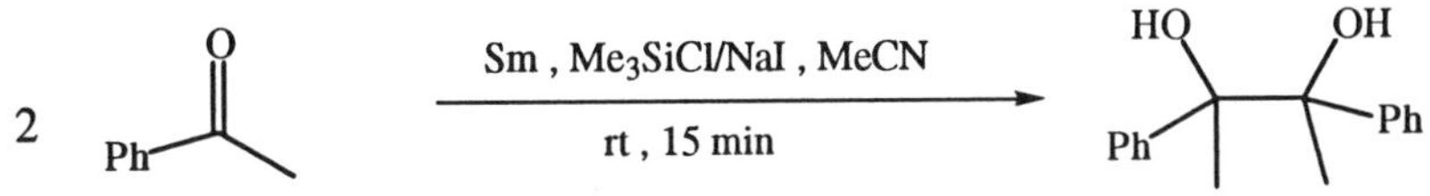

2 Ph (ketone) → Sm, Me_3SiCl/NaI, MeCN; rt, 15 min → HO OH Ph Ph

72%

Akane, N.; Hatano, T.; Kusui, H.; Nishiyama, Y.; Ishii, Y. *J. Org. Chem.*, ***1994***, *59*, 7902

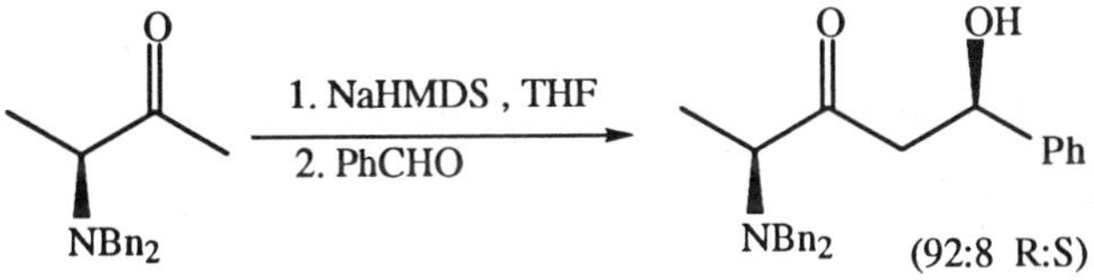

NBn_2 → 1. NaHMDS, THF; 2. PhCHO → O OH Ph NBn_2 (92:8 R:S) 90%

with LDA (63:7 R:S) 90%

Lagu, B.R.; Liotta, D.C. *Tetrahedron Lett.*, ***1994***, *35*, 4485

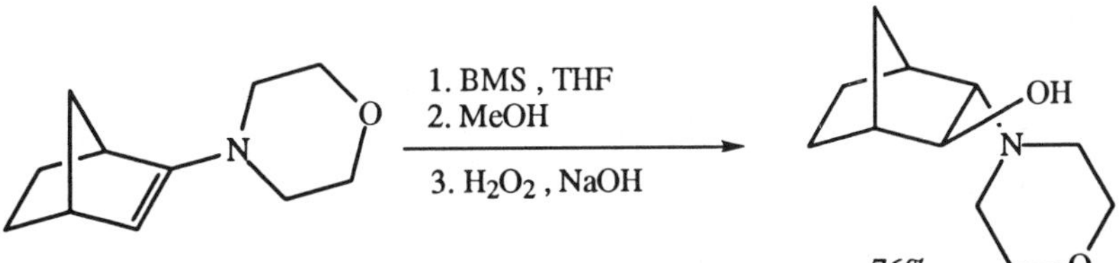

1. BMS, THF
2. MeOH
3. H_2O_2, NaOH

OH N O 76%

Goralski, C.T.; Hasha, D.L.; Nicholson, L.W.; Singaram, B. *Tetrahedron Lett.*, ***1994***, *35*, 5165

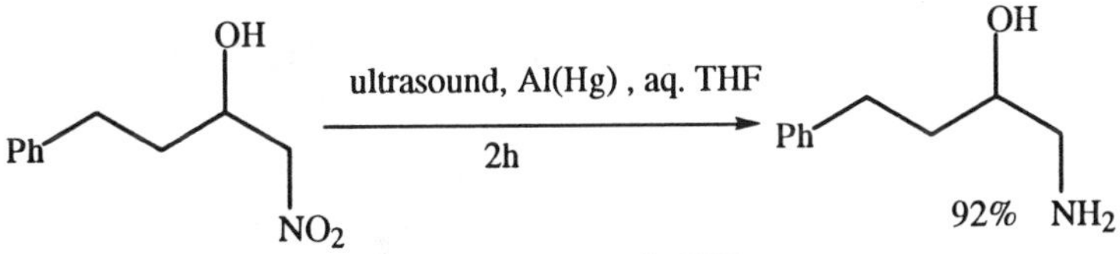

OH Ph NO_2 → ultrasound, Al(Hg), aq. THF; 2h → OH Ph 92% NH_2

Fitch, R.W.; Luzzio, F.A. *Tetrahedron Lett.*, ***1994***, *35*, 6013

t-BuNH$_2$, CH$_2$Cl$_2$
5% Yb(OTf)$_3$, rt , 16h

C_8H_{17} ; NHt-Bu ; OH ; + ; OH ; C_8H_{17} ; NHt-Bu

(>99 : <1) 99%

Crotti, P.; Favero, L.; Macchia, F.; Pineschi, M. *Tetrahedron Lett.*, ***1994***, *35*, 7089

Bn ; N ; O ; OMe

1. Cp$_2$Zr(Bu)$_2$, THF
-78°C → rt , 4h
2. BF$_3$•OEt$_2$, THF , 0°C
3. aq. NaOH

Bn ; N ; HO

82% (>98:2 *syn:anti*)

Ito, H.; Ikeuchi, Y.; Taguchi, T.; Hanzawa, Y.; Shiro, M. *J. Am. Chem. Soc.*, ***1994***, *116*, 5469

O ; NH$_2$

Na , iPrOH , THF

OH ; NH$_2$

76% (*erythro/threo* = 0.8)

Bartoli, G.; Cimarelli, C.; Palmieri, G. *J. Chem. Soc., Perkin Trans. 1.*, ***1994***, 537

N ; I ; O ; , SmI$_2$

THP , HMPA

N ; OH

85%

Booth, S.E.; Benneche, T.; Undheim, K. *Tetrahedron*, ***1995***, *51*, 3665

MeO$_2$C ; Ph ; O

PhNH$_2$, TiCl$_3$, PhCHO
THF/CH$_2$Cl$_2$, Py , rt

NHPh ; MeO$_2$C ; Ph ; HO ; Ph

76%

Clerici, A.; Clerici, L.; Porta, O. *Tetrahedron Lett.*, ***1995***, *36*, 5955

$BnNH_2$, 10kbar ,CH_2Cl_2 ; cat. $Yb(OTf)_3$, rt , 6d

OH, NHBn 90%

Meguro, M.; Asao, N.; Yamamoto, Y. *J. Chem. Soc., Perkin Trans. 1.*, ***1994***, 2597

Ph, Ph; $O_3Os{=}Nt\text{-}Bu$, DME , 25°C ; quinuclidine , 36h

NH*t*-Bu, Ph, Ph, OH + OH, Ph, Ph, OH

(97 : 3)

Rubinstein, H.; Svendsen, J.S. *Acta Chem. Scand. B.*, ***1994***, *48*, 439

⊕N, ⊖O

1. TFAA , CH_2Cl_2 rt , 4h
2. K_2CO_3

N, OH 95%

Fontenas, C.; Bejan, E.; Haddon, H.A.; Belavoine, G.G.A. *Synth. Commun.*, ***1995***, *25*, 629

SECTION 327: ALCOHOL, THIOL - ESTER

OBn, CHO; $OSiMe_2t\text{-}Bu$, OMe; 3% $LiClO_4$, -30°C ; CH_2Cl_2 , 4h

OBn, CO_2Me, $OSiMe_2t\text{-}Bu$ + OBn, CO_2Me, $OSiMe_2t\text{-}Bu$

(92 : 8) >84%

Reetz, M.T.; Fox, D.N.A. *Tetrahedron Lett.*, ***1993***, *34*, 1119

O; EtOH , CO , NaBr , iPr_2NEt ; $Pd_2(C_4H_7)_2Cl_2$, maleic anhydride

HO, CO_2Et 86%

Shimizu, I.; Maruyama, T.; Makuta, T.; Yamamoto, A. *Tetrahedron Lett.*, ***1993***, *34*, 2135

OMe, OTMS; PhCHO , 5% SmI_2 , 5 min ; CH_2Cl_2 , -78°C

Ph, CO_2Me, OTMS 95%

Van de Weghe, P.; Collin, J. *Tetrahedron Lett.*, ***1993***, *34*, 3881

Yeast , pet. ether , H_2O

rt , 1d

CO_2Et → OH, CO_2Et 58% (94% ee)

Yeast = *Saccharomyces cerevisiae*

Jayasinghe, L.Y.; Smallridge, A.J.; Trewhella, M.A. *Tetrahedron Lett.*, ***1993***, *34*, 3949

CO_2Me —[$Mn(OAc)_3$•2 H_2O; $Cu(OAc)_2$•H_2O, 25°C]→ CO_2Me, OH 78% + OH, OMe 9%

Oshima, T.; Sodeoka, M.; Shibasaki, M. *Tetrahedron Lett.*, ***1993***, *34*, 8509

$NHCO_2t$-Bu, CO_2Me, $SiMe_2Ph$ —[1. cat. OsO_4 , TMNO, acetone , water; 2. 5% aq. HCl]→ t-BuO_2CHN, $SiMe_2Ph$, O, O, OH

Panek, J.S.; Zhang, J. *J. Org. Chem.*, ***1993***, *58*, 294

$OSiMe_3$, OMe —[PhCHO; HgI_2]→ $OSiMe_3$, Ph, CO_2Me 92%

Dicker, I.B. *J. Org. Chem.*, ***1993***, *58*, 2324

OTMS, OMe —[10% $[Cp_2Zr(Ot$-Bu)THF][BPh_4]; CH_2Cl_2 , PhCHO , rt , 1h]→ OTMS, CO_2Me + OTMS, CO_2Me

(1.1 : 1) 95%

Hong, Y.; Norris, D.J.; Collins, S. *J. Org. Chem.*, ***1993***, *58*, 3591

CO_2t-Bu —[1. LDA; 2. Et_2AlCl; 3. (epoxide) O]→ OH, CO_2t-Bu 56%

(84:16 *syn:anti*)

Taylor, S.K.; Fried, J.A.; Grassl, Y.N.; Marolewski, A.E.; Pelton, E.A.; Poel, T.-J.; Rezanka, D.S.; Whittaker, M.R. *J. Org. Chem.*, ***1993***, *58*, 7304

$BrCH_2CO_2Et$, PhTeLi/$CeCl_3$

ether, 0°C → rt

95%

Fukuzawa, S.-i.; Hirai, K. *J. Chem. Soc., Perkin Trans. 1.*, ***1993***, 1963

oxido reductase from *Geotrichum candidum*

500 mM *bis*-TRIS-HCl
24h, NaDPH, ADH, glucose

93%

(>99% de; 92% ee)

Kawai, Y.; Takanobe, K.; Tsujimoto, M.; Ohno, A. *Tetrahedron Lett.*, ***1994***, *35*, 147

PhCHO +

$Sn(OTf)_2$

$Sn(OAc)_4$
chiral diamine
CH_2Cl_2, -78°C

(6 : 94) 95%

92%ee

PMP = *p*-methoxyphenyl

Kobayashi, S.; Kawasuji, T. *Tetrahedron Lett.*, ***1994***, *35*, 3329

3 eq. $CH_2=CHCO_2Me$

1.2 $MgBr_2$, CH_2Cl_2
0°C, 1h

58% (> 20:1 *endo:exo*)

Woo, S.H. *Tetrahedron Lett.*, ***1994***, *35*, 3975

CH_2Cl_2, -78°C
PhCHO

13% (26% ee)

Kiyooka, S.; Kido, Y.; Kaneko, Y. *Tetrahedron Lett.*, ***1994***, *35*, 5243

TMSO / EtS (vinyl sulfide) → [$BnOCH_2CHO$, 5% binol$TiCl_2$; toluene, 0°C, 2h] → EtS(C=O)CH$_2$CH(OTMS)CH$_2$OBn

81% (94% ee, R)

Mikami, K.; Matsukawa, S. *J. Am. Chem. Soc.*, ***1994***, *116*, 4077

γ-butyrolactone → [MeOH, Amberlyst-15, 20h] → CO_2Me ... OH 96%

Anand, R.C.; Selvapalam, N. *Synth. Commun.*, ***1994***, *24*, 2743

OHC (p-tolualdehyde) → [1. diketene, $Ti(OiPr)_4L$; CH_2Cl_2, -20°C, 2d; 2. iPrOH; 3. 1M HCl] → CO_2iPr, O, OH

90% (81% ee)

L = chiral imine ligand

Hayashi, M.; Inoue, T.; Oguni, N. *J. Chem. Soc. Chem. Commun.*, ***1994***, 341

cyclohexanone → [BrCH_2CO_2Et, THF, 2h; LVT, rt] → 1-hydroxy-1-(CH_2CO_2Et)cyclohexane

LVT = low valent titanium

67%

Aoyagi, Y.; Tanaka, W.; Ohta, A. *J. Chem. Soc. Chem. Commun.*, ***1994***, 1225

OMe (methoxyallene) → [1. BuLi, ether, -40°C; 2. PhCHO; 3. O_3, CH_2Cl_2, -78°C; 4. Me_2S] → Ph-CH(OH)-CO_2Me

51%

Hormuth, S.; Reißig, H.-U.; Dorsch, D. *Liebigs Ann. Chem.*, ***1994***, 121

$CH_3COCH_2CO_2Me$ → [$RuBr_2$(chiral bis-phosphine); 60 psi H_2, aq. MeOH] → OH ... CO_2Me

quant. (99% ee, S)

Burk, M.J.; Harper, T.G.P.; Kalberg, C.S. *J. Am. Chem. Soc.*, ***1995***, *117*, 4423

Ph-(1,3-dioxolan-2-yl) → $VO(OAc)_2$, PhH, TBHP → PhC(=O)O-CH$_2$CH$_2$-OH 98%

Choudary, B.M.; Reddy, P.N. *Synlett*, ***1995***, 959

Also via: Section 313 (Alcohol - Carboxylic Acid).

SECTION 328: ALCOHOL, THIOL - ETHER, EPOXIDE, THIOETHER

1. 5 NaH, THF, 50°C
2. NaOAc, Ac_2O, 100°C
$LiAlH_4$, THF

E+Z (18 : 1) 78%

Mandai, T.; Ueda, M.; Kashiwagi, K.; Kawada, M. *Tetrahedron Lett.*, ***1993***, ***34***, 111

1. 0.1% $[PdCl(\pi C_3H_5)]_2$
0.2% R-MOP, $HSiCl_3$
40°C, 24h
2. KF, $KHCO_3$, H_2O_2

HO

65 x 83% (95% ee)

RMOP = OMe PPh$_2$

Uozumi, Y.; Hayashi, T. *Tetrahedron Lett.*, ***1993***, ***34***, 2335

Me_2HC $CHMe_2$ O O O Cl

1. $BH_3{\bullet}SMe_2$, Me_2S, 0°C
2. aq. NaOH, H_2O_2
3. NaOH

HO O

97% (93% ee)

Molander, G.A.; Bobbitt, K.L. *J. Am. Chem. Soc.*, ***1993***, ***115***, 7517

EtO⁀Cl → 1. BuCHO, $Li_{(powder)}$, 5% DTBB, THF, 0°C; 2. H_2O → EtO⁀CH(OH)Bu 87%

Guijarro, A.; Yus, M. *Tetrahedron Lett.*, ***1993***, *34*, 3487

Ph⁀CHO → TMS⁀C(=CH₂)⁀TMS, cat. $BF_3 \cdot OEt_2$ → Ph, OH, Ph 42%

Markó, I.E.; Bayston, D.J. *Tetrahedron Lett.*, ***1993***, *34*, 6595

Ph, O, O, Ph → 2 eq. $BH_3 \cdot SMe_2$, 2 eq. TMSOTf → Ph, O, OH, Ph 95%

Bartels, B.; Hunter, R. *J. Org. Chem.*, ***1993***, *58*, 6756

Me_3Si, O, C_6H_{13} → 7 PhSH, SiO_2, 20°C, 20h → Me_3Si, PhS, OH, C_6H_{13} 89% + Me_3Si, HO, SPh, C_6H_{13} 14%

Raubo, P.; Wicha, J. *Synlett*, ***1993***, 25

cyclohexanone (=O) → $ClCH_2SPh$, SmI_2 → OH, SPh 73%

Yamashita, M.; Kitagawa, K.; Ohhara, T.; Iida, Y.; Masumi, A.; Kawasaki, I.; Ohta, S. *Chem. Lett.*, ***1993***, 653

O, S, O, O, OBn → 1. PhONa, DMF; 2. HCl-H_2O → PhO, OBn, OH 81%

Carlsen, P.H.J.; Aase, K. *Acta Chem. Scand. B.*, ***1993***, *47*, 617

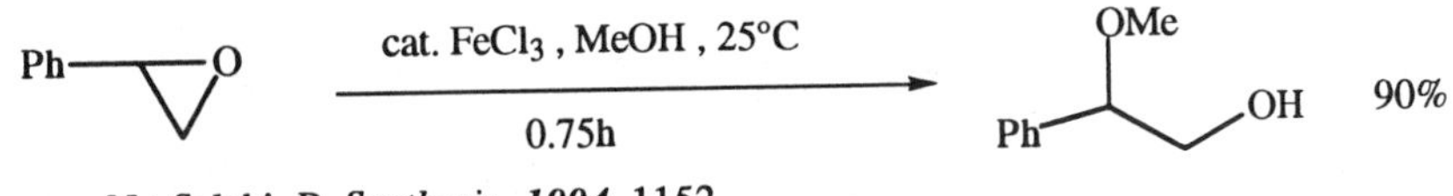

Ph–epoxide → cat. $FeCl_3$, MeOH, 25°C; 0.75h → Ph–CH(OMe)–CH$_2$OH 90%

Iranpoor, N.; Salehi, P. *Synthesis*, ***1994***, 1152

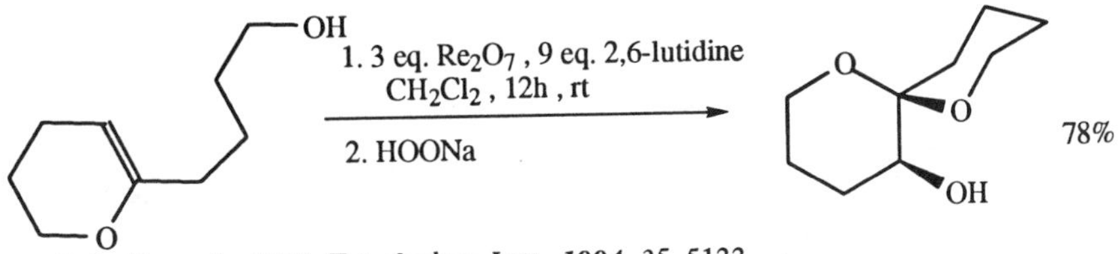

1. 3 eq. Re_2O_7, 9 eq. 2,6-lutidine, CH_2Cl_2, 12h, rt
2. HOONa

78%

Boyce, R.S.; Kennedy, R.M. *Tetrahedron Lett.*, ***1994***, *35*, 5133

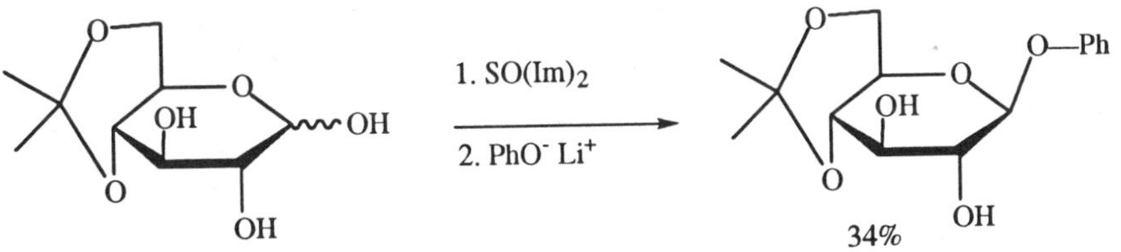

1. $SO(Im)_2$
2. PhO^- Li^+

34%

El Arabi Aouad, M.; El Meslouti, A.; Uzan, R.; Beaupere, D. *Tetrahedron Lett.*, ***1994***, *35*, 6279

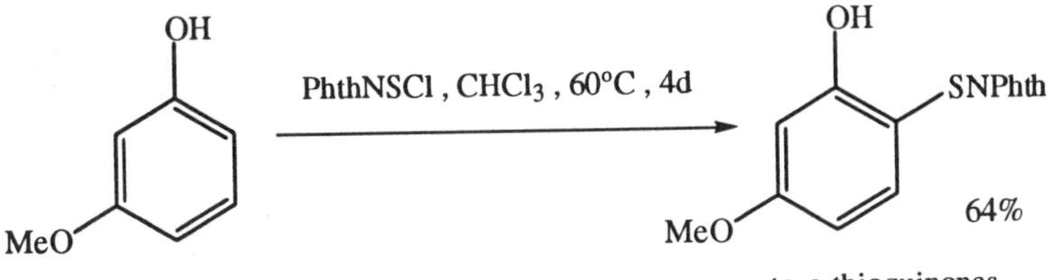

PhthNSCl, $CHCl_3$, 60°C, 4d

64%

precursor to *o*-thioquinones

Capozzi, G.; Menichetti, S.; Nativi, C.; Simonti, M.C. *Tetrahedron Lett.*, ***1994***, *35*, 9451

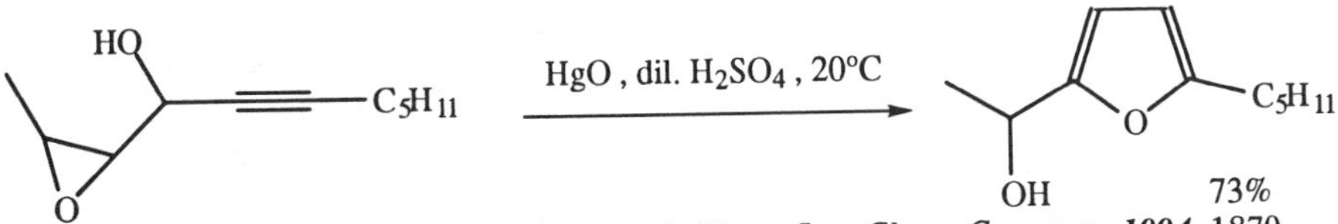

HgO, dil. H_2SO_4, 20°C

73%

Marson, C.M.; Harper, S.; Wrigglesworth, R. *J. Chem. Soc. Chem. Commun.*, ***1994***, 1879

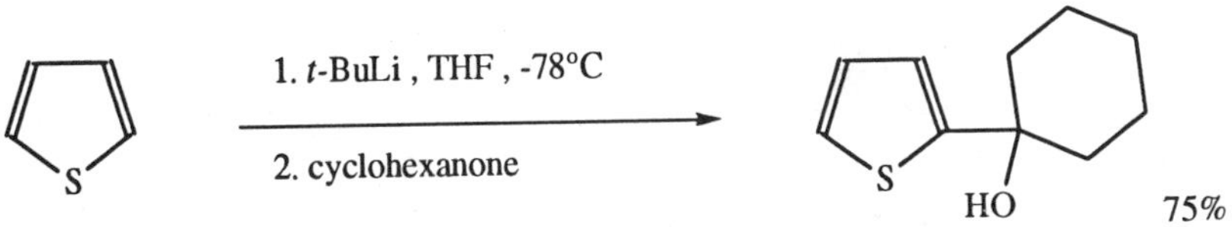

1. *t*-BuLi, THF, -78°C
2. cyclohexanone

75%

Paquette, L.A.; Dullweber, U.; Branan, B.M. *Heterocycles*, ***1994***, *37*, 187

PhCHO → 1. $Bu_3SnCH(OEt)_2$, BuLi, THF, -78°C; 2. aq. NH_4Cl → Ph–CH(OH)–$CH(OEt)_2$ 65%

Parrain, J.-L.; Beaudet, I.; Cintrat, J.-C.; Duchêne, A.; Quintard, J.-P. *Bull. Soc. Chim. Fr.*, ***1994***, *131*, 304

$(CF_3CO_2)ReO_3$, 0°C; 2,6-lutidine, CH_2Cl_2 → OH, H, O, H; 66% (6:1 *syn:anti*)

McDonald, F.E.; Towne, T.B. *J. Org. Chem.*, ***1995***, *60*, 5750

OH → PhSeCl, $CHCl_3$ → OH, SePh, OH + SePh, OH, OH (64 : 36) quant.

Cooper, M.A.; Ward, A.D. *Tetrahedron Lett.*, ***1995***, *36*, 2327

OH, OH → 1. O, O, I, OH, O, 23°C, 2h; 2. H_2O → OH, O 88%

Corey, E.J.; Palani, A. *Tetrahedron Lett.*, ***1995***, *36*, 3485

OTBS, O, Me, N, N, Ph → toluene, reflux, 3h → O, TBS, HO 66%

Kim, S.; Cho, C.M. *Tetrahedron Lett.*, ***1995***, *36*, 4845

1. 2% $Cp_2Ti(O4\text{-}ClC_6H_4)_2$, toluene
1% TBAF, Al_2O_3, 5 PMHS, rt
2. 1M NaOH, THF

PMHS = polymethylhydrosiloxane

94% (70:1)

Verdagauer, X.; Berk, S.C.; Buchwald, S.L. *J. Am. Chem. Soc.*, ***1995***, *117*, 12641

$BF_3 \cdot OEt_2$, $NaBH_3CN$
6h

87%

Srikrishna, A.; Viswajanani, R. *Tetrahedron*, ***1995***, *51*, 3339

e^-, MeOH, aq. MeCN, Bu_4NClO_4
1h

OMe
OH

96%

Safavi, A.; Iranpoor, N.; Fotuhi, L. *Bull. Chem. Soc. Jpn.*, ***1995***, *68*, 2591

REVIEW:

"Metal-Catalyzed Direct Hydroxy-Epoxidation of Olefins," Adam, W.; Richter, M.J. *Accts. Chem. Res.*, ***1994***, *27*, 57

SECTION 329: ALCOHOL, THIOL - HALIDE, SULFONATE

Cl

1. LDA. $ZnCl_2$, THF. -78°C
2. PhCHO

OH
Cl

80%

Mallaiah, K.; Satyanarayana, J.; Ila, H.; Junjappa, H. *Tetrahedron Lett.*, ***1993***, *34*, 3145

EtO_2C
Cl
C_3H_7

Baker's yeast

EtO_2C
Cl
OH
C_3H_7

+

EtO_2C
OH
Cl
C_3H_7

(23 : 77) 74%

Tsuboi, S.; Furutani, H.; Ansari, M.H.; Sakai, T.; Utaka, M.; Takeda, A. *J. Org. Chem.*, ***1993***, *58*, 486

CF_3CO_2Et → 1. Dibal, CH_2Cl_2, -78°C; 2. $ZnBr_2$, Bu_3Sn, CH_2Cl_2, 40°C → OH, F_3C 83%

Ishihara, T.; Hayashi, H.; Yamanaka, H. *Tetrahedron Lett.*, ***1993***, *34*, 5777

Ph, O, O, S, TsN, Ph → $MgBr_2$, Bu_4NBH_4; ether/CH_2Cl_2 → Br, HO, Ph 72% + HO, Ph 22%

Bailey, P.L.; Briggs, A.D.; Jackson, R.F.W.; Pietruszka, J. *Tetrahedron Lett.*, ***1993***, *34*, 6611

O → $Mg(NO_3)_2$, Bu_4NBr, 5h; $CHCl_3$ → Br, OH

93%

Suh, Y.-G.; Koo, B.-A.; Ko, J.-A.; Cho, Y.-S. *Chem. Lett.*, ***1993***, 1907

→ $NaIO_4$, $NaHSO_3$ → I, OH

90%

Masuda, H.; Takase, K.; Nishio, M.; Hasegawa, A.; Nishiyama, Y.; Ishii, Y. *J. Org. Chem.*, ***1994***, *59*, 5550

Ph, O, S, O, O → LiCl, DMF, 70°C, 1d → Cl, OH, Ph + OH, Cl

(35 : 65) 90%

Nymann, K.; Svendsen, J.S. *Acta Chem. Scand. B.*, ***1994***, *48*, 183

PhCHO → Br, CN, In → OH, CN, Ph

84% (56:44 *erythro:threo*)

Araki, S.; Yamada, M.; Butsugan, Y. *Bull. Chem. Soc. Jpn.*, ***1994***, *67*, 1126

REVIEW:

"Regioselective anbd Chemoselective Synthesis of Halohydrins by Cleavage of Oxiranes with Metal Halides," Bonini, C.; Righi, G. *Synthesis,* ***1994***, 225

SECTION 330: ALCOHOL, THIOL - KETONE

Bu_3SnH , AIBN , PhH

syringe pump addn reflux

72% (22:1 , β : α)

Rawal, V.H.; Krishnamurthy, V.; Fabre, A. *Tetrahedron Lett.*, ***1993***, *34*, 2899

15 min

96%

Curci, R.; D'Accolti, L.; Detomaso, A.; Fusco, C.; Takeuchi, K.; Ohga, Y.; Eaton, P.E. *Tetrahedron Lett.*, ***1993***, *34*, 4559

PhCHO , Me_3Al

88%

Shoda, H.; Nakamura, T.; Tanino, K.; Kuwajima, I. *Tetrahedron Lett.*, ***1993***, *34*, 6281

1. $Me_3Si(OTf)_2$, NEt_3

2. PhCHO

78% (91:9 *syn:anti*)

Kobayashi, S.; Nishio, K. *J. Org. Chem.*, ***1993***, *58*, 2647

$RuCl_3$, $MeCO_3H$

$MeCN - H_2O - CH_2Cl_2$

78%

Murahashi, S.-I.; Saito, T.; Hanaoka, H.; Murakami, Y.; Naota, T.; Kumobayashi, H.; Akutagawa, S. *J. Org. Chem.*, ***1993***, *58*, 2929

OH, OH → dimethyldioxirane, acetone, 0°C → O, OH >96% (>98% ee, R)

D'Accolti, L.; Detomaso, A.; Fusco, C.; Rosa, A.; Curci, R. *J. Org. Chem.*, ***1993***, *58*, 3600

O → 1. BuLi, CO, -110°C THF/pentane/ether 2. rt, H_3O^+ → Ph, OH, O 75%

Seyferth, D.; Hui, R.C.; Wang, W.-L. *J. Org. Chem.*, ***1993***, *58*, 5843

$OSiMe_3$ → PhCHO, 5% $Sc(OTf)_3$, aq. THF → OH, O, Ph 83%

Kobayashi, S.; Hachiya, I.; Ishitani, H.; Araki, M. *Synlett*, ***1993***, 472

PhCHO → OTMS, Ph, CH_2Cl_2, 2% $B(C_6F_5)_3$ → OH, O, Ph, Ph 84%

Ishihara, K.; Hananki, N.; Yamamoto, H. *Synlett*, ***1993***, 577

O, O → Ph, CHO, SmI_2, HMPA, rt, THF → OH, O, Ph, OH 83% (83:17 *anti:syn*)

Miyoshi, N.; Takeuchi, S.; Ohgo, Y. *Chem. Lett.*, ***1993***, 959

Ph → $MeCO_3H$, cat. $OsCl_3$, $MeCN\text{-}H_2O\text{-}CH_2Cl_2$ → Ph, OH, O 62%

Murahashi, S.-I.; Naota, T.; Hanaoka, H. *Chem. Lett.*, ***1993***, 1767

OSiMe$_3$ → PhCHO, Al_2O_3, ultrasound, 22h

(77 : 23) 75%

Ranu, B.C.; Chakraborty, R. *Tetrahedron*, ***1993***, *49*, 5333

PhCH$_2$CHO, H_2O, SmI_2

(39 : 61) 64%

Miyoshi, N.; Takeuchi, S.; Ohgo, Y. *Chem. Lett.*, ***1993***, 2129

PhCHO → 1. OTMS, Bu, EtCN, -78°C; 20% chiral acyloxyborane; 2. aq. HCl

81% (85% ee, R)

Ishihara, K.; Maruyama, T.; Mouri, M.; Gao, Q.; Furuta, K.; Yamamoto, H. *Bull. Chem. Soc. Jpn.*, ***1993***, *66*, 3483

1. 2.2 eq. SmI_3, tetrahydropyran
2. propanal
3. H_3O^+

95%

Namy, J.-L.; Colomb, M.; Kagan, H.B. *Tetrahedron Lett.*, ***1994***, *35*, 1723

TFA, $CDCl_3$, 2d

95:5 (*trans:cis*)

Dechoux, L.; Doris, E. *Tetrahedron Lett.*, ***1994***, *35*, 2017

Ph(C=O)CO$_2$Me → [PhCOCl, 3 eq. pyridine] → PhCOC(Ph)(OH)CO$_2$Me 86%

Araneo, S.; Clerici, A.; Porta, O. *Tetrahedron Lett.*, ***1994***, *35*, 2213

PhOH → [PhCOCl, cat. HZSM-5, DCE, reflux] → PhCO-C$_6$H$_4$-OH 67%

HZSM-5 = acidic zeolite [SiO_2/Al_2O_3 = 40]

Paul, V.; Sudalai, A.; Daniel, T.; Srinivasan, K.V. *Tetrahedron Lett.*, ***1994***, *35*, 2601

PhCOEt → [1. LDA, THF, -78°C; 2. (iPrO)$_3$TiCl; 3. acetone, dimethyldioxirane, -78°C; 4. aq. NH_4F] → PhCOCH(OH)CH$_3$

51% conversion (88% yield)

Adam, W.; Müller, M.; Prechtl, F. *J. Org. Chem.*, ***1994***, *59*, 2358

ClCH$_2$COC$_5$H$_{11}$ → [PhCHO, 10% Sm(hmds)$_3$, -30°C, 18h] → PhCH(OH)CH(Cl)COC$_5$H$_{11}$ 90%

Sasai, H.; Arai, S.; Shibasaki, M. *J. Org. Chem.*, ***1994***, *59*, 2661

1-(trimethylsilyloxy)cyclohexene (OSiMe$_3$) → [PhCHO, Yb(OTf)$_3$, aq. THF, rt, 19h] → 2-(hydroxy(phenyl)methyl)cyclohexanone

91% (73:27 *syn:anti*)

Kobayashi, S.; Hachiya, I. *J. Org. Chem.*, ***1994***, *59*, 3590

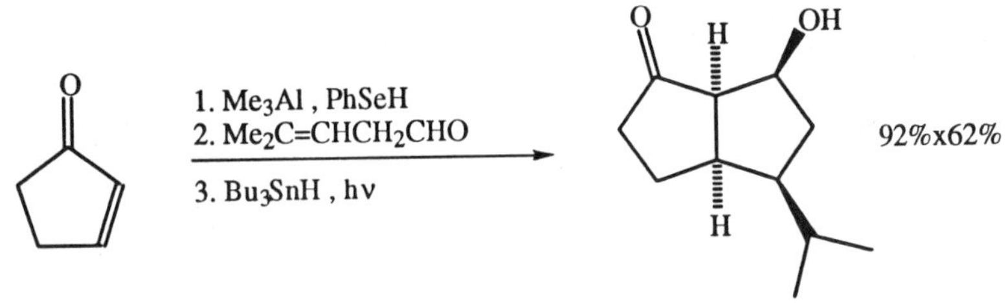

1. Me_3Al , PhSeH
2. $Me_2C=CHCH_2CHO$
3. Bu_3SnH , hν

O H OH H

92%x62%

Lee, I.-Y.C.; Lee, J.H.; Lee, H.W. *Tetrahedron Lett.*, ***1994***, *35*, 4173

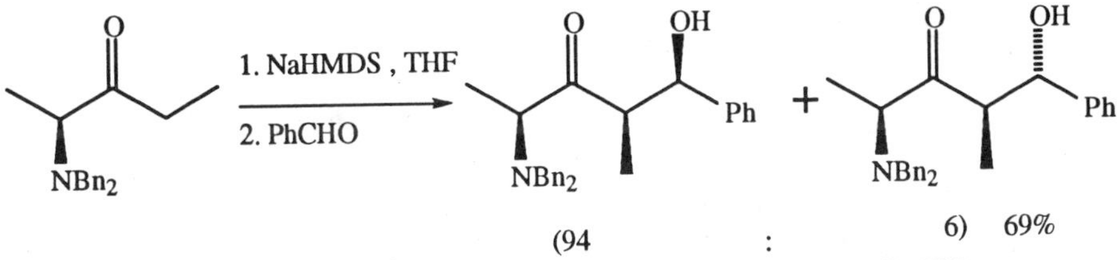

1. NaHMDS , THF
2. PhCHO

O NBn$_2$ O OH Ph + O OH Ph NBn$_2$ NBn$_2$

(94 : 6) 69%

Goh, J.B.; Lagu, B.R.; Wurster, J.; Liotta, D.C. *Tetrahedron Lett.*, ***1994***, *35*, 6029

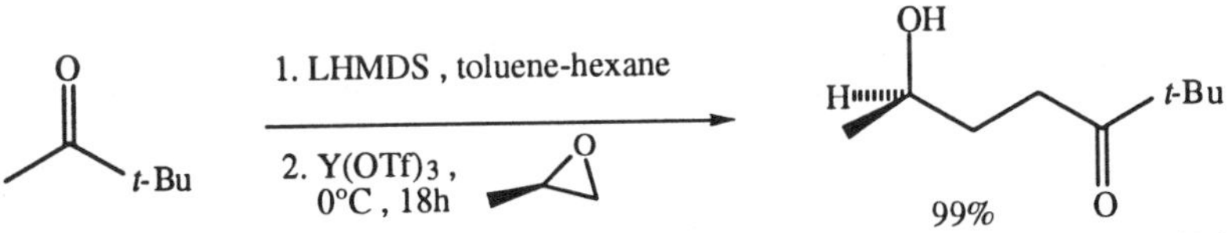

1. LHMDS , toluene-hexane
2. $Y(OTf)_3$, 0°C , 18h

O *t*-Bu O OH H *t*-Bu O

99%

Crotti, P.; Di Bussolo, V.; Favero, L.; Macchia, F.; Pineschi, M. *Tetrahedron Lett.*, ***1994***, *35*, 6537

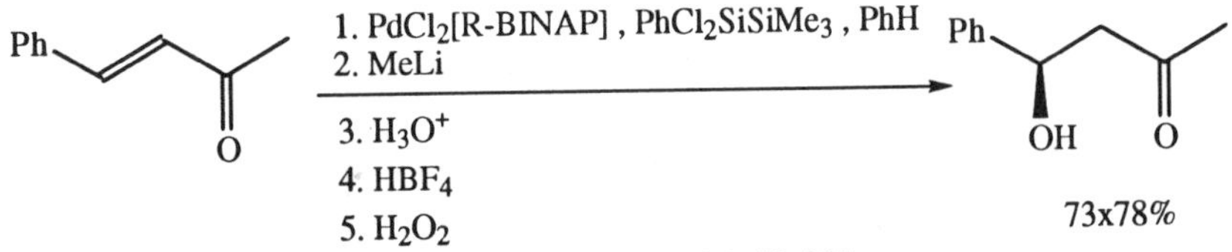

1. $PdCl_2$[R-BINAP] , $PhCl_2SiSiMe_3$, PhH
2. MeLi
3. H_3O^+
4. HBF_4
5. H_2O_2

Ph O Ph OH O

73x78%

Matsumoto, Y.; Hayashi, T.; Ito, Y. *Tetrahedron*, ***1994***, *50*, 335

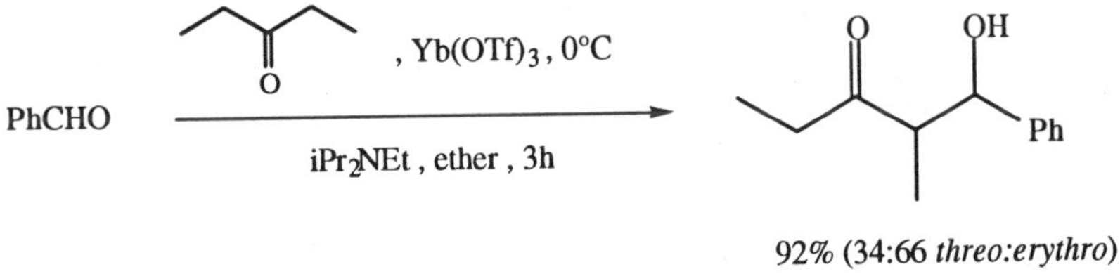

PhCHO , $Yb(OTf)_3$, 0°C
iPr$_2$NEt , ether , 3h

O O OH Ph

92% (34:66 *threo:erythro*)

Fukuzawa, S.-i.; Tsuchimoto, T.; Kanai, T. *Bull. Chem. Soc. Jpn.*, ***1994***, *67*, 2227

OTMS / Me

Bu_4NClO_4 , e^-

94%

Yoshida, J.; Morita, Y.; Ishichi, Y.; Isoe, S. *Tetrahedron Lett.*, ***1994***, *35*, 5247

SmI_2, $PhCH_2Br$

HMPA

86%

Enholm, E.J.; Schreier, J.A. *J. Org. Chem.*, ***1995***, *60*, 1110

PhCHO , MS 4Å , 13h

5% $PdCl_2$[R-DINAP]-AgOTf

23°C

87% + 9%

Sodeoka, M.; Ohrai, K.; Shibasaki, M. *J. Org. Chem.*, ***1995***, *60*, 2648

, acetone , rt , 1d

92%

Bovicelli, P.; Lupattelli, P.; Sanetti, A.; Mincione, E. *Tetrahedron Lett.*, ***1995***, *36*, 3031

$Zn(NO_3)_2$, acetone

$Me_2NCH_2CH_2OH$

30°C , 6h

pH 10.1	98%	0%
pH 11.0	0%	100%

Buonora, P.T.; Rosauer, K.G.; Dai, L. *Tetrahedron Lett.*, ***1995***, *36*, 4009

1. Bu_3SnH , AIBN , PhH relux
2. PhCHO , 10°C , 12h

73% (6:1 *erytrho:threo*)

Enholm, E.J.; Whitley, P.E. *Tetrahedron Lett.*, ***1995***, *36*, 9157

1. SmI_2 ,
2. H_2O

77%

Arime, T.; Takahashi, H.; Kobayashi, S.; Yamaguchi, S.; Mori, N. *Synth. Commun.*, ***1995***, *25*, 389

DMSO , KSF , microwave , 3 min

55%

Villemin, D.; Hammadi, M. *Synth. Commun.*, ***1995***, *25*, 3141

1. LDA
2. $Ti(OiPr)_3Cl$
3. TBHP , BaO
4. NH_4F , H_2O

53%

Schulz, M.; Kluge, R.; Schüßer, M.; Hoffmann, G. *Tetrahedron*, ***1995***, *51*, 3175

Cl_3CCO_2H , 11°C , Cl_2 , MeOH
EtOH , $NaHCO_3$

98%

Tassignon, P.S.G.; De Wit, D.; De Rijk, T.C.; De Buyck, L.F. *Tetrahedron*, ***1995***, *51*, 11863

1. 2 eq. SmI_2 , -60°C → -50•C
2. cyclohexanone , -50°C → 0°C

98%

Utimoto, K.; Matsui, T.; Takai, T.; Matsubara, S. *Chem. Lett.*, ***1995***, 197

1. $AlMe_3$
2. PhCHO

79% (13:87 *erythro:threo*)

Aoki, Y.; Oshima, K.; Utimoto, K. *Chem. Lett.*, ***1995***, 463

PhCHO

OTMS, Ph, $(C_6F_5)_3B$, CH_2Cl_2

-78°C, 6.5h

OH O Ph Ph 99%

Ishihara, K.; Hanaki, N.; Funawashi, M.; Miyata,M.; Yamamoto, H. *Bull. Chem. Soc. Jpn.*, ***1995***, *68*, 1721

SECTION 331: ALCOHOL, THIOL - NITRILE

C_6H_{13}—CHO

3:1 EtCN-CH_2Cl_2, 10h
20% Mg-dihydrooxazolidine catalyst

12% bis-dihydro oxazolidine derivative

OTMS, C_6H_{13}, CN

88%
(95% ee, S)

Corey, E.J.; Wang, Z. *Tetrahedron Lett.*, ***1993***, *34*, 4001

CHO

HO CN, 2h, rt

100 units enzyme
0.1M citrate buffer

H OH CN

93% conversion
(80%ee)

enzyme = hydroxynitrile lyase from *Hevea brasiliensis*

Klempier, N.; Griengl, H.; Hayn, M. *Tetrahedron Lett.*, ***1993***, *34*, 4769

Bu O

$Me_2C(CN)OH$, 1h

$La(OiPr)_3$, 50°C

OH Bu CN

80%

Ohno, H.; Mori, A.; Inoue, S. *Chem. Lett.*, ***1993***, 975

O_2 S O O $C_{10}H_{21}$

1. $LiCH_2CN$
2. H_3O^+

OH $C_{10}H_{21}$ CN

73%

Hoye, T.R.; Crawford, K.B. *J. Org. Chem.*, ***1994***, *59*, 520

1. TMSCN, CH_2Cl_2, rt; KCN/18-crown-6
2. MeOH/THF, SmI_2

HO—C6H4—CN 80%

Olson, S.H.; Danishefsky, S.J. *Tetrahedron Lett.*, ***1994***, *35*, 7901

Ph—CH2—CN → 1. LDA, THF, -78°C; 2. PhCHO

(5.5 : 1) 70%

Carlier, P.R.; Lo, K.M.; Lo, M.M.-C.; Williams, I.D. *J. Org. Chem.*, ***1995***, *60*, 7511

PhCHO → [LDA, $PhCH_2CN$], TMSCl 97%

Zhou, J.J.P.; Zhong, B.; Silverman, R.B. *J. Org. Chem.*, ***1995***, *60*, 2261

PhCHO → Me_3SiCN, $Ti(OiPr)_4$

72%
(91% ee, S)

Bolm, C.; Müller, P. *Tetrahedron Lett.*, ***1995***, *36*, 1625

ICH_2CN, $PdCl_2$/Ga-THF
2 eq. Cl_3CCO_2Et
reflux, 10h

60%

Zhang, X.-L.; Han, Y.; Tao, W.-T.; Huang, Y.-Z. *J. Chem. Soc., Perkin Trans. 1.*, ***1995***, 189

Review:

"Catalytic Asymmetric Cyanohydrin Synthesis," North, M. *Synlett*, ***1993***, 807

SECTION 332: ALCOHOL, THIOL - ALKENE

Allylic and benzylic hydroxylation (C=C-C-H → C=C-C-OH, etc.) is listed in Section 41 (Alcohols from Hydrides).

BnO ... 2 MeMgBr, CuCN, BF_3•OEt_2, THF, -78°C → BnO ... OH ... Me 88%

Kang, S-K.; Lee, D-H.; Sim, H-S.; Lim, J-S. *Tetrahedron Lett.*, ***1993***, *34*, 91

O, NEt_2; PhCHO, TMSI, THF → O, NEt_2 81%; 1. TIPSCl 2. MeLi 3. MeI → OTIPS O 47%; MeMgBr → OTIPS 72%

Tsukazaki, M.; Snieckus, V. *Tetrahedron Lett.*, ***1993***, *34*, 411
Lee, J.; Tsukazaki, M.; Snieckus, V. *Tetrahedron Lett.*, ***1993***, *34*, 415

MeO_2C ... CHO; 2 SmI_2, rt, 2 *t*-BuOH, THF → MeO_2C ... OH + MeO_2C ... OH (2 : 1) 66%

Gillmann, T. *Tetrahedron Lett.*, ***1993***, *34*, 607

H, H, C_5H_{11}, O, SPh; BuLi, THF, -78°C → C_5H_{11}, H, OH, SPh 83% 10.8:1, *E:Z*

Takano, S.; Sugihara, Y.; Ogasawara, K. *Tetrahedron Lett.*, ***1993***, *34*, 845

Cl, Cl; 1. xs Li, 2 iPrCHO 5% DTBB, THF, 0°C 2. H_2O-HCl → OH, OH 71% (1:1 diastereomers)

Guijarro, A.; Yus, M. *Tetrahedron Lett.*, ***1993***, *34*, 2011

1. Me, BBr_2 (2-propenyl), 25°C, 2h
2. NaOH, H_2O_2

Ph Me OH + Ph Me OH

(87 : 13) 90%

after 24 h at rt Ph Me OH only product (90%)

Singleton, D.A.; Kim, K.; Martinez, J.P. *Tetrahedron Lett.*, ***1993***, *34*, 3071

$SiCl_3$

1. PhCHO, DMF, 0°C
2. H_2O

Ph OH + Ph OH

(3 : 97) 89%

Kobayashi, S.; Nishio, K. *Tetrahedron Lett.*, ***1993***, *34*, 3453

O H OH

10 eq. BuLi, 10 min

OH Bu 71%

Lautens, M.; Gajda, C. *Tetrahedron Lett.*, ***1993***, *34*, 4591

Br Br Br

C_3H_7CHO, EG-Zn
DMF, 0°C, 2h

OH C_3H_7

72%

EG-Zn = electrogenerated reactive zinc

Tokuda, M.; Mimura, N.; Karasawa, T.; Fujita, H.; Suginome, H. *Tetrahedron Lett.*, ***1993***, *34*, 7607

Et $SnBu_3$

PhCHO, hν (>320 nm)
EtCN, -78°C, 5h

Et Ph OH + Et OH Ph

(30 : 70)
(86:14 *E:Z*)

Takuwa, A.; Shiigi, J.; Nishigaichi, Y. *Tetrahedron Lett.*, ***1993***, *34*, 3457

OTIPS

CHO

Et_2Zn , $Ti(OiPr)_4$, 12h
-60°C → -55°C

8% Tf N $Ti(OiPr)_2$ N Tf

OTIPS

OH

Et

68%

(93% ee)

Rozema, M.J.; Eisenberg, C.; Lütjens, H.; Ostwald, R.; Belyk, K.; Knochel, P. *Tetrahedron Lett.*, ***1993***, *34*, 3115

OH

1,2-dichlorobenzene , 12h
reflux , <0.5 torr

OH

90%

Janardhanam, S.; Devan, B.; Rajogopalan, K. *Tetrahedron Lett.*, ***1993***, *34*, 6761

Ph

OH

1. Sharpless epoxidation
2. TsCl
3. Te , $LiBHEt_3$, THF
TBAF , rt

Ph OH

78% (>90% ee)

Dittmer, D.C.; Discordia, R.P.; Zhang, Y.; Murphy, C.K.; Kumar, A.; Pepito, A.S.; Wang, Y. *J. Org. Chem.*, ***1993***, *58*, 718

O Bu

iPrMgBr , THF
$(tbp)_2Ni_2Cl_2$

Bu

$(CH_2)_2CH_2OH$

+

Bu $(CH_2)_2CH_2OH$

(>25 : 1) 60

Ducoux, J.-P.; LeMénez, P.; Kunesch, N.; Wenkert, E. *J. Org. Chem.*, ***1993***, *58*, 1290

O

O I

$Me_3SiSnBu_3$, DMF
CsF , rt , 1h

O

OH

86%

Mori, M.; Isono, N.; Kaneta, N.; Shibasaki, M. *J. Org. Chem.*, ***1993***, *58*, 2972

$Cr(CO)_5$ NMe$_2$ → hv , CO , THF , 8h → NMe$_2$ OH 75

Merlic, C.A.; Xu, D.; Gladstone, B.G. *J. Org. Chem.*, ***1993***, *58*, 538

EtMgBr , 10% Zr complex / THF , 25°C → H, HO 65% (>97% ee)

Morken, J.P.; Didiuk, M.T.; Hoveyda, A.H. *J. Am. Chem. Soc.*, ***1993***, *115*, 6997

OTMS 5% (binol)$TiCl_2$, BnO_2C—CHO → OTMS OH CO_2Bn Me

63% (99:1 *syn:anti*; 99:1 *Z:E*; 99% ee, R)

Mikami, K.; Matsukawa, S. *J. Am. Chem. Soc.*, ***1993***, *115*, 7039

PhCHO 1. $SiMe_3$, EtCN 20% chiral ligand 2. TBAF → OH Ph OH

63% (92% ee)(96:4 *erythro:threo*)

Ishihara, K.; Mouri, M.; Gao, Q.; Maruyama, T.; Furuta, K.; Yamamoto, H. *J. Am. Chem. Soc.*, ***1993***, *115*, 11490

O I , 10% Pd(OAc)$_2$ / 2 NaO_2CH , 2 Bu_4NCl , 2 iPr_2NEt Me_2NCOMe , 80°C , 1d → OH 70% (56:44 *E:Z*)

Larock, R.C.; Ding, S.; Tu, C. *Synlett*, ***1993***, 145

1. xylene, NEt_3 100°C, 20h
2. H_2O_2, NaOH

69%

X = catecholboryl

Singleton, D.A.; Redman, A.M. *Tetrahedron Lett.*, ***1994***, *35*, 509

$(Ipc)_2B$-allyl

Buynak, J.D.; Geng, B.; Uang, S.; Strickland, J.B. *Tetrahedron Lett.*, ***1994***, *35*, 985

1. $Cp_2Zr(H)Cl$, CH_2Cl_2
2. Me_2Zn
3. PhCHO, 0°C

94%

Wipf, P.; Xu, W. *Tetrahedron Lett.*, ***1994***, *35*, 5197

2 eq. $Me_2S{=}CH_2$, THF
-10°C → rt

80%

Alcaraz, L.; Harnett, J.J.; Mioskowski, C.; Martel, J.P.; Le Gall, T.; Shin, D.-S.; Falck, J.R. *Tetrahedron Lett.*, ***1994***, *35*, 5449

1. $(\eta^5\text{-}C_5H_5)_2TiCl$ iPrMgCl, THF, -50°C
2. PhCHO, -50°C → -20°C

(1.44 : 1) 42%

Szymoniak, J.; Felix, D.; Moïse, C. *Tetrahedron Lett.*, ***1994***, *35*, 8613

1. $HSiCl_3$, cat. $Pd(PPh_3)_4$
2. $PhCH_2CH_2CHO$, DMF

87% (>99:1 *syn:anti*)

Kobayashi, S.; Nishio, K. *Synthesis*, ***1994***, 457

Toluene , reflux , 3h

Me

Ph

OH

Me

73%

Ph (3:2 *syn:anti*)

Kim, S.; Cho, C.M. *Tetrahedron Lett.*, ***1994***, *35*, 8405

Br

CHO

$Pd(OAc)_2$, 3 PPh_3 , PhH

70•C , 16h

OH

47%

Watanabe, T.; Sakai, M.; Miyaura, N.; Suzuki, A. *J. Chem. Soc. Chem. Commun.*, ***1994***, 467

C_7H_{15} O N-NHTs

BuMgBr

OH

C_7H_{15} Bu

Chandrasekhar, S.; Takhi, M.; Yadav, J.S. *Tetrahedron Lett.*, ***1995***, *36*, 307

O

$SiMe_2Ph$

1. Li , NH_3 , THF—EtOH , -78°C
2. TBAF , THF
3. 30% aq. H_2O_2 , $KHCO_3$, MeOH

O

OH

60% overall

Taber, D.F.; Yet, L.; Bhamidipati, R.S. *Tetrahedron Lett.*, ***1995***, *36*, 351

MPMO

O

O=

5% $Pd_2(dba)_3$•$CHCl_3$, THF
20% PPh_3 , heat

OH

MPMO

93%

Kang, S.-K.; Park, D.-C.; Park, C.-H.; Hong, R.-K. *Tetrahedron Lett.*, ***1995***, *36*, 405

PhS

1. Cp_2TiCl_2/2 BuLi , 0°C , THF
2. Ph CHO

OH

Ph

+

OH

Ph

(97 : 3) 69%

Takeda, T.; Miura, I.; Horikawa, Y.; Fujiwara, T. *Tetrahedron Lett.*, ***1995***, *36*, 1495

1. Co(tdcpp), O_2-Et_3SiH, 1h
2. $(MeO)_3P$, iPrOH, CH_2Cl_2

C_3H_7 ... CO_2Et → HO, C_3H_7, CO_2Et 82%

Co(tdcpp) = [5,10,15,20-tetrakis(2,6-dichlorophenyl) porphinato] Cobalt (II)

Matsushita, Y.; Sugamoto, K.; Nakama, T.; Sakamoto, T.; Matsui, T.; Nakayama, M. *Tetrahedron Lett.*, ***1995***, *36*, 1879

3 *t*-BuOK, BuLi
THF, -78°C

RO, OBn → Ph, OH, RO + Ph, OH, RO

R = TBS	(79	:	21) 48%
R = H	(20	:	80) 90%

Tomooka, K.; Keong, P.-H.; Nakai, T. *Tetrahedron Lett.*, ***1995***, *36*, 2789

1. BuLi, THF, -78°C
2. PhCHO, -78°C → rt

BuTe → Ph, OH 85%

Mo, X.-S.; Huang, Y.-Z. *Tetrahedron Lett.*, ***1995***, *36*, 3589

Dibal, 14% $Ni(cod)_2$
20% R-BINAP, 1h

O, OMe, OMe, H, H → OMe, OMe, OH

78% (97% ee)

Lautens, M.; Chiu, P.; Ma, S.; Rovis, T. *J. Am. Chem. Soc.*, ***1995***, *117*, 532

$SnCl_4$-NBu_3 , MeCN reflux

65%

Yamaguchi, M.; Hayashi, A.; Hirama, M. *J. Am. Chem. Soc.*, ***1995***, *117*, 1151

2 eq. HCO_2NH_4 , THF
5% $Pd(PPh_3)_4$, reflux

95%

Kang, S.-K.; Park, D.-C.; Rho, H.-S.; Yu, C.-M.; Hong, J.-H. *Synth. Commun.*, ***1995***, *25*, 203

PhCHO , 10% $Ti(OiPr)_4$, 20% chiral binaphthylamine

93% (66% ee)

Berrisford, D.J.; Bolm, C. *Angew. Chem. Int. Ed. Engl.*, ***1995***, *34*, 1717

Also via: Section 302 (Alkyne - Alcohol).

SECTION 333: ALDEHYDE - ALDEHYDE

NO ADDITIONAL EXAMPLES

SECTION 334: ALDEHYDE - AMIDE

CO (1 atm) , Ph_2MeSiH
1% $Pd(OAc)_2$, 20% PPh_3
toluene , 110°C , Et_4NCl

61%

Brown, S.; Clarkson, S.; Grigg, R.; Sridharan, V. *J. Chem. Soc. Chem. Commun.*, ***1995***, 1135

SECTION 335: ALDEHYDE - AMINE

DMF , $POCl_3$

8%

Balasundaram, B.; Venugopal, M.; Permumal, P.T. *Tetrahedron Lett.*, ***1993***, *34*, 4249

SECTION 336: ALDEHYDE - ESTER

NO ADDITIONAL EXAMPLES

SECTION 337: ALDEHYDE - ETHER, EPOXIDE, THIOETHER

$Rh_2(\mu\text{-}OMe)_2)COD)_2$
10 eq. PPh_3 , CO , H_2

CHO + OEt EtO

+ $HC(OEt)_3$, 2d (96 : 4) quant.
+ $HC(OEt)_3$ + PPTS , 12h (0 : 100) quant.

Fernández, E.; Castillón, S. *Tetrahedron Lett.*, ***1994***, *35*, 2361

C_5H_{11}

2% $Pd(PPh_3)_4$, 5 atm CO , THF
PhSH , 70°C , 16h

C_5H_{11} =CHCHO PhS 82% (13:87 *E:Z*)

Ogawa, A.; Takeba, M.; Kawakami, J.; Ryu, I.; Kambe, N.; Sonoda, N. *J. Am. Chem. Soc.*, ***1995***, *117*, 7564

SECTION 338: ALDEHYDE - HALIDE, SULFONATE

NO ADDITIONAL EXAMPLES

SECTION 339: ALDEHYDE - KETONE

NO ADDITIONAL EXAMPLES

SECTION 340: ALDEHYDE - NITRILE

NO ADDITIONAL EXAMPLES

SECTION 341: ALDEHYDE - ALKENE

For the oxidation of allylic alcohols to alkene aldehydes, also see Section 48 (Aldehydes from Alcohols).

CHO

1. $Cp_2Zr(Cl)CH\text{=}CHOEt$
cat. $AgClO_4$
2. H_3O^+

CHO 89%

Maeta, H.; Suzuki, K. *Tetrahedron Lett.*, ***1993***, *34*, 341

PhCHO → 1. Li-(trienyl)-OMe (E+Z), -78°C → 0°C; 2. 1N HCl, 0•C → Ph-(tetraenyl)-CHO 75%

Duhamel, L.; Duhamel, P.; Le Gallic, Y. *Tetrahedron Lett.*, ***1993***, *34*, 319

OTMS-diene → (cyclohexenyl OTMS), CAN, MeCN, 0°C → 2-substituted cyclohexanone-CHO 62%

Paolobelli, A.B.; Latini, D.; Ruzziconi, R. *Tetrahedron Lett.*, ***1993***, *34*, 721

2,3-dimethylbutadiene → acrolein (CHO), rt; 25% polymeric Fe catalyst → CHO 97%

Saha, A.K.; Hossain, M.M. *Tetrahedron Lett.*, ***1993***, *34*, 3833

furan → 2-bromoacrolein (Br, CHO), CH_2Cl_2; cyclic B catalyst, -78°C, 8h → CHO, Br 79%

(99:1 *exo:endo*), 92% ee

Corey, E.J.; Loh, T.-P. *Tetrahedron Lett.*, ***1993***, *34*, 3979

PhCHO → 1. Me_3Si-(dienyl)=N-tBu, CsF-DMSO, rt → 100°C; 2. $ZnCl_2$, rt, 2h → Ph-(dienyl)-CHO 70%

Bellassoued, M.; Salemkour, M. *Tetrahedron Lett.*, ***1993***, *34*, 5281

Me_3Si, $OSiMe_3$ → PhCHO, 5% $TfSiMe_3$, CH_2Cl_2, rt, 15h → Ph, CHO 80%

Duhamel, L.; Gralek, J.; Bouyanzer, A. *Tetrahedron Lett.*, ***1993***, *34*, 7745

$(Me_3Si)_2HC$–CH=N-t-Bu → PhCHO, 10% $ZnBr_2$, THF → Ph–CH=CH–CHO 86%

Bellassoued, M.; Majidi, A. *J. Org. Chem.*, ***1993***, *58*, 2517

CHO, allyl-$SnBu_3$, $C_8H_{17}Br$, AIBN; 20 atm. CO, PhH, 80°C, 8h → C_8H_{17} O CHO

74%

Ryu, I.; Yamazaki, H.; Ogawa, A.; Kambe, N.; Sonoda, N. *J. Am. Chem. Soc.*, ***1993***, *115*, 1187

O → 1. dimethyldioxirane, acetone; 2. PH_3P=CHCHO, CH_2Cl_2 → OHC CHO

84%

Adger, B.J.; Barrett, C.; Brennan, J.; McGuigan, P.; McKervey, M.A. *J. Chem. Soc. Chem. Commun.*, ***1993***, 1220

CHO, SmI_2, 25°C → CHO

(95:5 [1,4:1,3]) 72%

Van de Weghe, P.; Collin, J. *Tetrahedron Lett.*, ***1994***, *35*, 2545

O → 1. Bu_3SH, $TiCl_4$, CH_2Cl_2; 2. $POCl_3$, DMF, 0-5°C → SBu CHO 67%

Asokan, C.V.; Mathews, A. *Tetrahedron Lett.*, ***1994***, *35*, 2585

Ph → allyl-OH, H_2O, 90°C; $[RhCl(cod)(C_5Me_5)]$, 1h → Ph CHO + Ph CHO

75 : 25) 85%

Dérien, S.; Dixneuf, P.H. *J. Chem. Soc. Chem. Commun.*, ***1994***, 2551

Ph–CH=CH–CH$_2$OH $\xrightarrow[O_2\text{ , heat}]{Pd(PPh_3)_4\text{ , }NH_4PF_6\text{ , toluene}}$ Ph–CH=CH–CHO 90%

Gómez-Bengoa, E.; Noheda, P.; Echavarren, A.M. *Tetrahedron Lett.*, ***1994***, *35*, 7097

PhCHO $\xrightarrow[BF_3\bullet OEt_2\text{ , }CH_2Cl_2\text{ , -78°C , 1h}]{EtO–CH=CH–SnBu_3}$ Ph–CH=CH–CHO 86%

Cabezas, J.A.; Oehlschlager, A.C. *Tetrahedron Lett.*, ***1995***, *36*, 5127

Ph–CH=CH–CH=N–OH $\xrightarrow{MnO_2\text{ , hexane , 15 min}}$ Ph–CH=CH–CHO 87%

Shinada, T.; Yoshihara, K. *Tetrahedron Lett.*, ***1995***, *36*, 6701

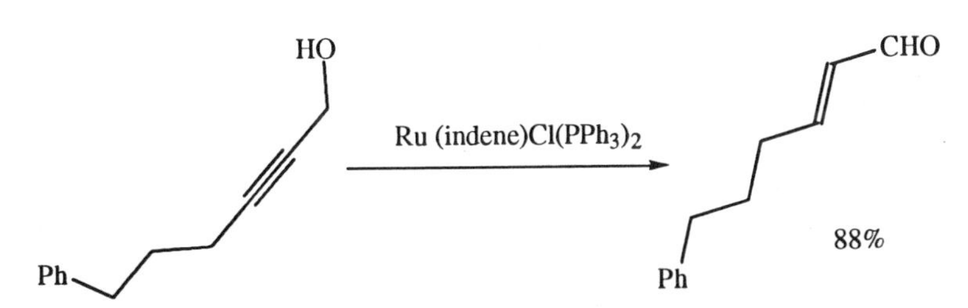

Trost, B.M.; Livingston, R.C. *J. Am. Chem. Soc.*, ***1995***, *117*, 9586

C_3H_7–C≡C–C_3H_7 $\xrightarrow[2.2\%\text{ biaryloxyl phophite}]{2\%\text{ [}Rh(CO)_2(acac)\text{]}}$ C_3H_7–CH=C(C_3H_7)–CHO 85%

Johnson, J.R.; Cuny, G.D.; Buchwald, S.L. *Angew. Chem. Int. Ed. Engl.*, ***1995***, *34*, 1760

Also via β-Hydroxy aldehydes: Section 324 (Alcohols - Aldehyde).

SECTION 342: AMIDE - AMIDE

NO ADDITIONAL EXAMPLES

Also via Dicarboxylic Acids:	Section 312 (Carboxylic Acid - Carboxylic Acid)
Diamines	Section 350 (Amines - Amines)

SECTION 343: AMIDE - AMINE

N-OH
Me C≡N
Me
$AcCl, Ac_2O, HCl$
100-130°C, 4h
Me
Me N NHAc
63%

Vijn, R.J.; Arts, H.J.; Maas, P.J.; Castelijns, A.M. *J. Org. Chem.*, ***1993***, *58*, 887

1. $TMSCl, NaI, I_2$
TMEDA, MeCN
2. NH_3, 100°C
3. racemization-resolution
N H O
NH_2
95x77x80%

Armstrong III, J.E.; Eng, K.K.; Keller, J.L.; Purick, R.M.; Hartner Jr., F.W.; Choi, W.-B.; Askin, D.; Volante, R.P. *Tetrahedron Lett.*, ***1994***, *35*, 3239

TsHN
$Pd(OAc)_2, Et_2NH$
1 atm CO, THF
CH_2Cl_2
N—Ts
O
NEt_2
+
N—Ts
NEt_2
O
(86 : 14) 72%

Andersson, P.G.; Aranyos, A. *Tetrahedron Lett.*, ***1994***, *35*, 4441

N
N Ph
Me
-70°C → -20°C
OZnCl
OMe
Me
H N Ph
MeO_2C
N
58%

van Maanen, H.L.; Kleijn, H.; Jastrezebski, T.B.H.; van Koten, G. *Recl. Trav. Chim. Pays-Bas*, ***1994***, *113*, 567

Ts, N (2-methylaziridine) → $BnNH_2$, 10% $Yb(OTf)_3$ / CH_2Cl_2, rt, 10h → NHTs / NHBn + NHBn / NHTs

(>99 : 1) 75%

Meguro, M.; Asao, N.; Yamamoto, Y. *Tetrahedron Lett.*, ***1994***, *35*, 7395

N, O, Cl → t-$BuNH_2$, NEt_3 / CH_2Cl_2, reflux, 1d → N, O, NHt-Bu 78%

Chen, P.; Suh, D.-J.; Smith, M.B. *J. Chem. Soc., Perkin Trans. 1.*, ***1995***, 1317

SECTION 344: AMIDE - ESTER

Me, H, BocHN, CO_2H → t-BuBr, K_2CO_3, DMAC / BTEAC, 55°C → Me, H, BocHN, CO_2t-Bu 92%

DMAC = dimethyl acetamide
BTEAC = $BnEt_3NCl$

Chevallet, P.; Garroust, P.; Malawska, B.; Martinez, J. *Tetrahedron Lett.*, ***1993***, *34*, 7409

CO_2Me, N, Ph, O → $Mn(OAc)_3$, $CH_2(CO_2Me)_2$ → CO_2Me, CO_2Me, CO_2Me, N 83%

Chuang, C.-P.; Wang, S.-F. *Tetrahedron Lett.*, ***1994***, *35*, 1283

O, N, H → CO_2Et / $Si(OEt)_4$, 0.1 CsF → O, N, CO_2Et 98%

Ahn, K.H.; Lee, S.J. *Tetrahedron Lett.*, ***1994***, *35*, 1875

1. 15% $ZnCl_2$, CH_2Cl_2, CpFe(CO)$_2$(allyl)
2. CAN, MeCN, CO

Chen, T.; Jiang, S.; Turos, E. *Tetrahedron Lett.*, ***1994***, *35*, 8325

Bu_3SnH, AIBN, PhH

52%

Goodall, K.; Parsons, A.F. *J. Chem. Soc., Perkin Trans. 1.*, ***1994***, 3257

1. $Sn(OTf)_2$, CH_2Cl_2, N-ethylpiperidine
2. N-CO_2Et-2-ethoxypyrrolidine

78%

Nagasaka, T.; Nishida, S.; Sugihara, S.; Kawahara, T.; Adachi, K.; Hamaguchi, F. *Heterocycles*, ***1994***, *39*, 171

Related Methods:

Section 315 (Carboxylic Acid - Amide)
Section 316 (Carboxylic Acid - Amine)
Section 351 (Amine - Ester)

SECTION 345: AMIDE - ETHER, EPOXIDE, THIOETHER

CO/H_2 (1600 psi), 100°C
$HC(OEt)_3$, 24h
$HRh(CO)(PPh)_3$

42%

(2:1 *syn:anti*)

Eguchi, M.; Zeng, Q.; Korda, A.; Ojioma, I. *Tetrahedron Lett.*, ***1993***, *34*, 915

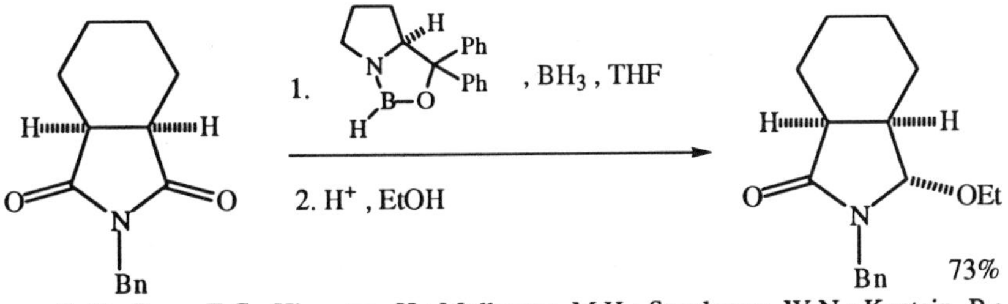
H
H
O
N
O
Bn
1.
H
Ph
N
Ph
B—O
H
, BH3 , THF
2. H+ , EtOH
H
H
O
N
OEt
Bn
73%

Romagnoli, R.; Roos, E.C.; Hiemstra, H.; Mollenaar, M.H.; Speckamp, W.N.; Kaptein, B.; Schoemaker, H.E. *Tetrahedron Lett.*, ***1994***, *35*, 1087

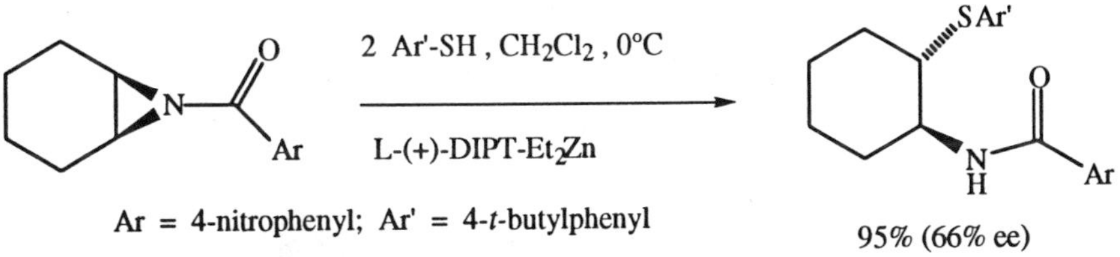
N
O
Ar
2 Ar'-SH , CH2Cl2 , 0°C
L-(+)-DIPT-Et2Zn
SAr'
O
N
H
Ar
Ar = 4-nitrophenyl; Ar' = 4-t-butylphenyl
95% (66% ee)

Hayashi, M.; Ono, K.; Hoshimi, H.; Oguni, N. *J. Chem. Soc. Chem. Commun.*, ***1994***, 2699

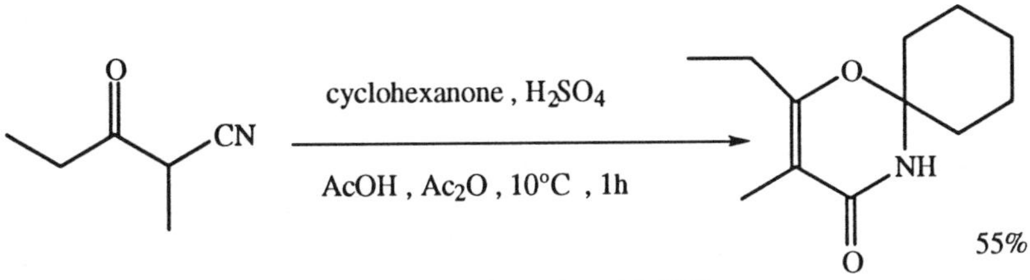
O
CN
cyclohexanone , H2SO4
AcOH , Ac2O , 10°C , 1h
O
NH
O
55%

Stambach, J.-F.; Jung, L.; Hug, R. *Heterocycles*, ***1994***, *38*, 297

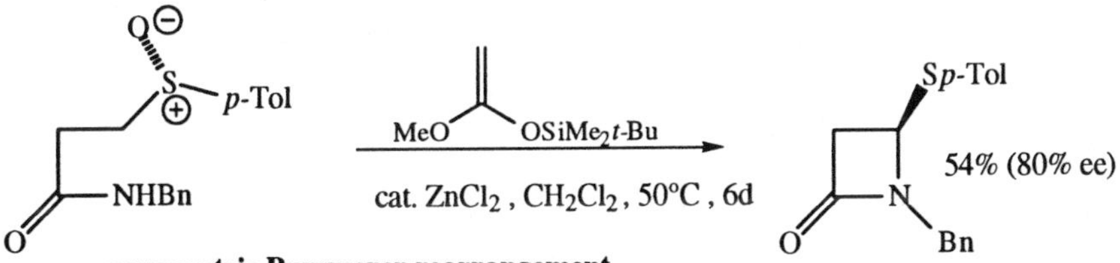
O
S
p-Tol
NHBn
O
MeO
OSiMe2t-Bu
cat. ZnCl2 , CH2Cl2 , 50°C , 6d
Sp-Tol
54% (80% ee)
N
O
Bn

asymmetric Pummerer rearrangement

Kita, Y.; Shibata, N.; Kawano, N.; Tohjo, T.; Fujimori, C.; Matsumoto, K. *Tetrahedron Lett.*, ***1995***, *36*, 115

1.5 NMO , CH_2Cl_2

90%

Andres, C.J.; Spetseris, N.; Norton, J.R.; Meyers, A.I. *Tetrahedron Lett.*, ***1995***, *36*, 1613

SECTION 346: AMIDE - HALIDE, SULFONATE

CuCl , cat. bipy

98%

Nagashima, H.; Ozaki, N.; Ishii, M.; Seki, K.; Washiyama, M.; Itoh, K. *J. Org. Chem.*, ***1993***, *58*, 464

$Et_3N \bullet 3$ HF , MeCN

e^-

77%

Narizuka, S.; Fuchigami, T. *J. Org. Chem.*, ***1993***, *58*, 4200

$Ph_2Se(OH)OAc$, I_2
hν , 6.5h

76%

Dorta, R.L.; Francisco, C.G.; Suárez, E. *Tetrahedron Lett.*, ***1994***, *35*, 1083

1. $[p\text{-}AnP(S)S]_2$, PhMe , 100°C
2. $Bu_4NH_2F_3$, NIS , CH_2Cl_2
0°C → rt

98x64%

Kuroboshi, M.; Hiyama, T. *Tetrahedron Lett.*, ***1994***, *35*, 3983

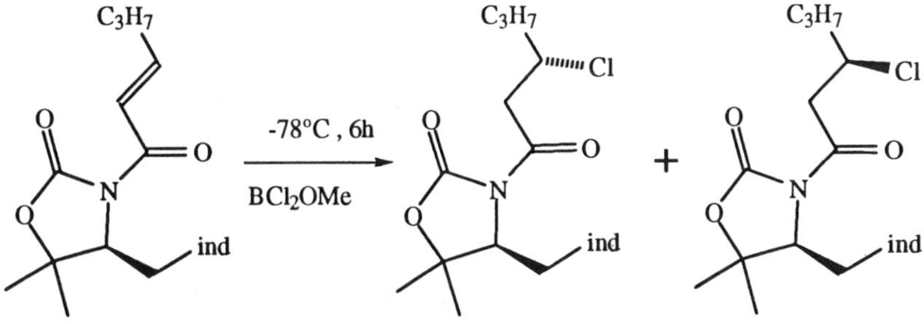

C_3H_7 C_3H_7 C_3H_7

-78°C , 6h

BCl_2OMe

O N O ind → O N Cl O ind + O N Cl O ind

(>99 : 1) 80%

Cardillo, G.; Di Martino, E.; Gentilucci, L.; Tomasini, C.; Tomasoni, L. *Tetrahedron Asymmetry*, ***1995***, *6*, 1957

SECTION 347: AMIDE - KETONE

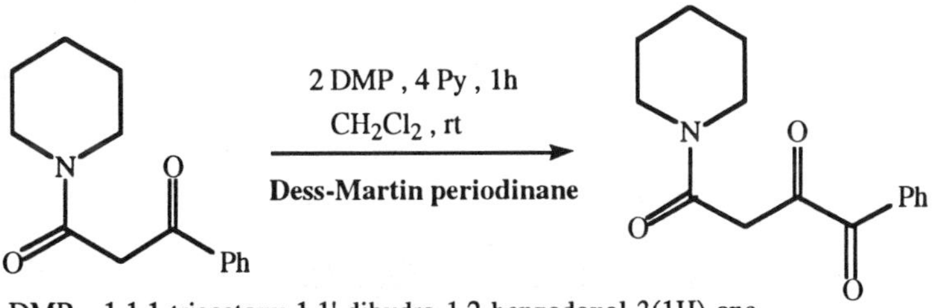

2 DMP , 4 Py , 1h

CH_2Cl_2 , rt

Dess-Martin periodinane

N O O Ph → N O O Ph O 75%

DMP = 1,1,1-triacetoxy-1,1'-dihydro-1,2-benzodoxol-3(1H)-one

Batchelor, M.J.; Gillespie, R.J.; Golec, J.M.C.; Hedgecock, C.J.R. *Tetrahedron Lett.*, ***1993***, *34*, 167

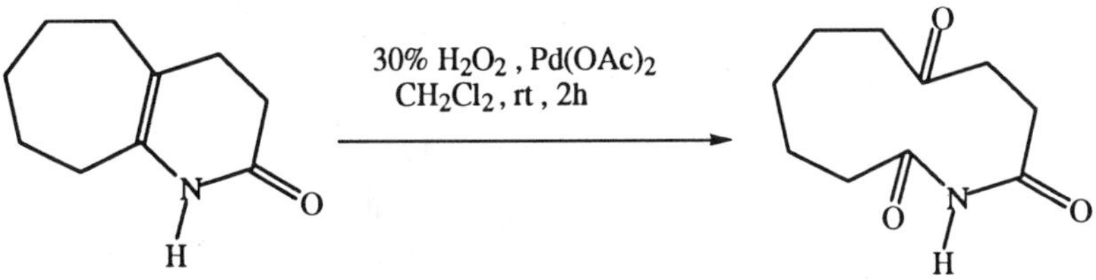

30% H_2O_2 , $Pd(OAc)_2$

CH_2Cl_2, rt , 2h

N H O → O O N H O 73%

Naota, T.; Sasao, S.; Tanaka, K.; Yamamoto, H.; Murahashi, S. *Tetrahedron Lett.*, ***1993***, *34*, 4843

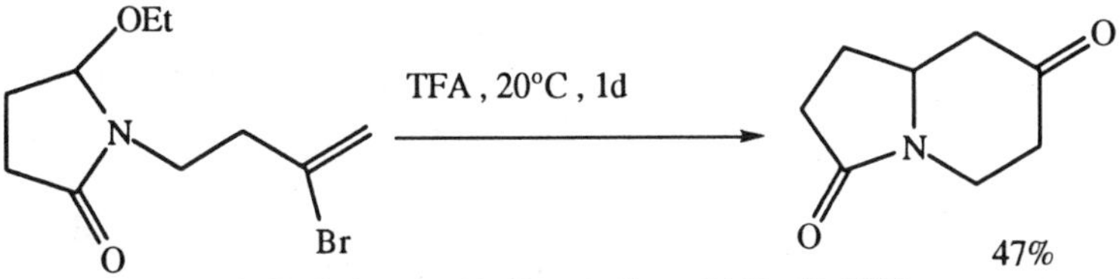

TFA , 20°C , 1d

OEt N O Br → O N O 47%

Gesson, J.-P.; Jacquesy, J.-C.; Rambaud, D. *Tetrahedron*, ***1993***, *49*, 2239

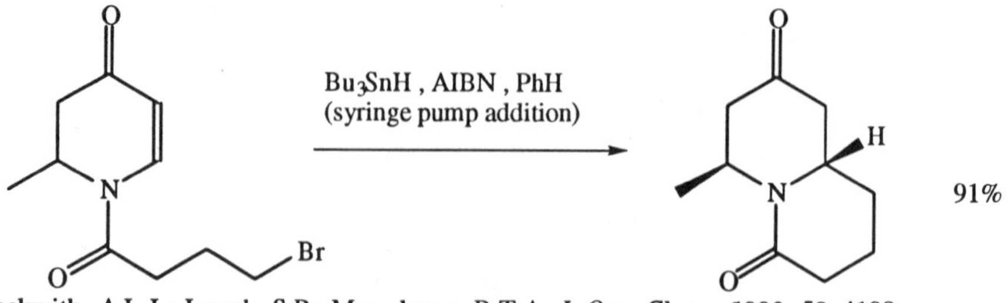

Bu_3SnH , AIBN , PhH
(syringe pump addition)

Br

H

91%

Beckwith, A.L.J.; Joseph, S.P.; Mayadunne, R.T.A. *J. Org. Chem.*, ***1993***, *58*, 4198

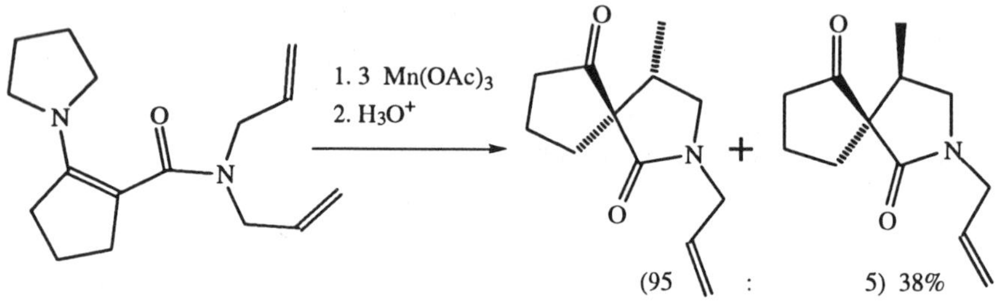

1. 3 $Mn(OAc)_3$
2. H_3O^+

(95 : 5) 38%

Cossy, J.; Bouzide, A.; Leblanc, C. *Synlett*, ***1993***, 202

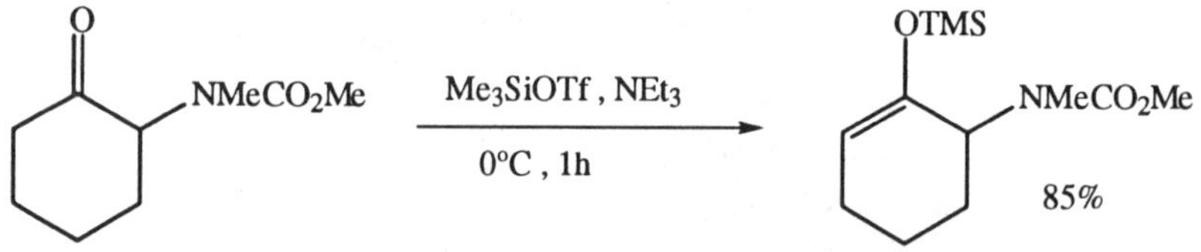

$NMeCO_2Me$

Me_3SiOTf , NEt_3

0°C , 1h

OTMS

$NMeCO_2Me$

85%

Rossi, L.; Pecunioso, A. *Tetrahedron Lett.*, ***1994***, *35*, 5285

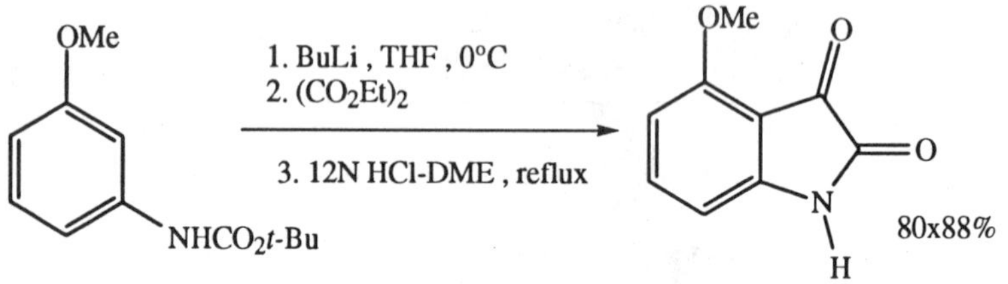

OMe

$NHCO_2t$-Bu

1. BuLi , THF , 0°C
2. $(CO_2Et)_2$

3. 12N HCl-DME , reflux

OMe

80x88%

Hewawasam, P.; Meanwell, N.A. *Tetrahedron Lett.*, ***1994***, *35*, 7303

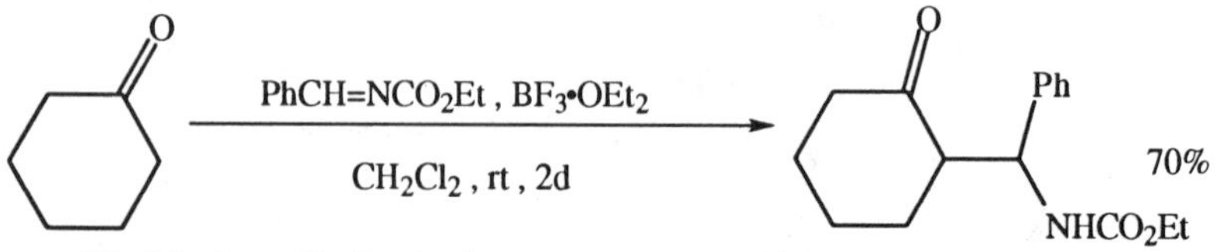

PhCH=NCO_2Et , BF_3•OEt_2

CH_2Cl_2 , rt , 2d

Ph

$NHCO_2Et$

70%

ten Hoeve, W.; Wynberg, H. *Synth. Commun.*, ***1994***, *24*, 899

1. MeNHOH
2. Cl(C=N-Me)Ph
3. H_2O

via hetero Cope

60%

Lantos, I.; Zhang, W.-Y. *Tetrahedron Lett.*, ***1994***, *35*, 5977

1. RCOCl
2. MeMgX
3. oxalic acid
4. TFA , $CHCl_3$, heat

R = (-)-8-phenylmenthyl

CO_2R

78x86%

Comins, D.L.; LaMunyon, D.H. *Tetrahedron Lett.*, ***1994***, *35*, 7343

1. *t*-BuLi , TMEDA , -78°C
2. (δ-valerolactone)

$NiPr_2$

80%

Brenstrum, T.J.; Brimble, M.A.; Stevenson, R.J. *Tetrahedron*, ***1994***, *50*, 4897

O_2N–C6H4–CHO

$CoCl_2$, MeCN , AcCl

NHAc

NO_2

56%

Bhatia, B.; Reddy, M.M.; Iqbal, J. *J. Chem. Soc. Chem. Commun.*, ***1994***, 713

N_2

NHMe (allyl)

4 eq. DMAP , toluene , reflux

95%

Cossy, J.; Belotti, D.; Bouzide, A.; Thelland, A. *Bull. Soc. Chim. Fr.*, ***1994***, *131*, 723

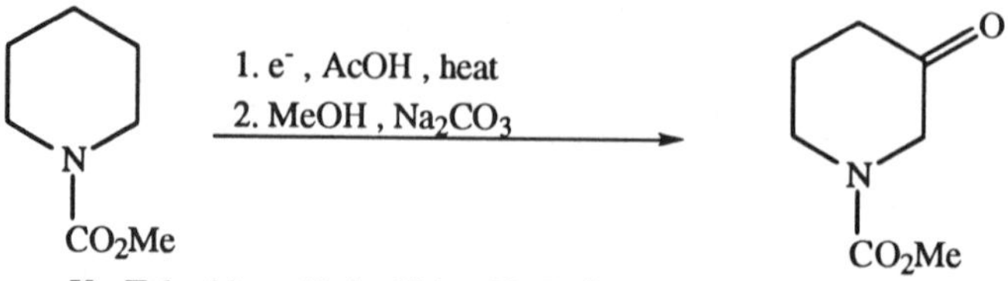

1. e^-, AcOH, heat
2. MeOH, Na_2CO_3

CO_2Me CO_2Me 75x95%

Matsumura, Y.; Takeshima, Y.-i.; Okita, H. *Bull. Chem. Soc. Jpn.*, ***1994***, *67*, 304

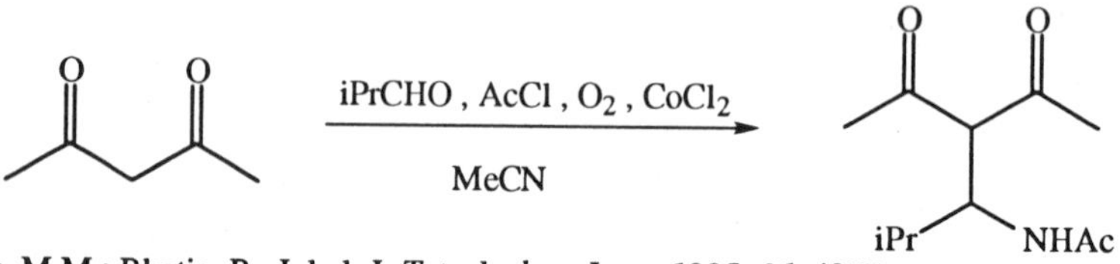

iPrCHO, AcCl, O_2, $CoCl_2$
MeCN

iPr NHAc 76%

Reddy, M.M.; Bhatia, B.; Iqbal, J. *Tetrahedron Lett.*, ***1995***, *36*, 4877

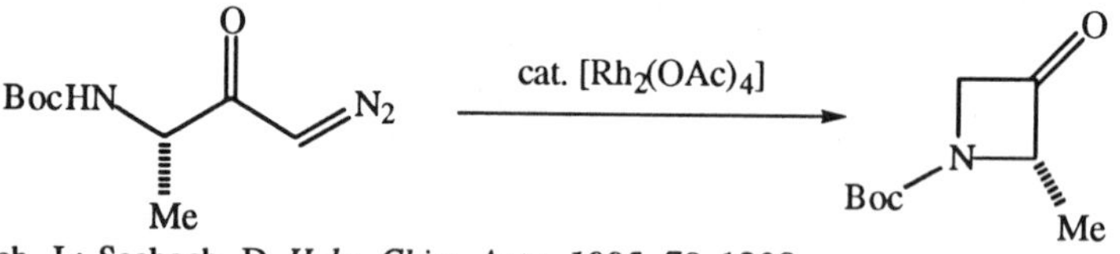

BocHN N_2 Me

cat. $[Rh_2(OAc)_4]$

Boc Me 59%

Podlech, J.; Seebach, D. *Helv. Chim. Acta*, ***1995***, *78*, 1238

SECTION 348: AMIDE - NITRILE

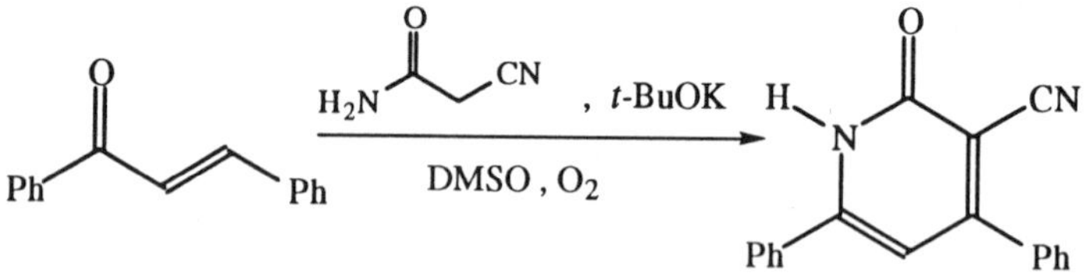

H_2N CN, *t*-BuOK
DMSO, O_2

Ph Ph H CN Ph Ph

88%

Jain, R.; Roschangar, F.; Ciufolini, M.A. *Tetrahedron Lett.*, ***1995***, *36*, 3307

SECTION 349: AMIDE - ALKENE

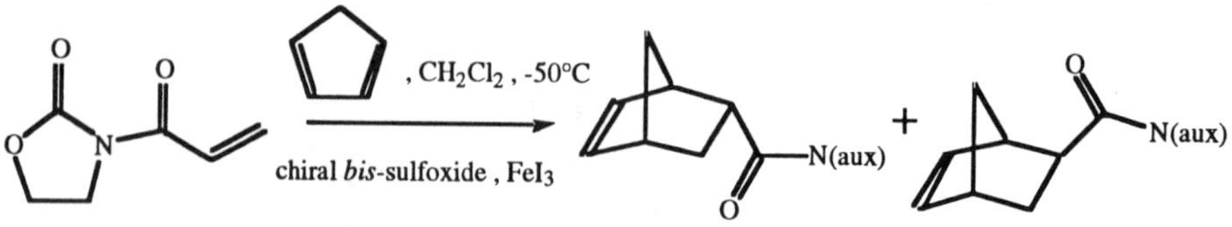

, CH_2Cl_2, -50°C
chiral *bis*-sulfoxide, FeI_3

N(aux) + N(aux)

(6 : 4) 78%

Khiar, N.; Fernández, I.; Alcudia, F. *Tetrahedron Lett.*, ***1993***, *34*, 111

1. LDA, THF, -78°C
2. PhCHO, THF reflux

92%
100% Z

Couture, A.; Deniau, E.; Grandclaudon, P. *Tetrahedron Lett.*, ***1993***, *34*, 1479

, MS4Å, CH_2Cl_2
20% chiral Yb triflate, 0°C

77%
(89:1 *endo:exo*; 95%ee)

Kobayashi, S.; Hachiya, I.; Ishitani, H.; Araki, M. *Tetrahedron Lett.*, ***1993***, *34*, 4535

4% $Pd_2(dba)_3$•$CHCl_3$
9.6% RS-BPPFOH
silver-exchanged zeolite
DMF:DMSO, 0°C, 5d

(1 : 1.4) 94%
(86%ee)

Nukui, S.; Sodeoka, M.; Shibasaki, M. *Tetrahedron Lett.*, ***1993***, *34*, 4965

, PPh_3, THF
2% $Pd_2(dba)_3$•$CHCl_3$
2 eq. iPr_2NEt

73%

Torii, S.; Okumoto, H.; Sadakane, M.; Hai, A.K.M.A.; Tanaka, H. *Tetrahedron Lett.*, ***1993***, *34*, 6553

, -20°C, 36h
bis-imine•cu(OTf)$_2$

(>97:3 *endo:exo*) 72%

Evans, D.A.; Lectka, T.; Miller, S.J. *Tetrahedron Lett.*, ***1993***, *34*, 7027

1. 0.2 eq.fe(acac)$_3$
3 eq. Et_3Al/L

2. ethylene glycol
cat. *p*-TsOH

BzO

(>20 : 1) 45%

Takacs, J.M.; Weidner, J.J.; Takacs, B.E. *Tetrahedron Lett.*, ***1993***, *34*, 6219

0.02 $PdCl_2$, 2 eq. $CuCL_2$
3 eq. $ClCH_2CO_2^-Na^+$

CO . MeOH

NHTs

CO_2Me

Ts

53%

Kimura, M.; Saeki, N.; Uchida, S.; Harayama, H.; Tanaka, S.; Fugami, K.; Tamaru, Y. *Tetrahedron Lett.*, ***1993***, *34*, 7611

2

1. EtO_2C-N=N-CO_2Et , $SnCl_4$
CH_2Cl_2 , -60°C , 5 min

2. Li , NH_3 , -33°C
3. 10 N HCl

$NHCO_2Et$

87 x 86% (11:1 *E:Z*)

Brimble, M.A.; Heathcock, C.H. *J. Org. Chem.*, ***1993***, *58*, 5261

Br

$Pd(OAc)_2$, P(*o*-Tol)$_3$, DMF
Na_2CO_3 , BU_4NCl , 65°C , 40h

sealed tube (*in vacuo*)

NHTs

N—Ts

(4 : 1) 78%

Harris Jr., G.D.; Herr, R.J.; Weinreb, S.M. *J. Org. Chem.*, ***1993***, *58*, 5452

$AcO\text{-}CH_2C(=CH_2)CH_2\text{-}TMS$

$Pd(PPh_3)_4$, PhMe ,100°C

N-Ts; t-Bu; 90%

Trost, B.M.; Marrs, C.M. *J. Am. Chem. Soc.*, ***1993***, *115*, 6636

CO_2Bn; N; Boc; O

$Ph_3P{=}CHCO_2Et$

toluene , reflux , 4h

EtO_2C; CO_2Bn; N; Boc; 80%

Baldwin, J.E.; Edwards, A.J.; Farthing, C.N.; Russell, A.T. *Synlett*, ***1993***, 49

O; CO_2Me

1. NaH , THF , BuLi

2. N—Ts (aziridine)

3. toluene , Amberlyst 15 , reflux

CO_2Me; N; Ts; >96% (>20:1 *E:Z*)

Lygo, B. *Synlett*, ***1993***, 764

$PhCH_2CO_2Et$

1. 2 eq. LDA

2. PhCH=NPh

Ph; Ph; PhHN; O; 58%

this reaction forms a β-lactam with 1 eq. of LDA

Manhas, M.S.; Chaudhary, A.G.; Raju, V.S.; Robbi, E.W.; Bose, A.K. *Heterocycles*, ***1993***, *35*, 635

Ph; Cl; N; THPO

allyl alcohol (OH)

NaH , DMF

Ph; O; N; THPO; 60%

1. xylenes , reflux

2. PPTS

Ph; O; N; HO; 68x75%

De la Torre, J.A.; Fernandez, M.; Morgans Jr., D.; Smith, D.B.; Talamas, F.X.; Trejo, A. *Tetrahedron Lett.*, ***1994***, *35*, 15

OMe; $NHCO_2Me$

Br (allyl bromide)

Zn , THF , rt , 1h

$NHCO_2Me$; 90%

Kise, N.; Yamazaki, H.; Mabuchi, T.; Shono, T. *Tetrahedron Lett.*, ***1994***, *35*, 1561

Ts; N; MeO_2CO

0.1 [Rh(CO)$_2$(acac) , dppb
AcOH , 70°C , 6h

Ts; N

79%

Oppolzer, W.; Fürstner, A. *Helv. Chim. Acta,* ***1993****, 76,* 2329

I; SPh; O

1. piperidine , CF_3CO_2Ag
THF , rt

2. $C_9H_{19}MgBr$, THF
$NiCl_2(dppe)$, 0°C → rt

C_9H_{19}; N; O

57x60%

Babudri, F.; Fiandanese, V.; Naso, F.; Punzi, A. *Tetrahedron Lett.,* ***1994****, 35,* 2067

N; O

$PhMe_2CCH$=Mo-N[2,6-(iPr)$_2C_6H_{13}$]
[OCMe(CF_3)$_2$]$_2$

PhH , rt

N; O

80-90%

Martin, S.F.; Liao, Y.; Chen, H.-J.; Pätzel, M.; Ramser, M.N. *Tetrahedron Lett.,* ***1994****, 35,* 6005

CO_2H; H; Me; $NHSO_2Ph$

1. MeLi
2. NaH
3. SPh; ⊕PPh_3

Me; SPh; Me; N; SO_2Ph

67%

Burley, I.; Hewson, A.T. *Tetrahedron Lett.,* ***1994****, 35,* 7099

O; N; CO_2Me; CO_2t-Bu

1. Dibal , toluene , -78°C
2. Ph_3P=CHPh , THF , 50°C
3. PCC , 25°C

O; N; Ph; CO_2t-Bu

52% (90:10 *E:Z*)

Wei, Z.Y.; Knaus, E.E. *Synlett,* ***1994****,* 345

O O N O Me

, MS 4Å , CH_2Cl_2

20% Yb-BINOL complex

Me O N O O + Me O N O O

(89 : 11) 77%

95% ee

Kobayashi, S.; Ishitani, H.; Araki, M.; Hachiya, I. *Tetrahedron Lett.*, ***1994***, *35*, 6325

O CO_2Et

1. NH_2 Me H Ph , BF_3•OEt_2, PhH , reflux

2. CH_2=$CHCO_2CO_2$, THF , reflux

CO_2Et O N Me H Ph

76% (97:3)

Barta, N.S.; Brode, A.; Stille, J.R. *J. Am. Chem. Soc.*, ***1994***, *116*, 6201

Ar NH_2

H_2CO , $[Rh(OAc)_2]_2$, PPh_3 , PhH

Ar O N H 40%

Campi, E.M.; Chong, J.M.; Jackson, W.R.; Van Der Schoot, M. *Tetrahedron*, ***1994***, *50*, 2533

NHTs I

C_8H_{17}—CH=C=CH_2
Ag_3PO_4 , 5% $Pd(OAc)_2$

bis-oxazoline catalyst
90°C , 1d , DMF

Ts N C_8H_{17}

94% (82% ee)

Larock, R.C.; Zenner, J.M. *J. Org. Chem.*, ***1995***, *60*, 482

Ph-N=C=O , $AlCl_3$, 40°C

CH_2Cl_2 , 12h

Me SnMe$_3$ → Me NHPh O 50% (4.9:1 *E:Z*)

Niestroj, M.; Neumann, W.P.; Thies, O. *Chem. Ber.*, ***1994***, *127*, 1131

Bu_4NF , 66°C

O NHBn → O N Bn 50%

Jacobi, P.A.; Brielmann, H.L.; Hauck, S.I. *Tetrahedron Lett.*, ***1995***, *36*, 1193

5% $Cl_2Pd(PPh_3)_2$, MeOH , 65°C

CO (1 atm) , 4 NEt_3

AcHN Bu I → Bu N O Ac 78%

Copéret, C.; Sugihara, T.; Negishi, E. *Tetrahedron Lett.*, ***1995***, *36*, 1771

TsNNaCl , EtOH , 20h

25°C

Ph TePh → Ph NHTs 84% + Ph OH 9%

Nishibayashi, Y.; Srivastava, S.K.; Ohe, K.; Uemura, S. *Tetrahedron Lett.*, ***1995***, *36*, 6725

1. LiHMDS , THF , HMPA , -78°C
2. $PhNTf_2$, -78°C → 25°C
3. Ph_2CuLi , THF-HMPA
-78°C → 25°C

N O Bz → N Ph Bz 94x69%

Tsushima, K.; Hirade, T.; Hasegawa, H.; Murai, A. *Chem. Lett.*, ***1995***, 801

0.1 AgNCO , 0.1 NEt_3 , PhH
50°C , 5h

H Me Me O NHTs O → Me Me O N Ts O 98%

Kimura, M.; Tanaka, S.; Tamaru, Y. *Bull. Chem. Soc. Jpn.*, ***1995***, *68*, 1689

Also via Alkenyl Acids: Section 322 (Carboxylic Acid -Alkene)

SECTION 350: AMINE - AMINE

In , aq. EtOH , 16h

NH_4Cl

Ph NHPh NHPh quant.

Kalyanam, N.; Rao, G.V. *Tetrahedron Lett.*, ***1993***, *34*, 1647

1. *m*-cpba , CH_2Cl_2 0°C → 25°C

2. H_2 (1 atm) , Pd/C MeOH , 15h , 25°C

42%

Plaquevent, J.-C.; Chichaoui, I. *Tetrahedron Lett.*, ***1993***, *34*, 5287

PhCN

quant.

Rousselet, G.; Capdevielle, P.; Maumy, M. *Tetrahedron Lett.*, ***1993***, *34*, 6395

$NHEt_2$, 0.8 GPa , CH_2Cl_2 , 2d

MeOH , 50°C

68%

Matsumoto, K.; Uchida, T.; Hashimoto, S.; Yonezawa, Y.; Iida, H.; Kakehi, A.; Otani, S. *Heterocycles*, ***1993***, *36*, 2215

1. *t*-BuLi , -78°C , THF
2. $ZnCl_2$
3. $PdCl_2(PPh_3)_2$, Dibal THF , reflux

4. TsOH , EtOH

93%

Amat, M.; Hadida, S.; Bosch., J. *Tetrahedron Lett.*, ***1994***, *35*, 793

2 Ph–CH=NPh —(Al, KOH, MeOH, rt, 10 min)→ Ph, Ph, PhHN, NHPh

90% (70:30 *dl:meso*)

Baruah, B.; Prajapati, D.; Sandhu, J.S. *Tetrahedron Lett.*, ***1995***, *36*, 6747

NAr, Ph, Ph, NAr —(BH_3•THF, THF; NH_2, TBDMSO, Ph, Ph, Me, OH)→ NAr, Ph, Ph, NAr + NAr, Ph, Ph, NAr

Ar = 4-OMe phenyl (95 : 5) 71%

Shimizu, M.; Kami, M.; Fujisawa, T. *Tetrahedron Lett.*, ***1995***, *36*, 8607

SECTION 351: AMINE - ESTER

CHO, CO_2Et —(1. N–H (pyrrolidine); 2. SmI_2, THF, 25°C)→ N, CO_2Et 70%

Aurrechoechea, J.M.; Fernánandez-Acebes, A. *Tetrahedron Lett.*, ***1993***, *34*, 549

Ph, N, Ph —(1. Im_2CO, BF_3•OEt_2 84%, THF, heat; 2. NaOMe, MeOH/THF 94%)→ PhHN, CO_2Me, Ph

Fustero, S.; Díaz, M.D.; Carlón, R.P. *Tetrahedron Lett.*, ***1993***, *34*, 725

TMS, N, TMS —(AgF, EtOAc; MeCN, rt, 15 min)→ H, CO_2Et, N + H, N, CO_2Et

(17 : 3) 98%

(7:3 de)

Pandey, G.; Lakshmaiah, G. *Tetrahedron Lett.*, ***1993***, *34*, 4861

N–Me pyrrole (N, Me) → 1. I, CO_2Et , BEt_3 , DMSO / $Fe_2(SO)_3{\bullet}H_2O$, hexane → N, Me, CO_2Et 54%

Baciocchi, E.; Muraglia, E. *Tetrahedron Lett.*, ***1993***, *34*, 5015

Ph, N, Ph + Me, $OSiMe_3$, OMe → rt , 1d / Montmorillonite K-10* → PhHN, CO_2Me, Ph, Me 95%

78:22 (*anti:syn*)

* dried under microwave conditions

Texier-Boullet, F.; Latouche, R.; Hamelin, J. *Tetrahedron Lett.*, ***1993***, *34*, 2123

EtO_2C, N=PPh_3, Me + O, H, Ph → Toluene , 160°C / sealed tube → EtO_2C, N, Me, Ph 58%

Molina, P.; Pastor, A.; Vilaplana, M.J. *Tetrahedron Lett.*, ***1993***, *34*, 3773

OAc → 1. Na^+ $^-NHSO_2Tol$, 60°C, THF/DMSO , 19h / 2. O_3 , NaOH/MeOH, -78°C , 3h → MeO_2C, Nphath 73x66%

Jumnah, R.; Williams, J.M.J. *Tetrahedron Lett.*, ***1993***, *34*, 6619

CO_2Me, CO_2Me → 1. $SnCl_4$, PhCN / 2. OH^- , Na_2CO_3 → Ph, CO_2Me, H_2N, CO_2Me 91%

Nicolini, M.; Citterio, A. *Org. Prep. Proceed. Int.*, ***1993***, *25*, 229

$OSiME_3$, OEt → PhCH=NPh , CH_2Cl_2 , -78°C , 2h / Fe-Montmorillonite → NHPh, Ph, CO_2Et

97% (96:e *threo:erythro*)

Onaka, M.; Ohno, R.; Yanigiya, N.; Izumi, Y. *Synlett*, ***1993***, 141

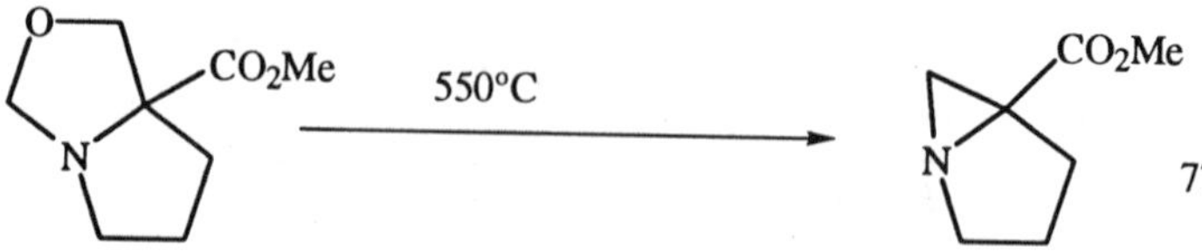
CO2Me
550°C
CO2Me
N
O
N

77%

Bureau, R.; Mortier, J.; Joucla, M. *Bull. Soc. Chim. Fr.*, ***1993***, *130*, 584

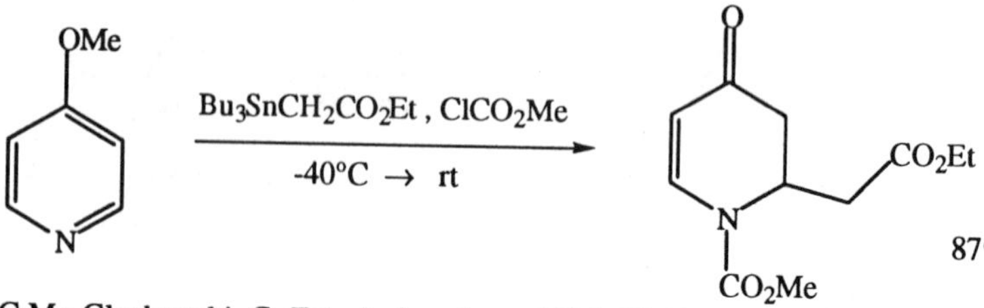
OMe
N
Bu3SnCH2CO2Et , ClCO2Me
-40°C → rt
O
CO2Et
N
CO2Me

87%

Dhar, T.G.M.; Gluchowski, C. *Tetrahedron Lett.*, ***1994***, *35*, 989

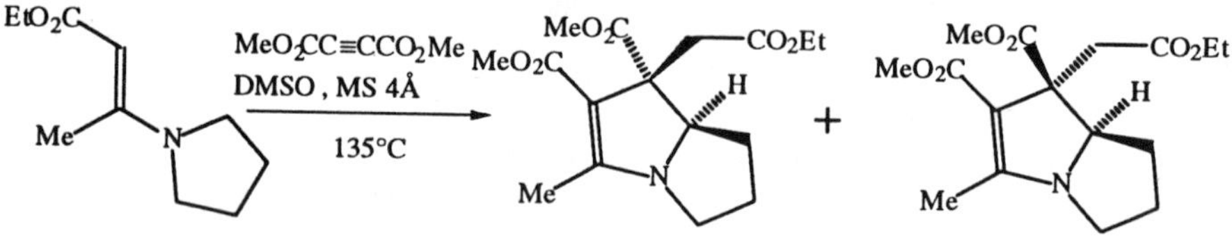
EtO2C
Me
N
MeO2CC≡CCO2Me
DMSO , MS 4Å
135°C
MeO2C
MeO2C
CO2Et
H
Me
N
+
MeO2C
MeO2C
CO2Et
H
Me
N

(36 : 64) 72%

Jiang, S.; Janousek, Z.; Viehe, H.G. *Tetrahedron Lett.*, ***1994***, *35*, 1185

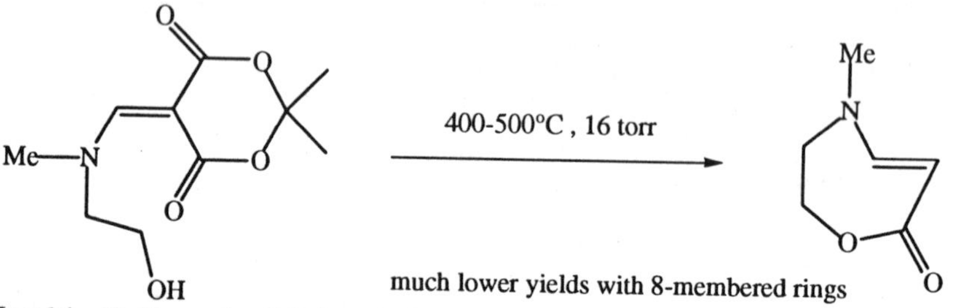
O
O
O
O
Me
N
OH
400-500°C , 16 torr
Me
N
O
O

60%

much lower yields with 8-membered rings

Jourdain, F.; Pommelet, J.C. *Tetrahedron Lett.*, ***1994***, *35*, 1545

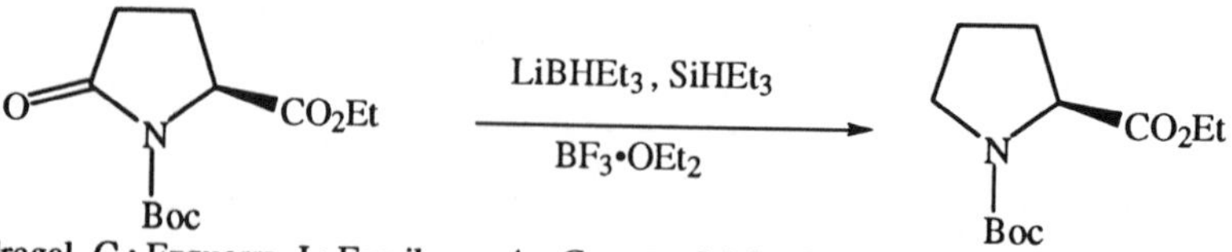
O
N
CO2Et
Boc
LiBHEt3 , SiHEt3
BF3•OEt2
N
CO2Et
Boc

85%

Pedregal, C.; Ezquerra, J.; Escribano, A.; Carreño, M.C.; García Ruano, J.L.G. *Tetrahedron Lett.*, ***1994***, *35*, 2053

CO_2Me / NO_2 — 2 eq. CO , $SnCl_2$ / $PdCl_2(PPh_3)_2$ → CO_2Me N H 62%

Akazome, M.; Kondo, T.; Watanabe, Y. *J. Org. Chem.*, ***1994***, *59*, 3375

CO_2Et N EtO_2C — 1. BBr_3 / 2. H_2 , Pd-C , AcOH → H N EtO_2C 88x75%

Jefford, C.W.; Thornton, S.R.; Sienkiewicz, K. *Tetrahedron Lett.*, ***1994***, *35*, 3905

Bn CbzHN CO_2H — BOP , MeOH , CH_2Cl_2 / -20°C → rt → Bn CbzHN CO_2Me 97%

BOP = benzotriazol-1-yloxy(dimethylamino)phosphonium hexafluorophosphate

Kim, M.H.; Patel, D.V. *Tetrahedron Lett.*, ***1994***, *35*, 5603

Bn TMS N TMS — MeO_2C CO_2Me / e$^-$, AcONa → MeO_2C N—Bn MeO_2C 47%

Torii, S.; Okumoto, H.; Genba, A. *Synlett*, ***1994***, 217

CO_2Et — $BnNH_2$, 10% $La(OTf)_3$, CH_2Cl_2 / rt → CO_2Et NHBn 95%

Matsubara, S.; Yoshioka, M.; Utimoto, K. *Chem. Lett.*, ***1994***, 827

N I — CO_2Me , Bu_3SnH / PhH , AIBN , heat → N CO_2Me 95%

Undheim, K.; Williams, L. *J. Chem. Soc. Chem. Commun.*, ***1994***, 883

$CH_2=CHCO_2Me$ —[$(iPr)_2NH$, 50°C, 3 KBar; cat. $Yb(OTf)_3$]→ $(Pri)_2N$–CH_2CH_2–CO_2Me

80% (13% with no catalyst
17% with catalyst at 1 Bar)

Jenner, G. *Tetrahedron Lett.*, ***1995***, *36*, 233

Br ... N ... CO_2Et, CO_2Me —[1.3 eq. bu_3SnH, 0.15 AIBN; PhH (0.025 M), reflux, 6h; syringe pump addition, 5h]→ N–CO_2Me pyrrolidine, CO_2Et

96% (59:41 - *trans:cisI*)

Lee, E.; Kang, T.S.; Joo, B.J.; Tae, J.S.; Li, K.S.; Chung, C.K. *Tetrahedron Lett.*, ***1995***, *36*, 417

H, N, Bn, CO_2Me —[1. DMSO, $(COCl)_2$; 2. NEt_3]→ N, Bn, CO_2Me 70%

Gentilucci, L.; Grijzen, Y.; Thijs, L.; Zwanenburg, B. *Tetrahedron Lett.*, ***1995***, *36*, 4665

PhCHO —[$Me_2C=C(OSiMe_3)(OMe)$, $PhNH_2$, rt; 10% $Yb(OTf)_3$, MS 4Å, CH_2Cl_2]→ NHPh, PhHN, CO_2Me

90%

Kobayashi, S.; Araki, M.; Yasuda, M. *Tetrahedron Lett.*, ***1995***, *36*, 5773

Ph–CH=N–Ph —[$Me_2C=C(OSiMe_3)(OMe)$; 5% $Yb(OTf)_3$, CH_2Cl_2, 0°C]→ NHPh, Ph, CO_2Me 97%

Kobayashi, S.; Araki, M.; Ishitani, H.; Nagayama, S.; Hachiya, I. *Synlett*, ***1995***, 233

piperidine N–H —[$CH_2=CHCO_2Me$, hν; anthraquinone, MeCN]→ N–CH_2CH_2–CO_2Me 93%

Das, S.; Kumar, J.S.D.; Shivaramayya, K.; George, M.V. *J. Chem. Soc., Perkin Trans. 1.*, ***1995***, 1797

Ph–CH=N–Ph → (1% $[Rh_2(OAc)_4]$, CH_2Cl_2; $MeO_2CCH=N_2$) → Ph–N aziridine (CO_2Me, Ph) 35%

Moran, M.; Bernardinelli, G.; Müller, P. *Helv. Chim. Acta*, ***1995***, *78*, 2048

Ph–CH=N–Ph → (N_2CHCO_2Et, CH_2Cl_2; $Cu(OTf)_2$, -25°C) → N-Ph aziridines (Ph, CO_2Et) + (Ph, CO_2Et)

(1.7 : 1) 80%

Rasmussen, K.G.; Jørgensen, K.A. *J. Chem. Soc. Chem. Commun.*, ***1995***, 1401

Ph–CH=N–Ph → (N_2CHCO_2Et, $CuPF_6(MeCN)_4$; CH_2Cl_2, bis-oxazoline ligand) → N-Ph aziridines (Ph, CO_2Et) + (Ph, CO_2Et)

(4 : 1) 37%

44% ee 35% ee

Hansen, K.B.; Finney, N.S.; Jacobsen, E.N. *Angew. Chem. Int. Ed. Engl.*, ***1995***, *34*, 676

Related Methods:

Section 315 (Carboxylic Acid - Amide)
Section 316 (Carboxylic Acid - Amine)
Section 344 (Amide - Ester)

SECTION 352: AMINE - ETHER, EPOXIDE, THIOETHER

Ph–CH(OTBDMS)–CN → (1. Dibal; 2. allyllithium; 3. H_2O) → Ph–CH(OTBDMS)–CH(NH_2)–CH_2CH=CH_2 + diastereomer

(91 : 9) 43%

Cainelli, G.; Panunzio, M.; Contento, M.; Giacomini, D.; Mezzina, E.; Giovagnoli, D. *Tetrahedron*, ***1993***, *49*, 3809

N-*t*-Bu

1. PhSeBr , CH_2Cl_2 , 0°C

2. $NaBH_4$, MeOH

t-Bu N SePh 90%

De Kimpe, N.; Boelens, M. *J. Chem. Soc. Chem. Commun.*, ***1993***, 916

N OH

hv (254 nm, Quartz) , acetone 2h

N O 88%

(48% conversion)

Cossy, J.; Guha, M. *Tetrahedron Lett.*, ***1994***, *35*, 1715

O*t*-Bu

1. NaN_3 , ICl
2. $LiAlH_4$
3. PhCOCl

ca. 30%

4. NaI , acetone
5. NiO_2

ca. 50%

O*t*-Bu N Ph O

Eastwood, F.W.; Perlmutter, P.; Yang, Q. *Tetrahedron Lett.*, ***1994***, *35*, 2039

OH NHCO$_2$Et

PhSeCl , CH_2Cl_2

-78°C → rt , 18h

OH N SePh CO_2Et + OH N SePh CO_2Et

(5 : 1) 61%

Cooper, M.A.; Ward, A.D. *Tetrahedron Lett.*, ***1994***, *35*, 5065

O Cl

1. NBS , $(BzO)_2$, CCl_4 , heat

2. $iPrNH_2$, ether , $TiCl_4$
3. $NaBH_4$, MeOH

NHiPr MeO OMe

52%

De Kimpe, N.; Stanoeva, E. *Synthesis*, ***1994***, 695

$PhCH_2NHSiMe_3$ → (propylene oxide, rt, 1d; $Co_2(CO)_8$, CO (1 atm)) → $PhCH_2NHC(O)CH_2CH(CH_3)OSiMe_3$ 84%

Watanabe, Y.; Nishiyama, K.; Zhang, K.; Okuda, F.; Kondo, T.; Tsuji, Y. *Bull. Chem. Soc. Jpn.*, ***1994***, *67*, 879

1. $NHEt_2$, CH_2Cl_2 (>90%)
2. PhOH, NaOH, THF

Et_2N ... OPh, C_8H_{17} 61% + PhO ... NEt_2, C_8H_{17} 30%

Richardson, P.F.; Nelson, L.T.J.; Sharpless, K.B. *Tetrahedron Lett.*, ***1995***, *36*, 9241

hν, PhSH, 2h

HS ... O, N ... + ... SH, O, N

(82 : 6:12) 66%
trans:cis

Naito, T.; Honda,Y.; Miyata, O.; Ninomiya, I. *J. Chem. Soc., Perkin Trans. 1.*, ***1995***, 19

SECTION 353: AMINE - HALIDE, SULFONATE

$PhNMe_2$ → (Br_2, CH_2Cl_2; 0.5 MeOH; 0.01 Et_3NHCl) → Br–C_6H_4–NMe_2 near quant.

Gervat, S.; Léonel, E.; Barraud, J-Y.; Ratovelomanana, V. *Tetrahedron Lett.*, ***1993***, *34*, 2115

2-iodonitrobenzene (NO_2, I) → ("Ni_2B", H_2O, 15°C, 30 min) → 2-iodoaniline (NH_2, I) 72%

"Ni_2B" = $Ni(OAc)_2$•4 $H_2O/NaBH_4$

Seltzman, H.H.; Berrang, B.D. *Tetrahedron Lett.*, ***1993***, *34*, 3083

CF_3SiMe_3 , THF
Et_4NF , 15h , rt

Ph Ph H N → Ph Ph F_3C H N H 86%

Félix, C.P.; Khatimi, N.; Laurent, A.J. *Tetrahedron Lett.*, ***1994***, *35*, 3303

CO_2Et O CN — iPrOH , $HCl_{(g)}$, rt → CO_2Et N H Cl 96%

Foley, L.H. *Tetrahedron Lett.*, ***1994***, *35*, 5989

SECTION 354: AMINE - KETONE

O O Ph — N—H , SiO_2 ; microwave (300W) 3 min → N O Ph 89%

Rechsteiner, B.; Texier-Boullet, F.; Hamelin, J. *Tetrahedron Lett.*, ***1993***, *34*, 5071

N F TfO — $O^- Li^+$, THF ; -78°C → 23°C → N O 35%

Kiselyov, A.S.; Strekowski, L. *J. Org. Chem.*, ***1993***, *58*, 4476

O TMS N Bn — $Pd(OAc)_4$, MeCN ; 25°C → O Bn—N 60%

hv , 9,10-dicyanoanthracene → 5% + O Bn—N 78%

Zhang, X.; Jung, Y.S.; Mariano, P.S.; Fox, M.A.; Martin, P.S.; Merkert, J. *Tetrahedron Lett.*, ***1993***, *34*, 5239

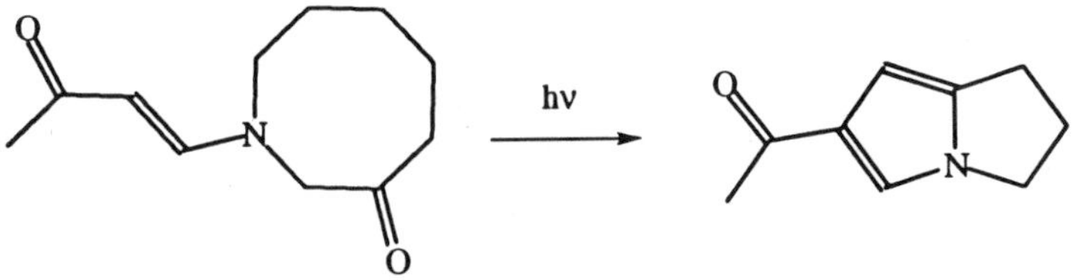

hν

Winkler, J.D.; Siegel, M.G. *Tetrahedron Lett.*, ***1993***, *34*, 7697

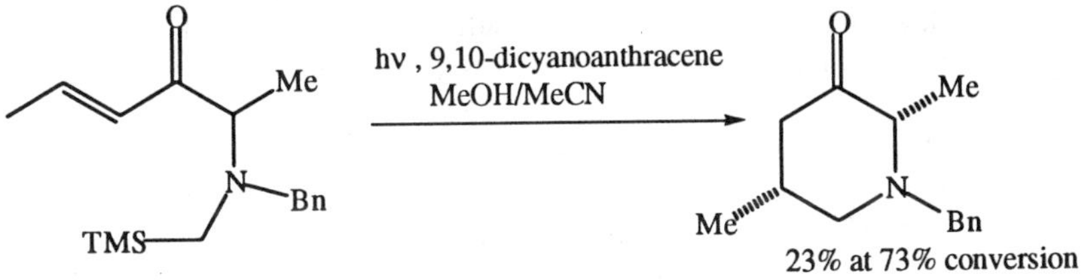

hν , 9,10-dicyanoanthracene
MeOH/MeCN

23% at 73% conversion

Khim, S.K.; Mariano, P.S. *Tetrahedron Lett.*, ***1994***, *35*, 999

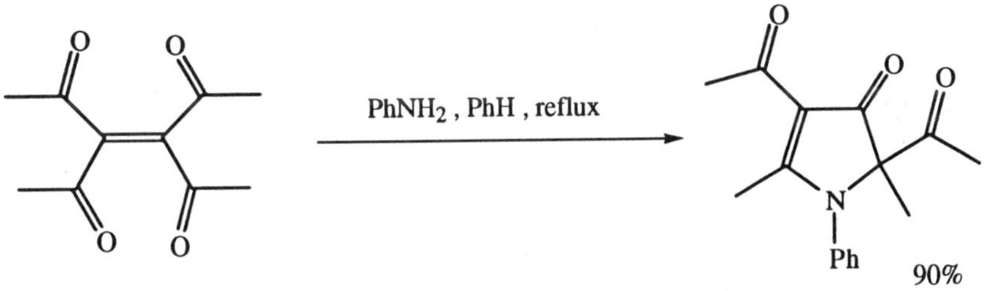

$PhNH_2$, PhH , reflux

90%

Adembri, G.; Celli, A.M.; Lampariello, L.R.; Scotton, M.; Sega, A. *Tetrahedron Lett.*, ***1994***, *35*, 4023

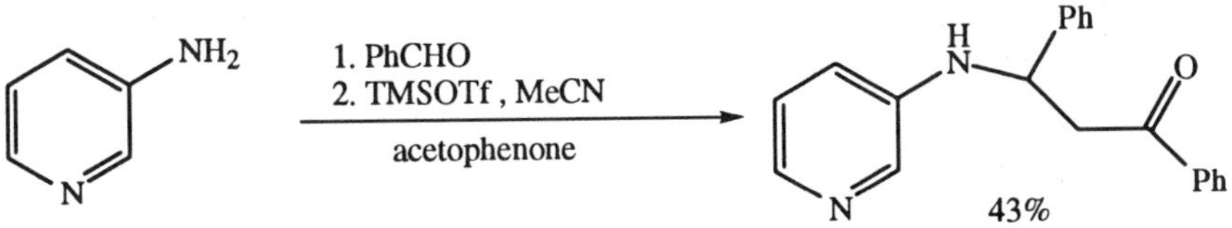

1. PhCHO
2. TMSOTf , MeCN
acetophenone

43%

Moutou, J.L.; Schmitt, M.; Wermuth, C.G.; Bourguignon, J.J. *Tetrahedron Lett.*, ***1994***, *35*, 6883

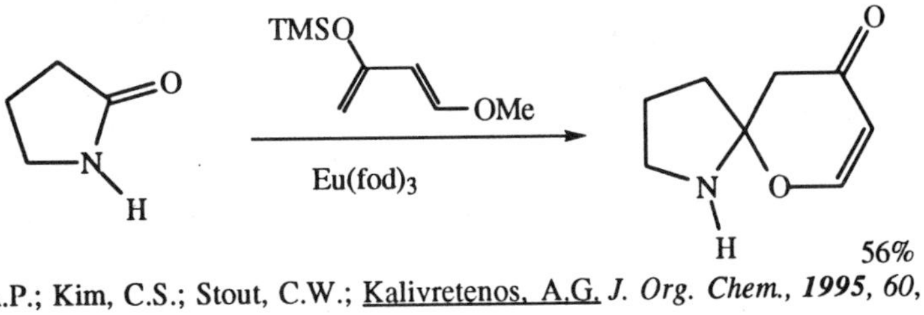

$Eu(fod)_3$

56%

Degnan, A.P.; Kim, C.S.; Stout, C.W.; Kalivretenos, A.G. *J. Org. Chem.*, ***1995***, *60*, 7724

hv , DCA , MeCN
DCA = 9,10-dicyanoanthracene

syn, 81%
anti, 18%

less selective with pyrrolidine derivatives

Hoegy, S.E.; Mariano, P.S. *Tetrahedron Lett.*, ***1994***, *35*, 8319

$Cu(acac)_2$, PhH , reflux

76%

Clark, J.S.; Hodgson, P.B. *J. Chem. Soc. Chem. Commun.*, ***1994***, 2701

20% Bu_4NReO_4 , CF_3SO_3H
DCE , reflux , 2h

76%

Kusama, H.; Uchiyama, K.; Yamashita, Y.; Narasaka, K. *Chem. Lett.*, ***1995***, 715

PhCHO → , $ArNH_2$, $THF-H_2O$; 10% $Yb(OTf)_3$

90%

Ar = 4-chlorophenyl

Kobayashi, S.; Ishitani, H. *J. Chem. Soc. Chem. Commun.*, ***1995***, 1379

CO_2Me ... Ph, 10% $NiBr_2(PPh_3)_2$, 1d → HN ... Ph, 75%

Me Me

Shida, N.; Kubota, Y.; Fukui, H.; Asao, N.; Kadoata, I.; Yamamoto, Y. *Tetrahedron Lett.*, ***1995***, *36*, 5023

SECTION 355: AMINE - NITRILE

$SiMe_3$

3% $BF_3•OEt_2$, 0°C, 4h

N—Ts

NHTs + NHTs

(2.6 : 1) 84%

Bergmeier, S.C.; Seth, P.P. *Tetrahedron Lett.*, ***1995***, *36*, 3793

SECTION 356: AMINE - ALKENE

Bn, Bn—N—H

1. $(CH_2O)_n$, 90°C

2. $(HO)_2B$ ... Ph

90°C, 30 min

Bn_2N ... Ph

81%

Petasis, N.A.; Akritopoulou, I. *Tetrahedron Lett.*, ***1993***, *34*, 583

TsO ... N ... O

1. *sec*-BuLi, -78°C
2. MeI
3. H_2O
4. SmI_2, THF-HMPA

Me ... N ... O

90 x 42%

Ibáñez, P.L.; Nájera, C. *Tetrahedron Lett.*, ***1993***, *34*, 2003

t-Bu
t-Bu N Li ⊖
THF, -78°C
N
t-Bu N
(72 : 28) 41%

Wegman, S.; Würthwein, E.U. *Tetrahedron Lett.*, ***1993***, *34*, 307

N
Me
PhCHO
cat. $PdCl_2(PPh_3)_2$
K_2CO_3
N
Me
(*E*+*Z*)
Ph
N
Me
Ph
(*E*+*Z*)

in Toluene 10% 85%
in DMF 72% 2%

Filippini, L.; Gusmeroli, M.; Riva, R. *Tetrahedron Lett.*, ***1993***, *34*, 1643

O
N Bu
1. LDA, -78°C
THF → xylene
2. xylene, reflux, 4h
O
NHBu
O
NHBu
(99.5 : 0.5) 92%

Tsunoda, T.; Tatsuki, S.; Shiraishi, Y.; Akasaka, M.; Itô, S. *Tetrahedron Lett.*, ***1993***, *34*, 3297

OMe
N
Ph H
1. BF_3•OEt_2, THF
2. $CrCl_2$
3. allyl bromide, rt
4. 10% aq. Na_2CO_3
OMe
H
N
Ph
65%

Giammaruco, M.; Taddei, M.; Ulivi, P. *Tetrahedron Lett.*, ***1993***, *34*, 3635

N_3 ... CO_2Et — 1. PPh_3; 2. EtO_2C–CO–CO_2Et → EtO_2C–CH=CH–N=C(CO_2Et)(CO_2Et) 92%

Palacios, F.; Perez de Heredia, I.; Rubiales, G. *Tetrahedron Lett.*, ***1993***, *34*, 4377

(R) Ph, N, MeO_2C, H — methylenecyclohexane, $SnCl_4$, 20°C, excess $Me_2C{=}CH_2$ → HN, Ph, MeO_2C + HN, Ph, MeO_2C (30 : 70) 61%

Mikami, K.; Kaneko, M.; Yajima, T. *Tetrahedron Lett.*, ***1993***, *34*, 4841

Ph, N, Ph, SePh — Bu_3SnH, AIBN, 0.05M (syringe pump), 2h → NHPh, Ph 23% + Ph, Ph, N 45%

Schwan, A.L.; Refvik, M.D. *Tetrahedron Lett.*, ***1993***, *34*, 4901

Me_3Si, Bn, N, H — $(CH_2O)_n$, *p*-tsOH•H_2O, MeCN, 60°C → N, Bn 67-69% (64-67%ee)

Castro, P.; Overman, L.E.; Zhang, X.; Mariano, P.S. *Tetrahedron Lett.*, ***1993***, *34*, 5243

Me, *t*-Bu, CHO — *t*-$BuNH_2$, MS 3Å, 2h, CH_2Cl_2, 22°C → *t*-Bu, N, H, Me, *t*-Bu 94%

Sigman, M.S.; Eaton, B.E. *Tetrahedron Lett.*, ***1993***, *34*, 5367

Ph N

1. LDA, THF
2. $ZnCl_2 \cdot OEt_2$
3. PhCN

Ph H N Ph NH

78%

Barluenga, J.; del Pozo Losada, C.; Olano, B. *Tetrahedron Lett.*, ***1993***, *34*, 5497

Ph N-Ph

Bi/Bu_4NBr, MeCN

Br

NHPh Ph

95%

Bhuyan, P.J.; Prajapati, D.; Sandhu, J.S. *Tetrahedron Lett.*, ***1993***, *34*, 7975

N I N CH_2OEt

1. Br Ni 2, toluene, hν
2. 23°C, 12h
3. TBAF

N N CH_2OEt

91%

Knapp, S.; Albaneze, J.; Schugar, H.J. *J. Org. Chem.*, ***1993***, *58*, 997

Me N MeO

$BF_3 \cdot OEt_2$, Tol
reflux, 2d

NHMe MeO + NHMe MeO

(66 : 34) 99%

Beholz, L.G.; Stille, J.R. *J. Org. Chem.*, ***1993***, *58*, 5095

NHBn

PhI, $Pd(PPh_3)_4$, K_2CO_3
DMF, 70°C

N Bn Ph

71%

Davies, I.W.; Scopes, D.I.C.; Gallagher, T. *Synlett*, ***1993***, 85

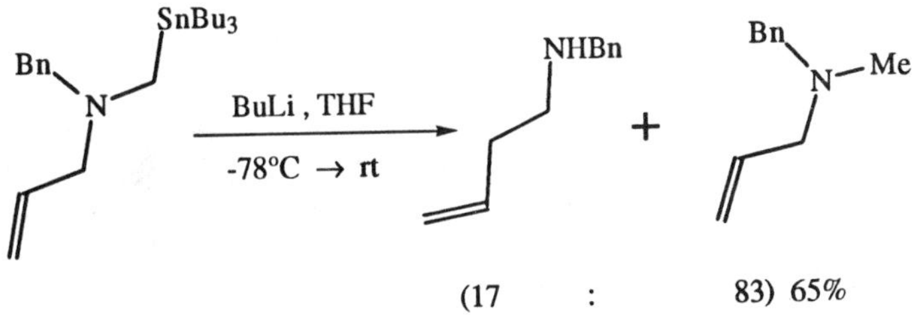
SnBu3
Bn
N
BuLi , THF
-78°C → rt
NHBn
+
Bn
N
Me
(17 : 83) 65%

Coldham, I. *J. Chem. Soc., Perkin Trans. 1.*, ***1993***, 1275

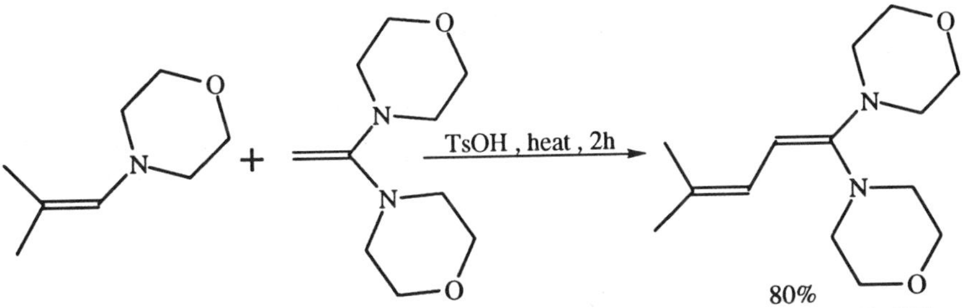
O
N
+
O
N
N
O
TsOH , heat , 2h
O
N
N
O
80%

Leurs, S.; Vanderbulcke-Coyette, B.; Viehe, H.G. *Bull. Soc. Chim. Belg.*, ***1993***, *102*, 645

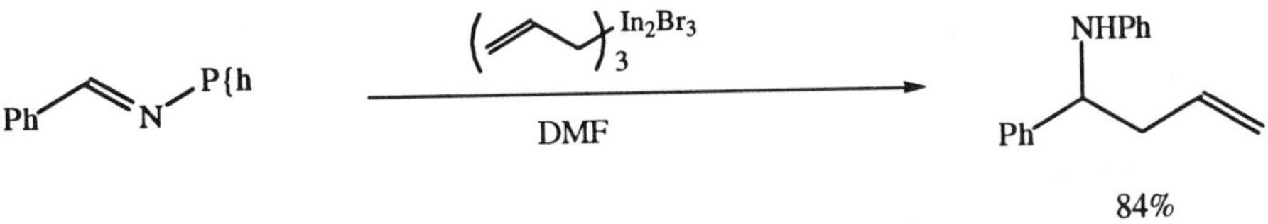
Ph
N
P{h
In2Br3
3
DMF
NHPh
Ph
84%

Jin, S.-J.; Araki, S.; Butsugan, Y. *Bull. Chem. Soc. Jpn.*, ***1993***, *66*, 1528

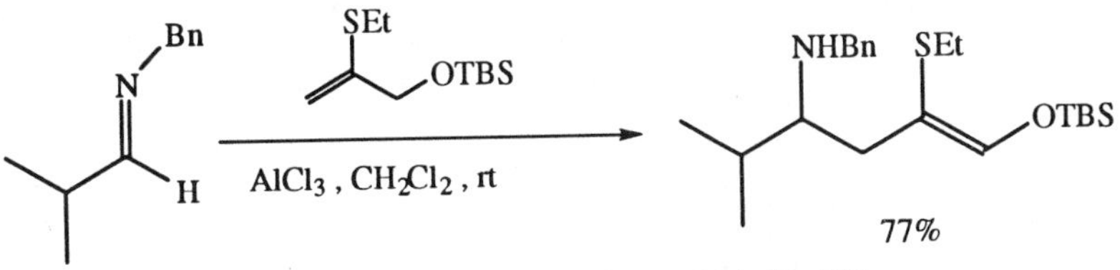
Bn
N
H
SEt
OTBS
AlCl3 , CH2Cl2 , rt
NHBn
SEt
OTBS
77%

Tohyama, Y.; Tanino, K.; Kuwajima, I. *J. Org. Chem.*, ***1994***, *59*, 518

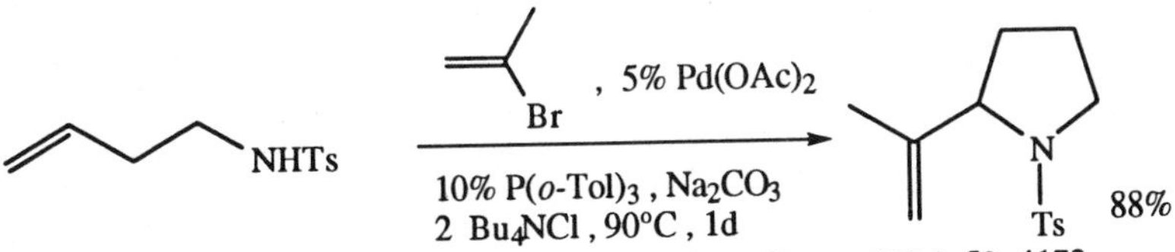
NHTs
Br
, 5% Pd(OAc)2
10% P(o-Tol)3 , Na2CO3
2 Bu4NCl , 90°C , 1d
N
Ts
88%

Larock, R.C.; Yang, H.; Weinreb, S.M.; Herr, R.J. *J. Org. Chem.*, ***1994***, *59*, 4172

Bu_3Sn

=N—NMe_2

2 eq. Bu_3SnH, AIBN, PhH

reflux

89% $NHNMe_2$

Bernard-Henriet, C.; Grimaldi, J.R.; Hatem, J.M. *Tetrahedron Lett.*, ***1994***, *35*, 3699

Bu_3Sn

N

1. 1.2 eq. BuLi

2 eq. Ph Ph

-78°C

2. MeI

Ph Ph + Ph Ph

N N

Me Me

(2.7 : 1) 65%

Pearson, W.H.; Jacobs, V.A. *Tetrahedron Lett.*, ***1994***, *35*, 7001

PhNHOH, $FeCl_2/FeCl_3$

NHPh 72%

Srivastava, R.S.; Nicholas, K.M. *Tetrahedron Lett.*, ***1994***, *35*, 8739

C_6H_{13}

N CO_2t-Bu

LDA

H

C_7H_{15} N CO_2t-Bu

98%

Ahman, J.; Somfai, P. *J. Am. Chem. Soc.*, ***1994***, *116*, 9781

Ph N

Br

1. $Ni(cod)_2$, MeCN

2. 16 $NaBH_4$, 6 MeOH

Ph N 45%

Solé, D.; Cancho, Y.; Llebaria, A.; Moretó, J.M.; Delgado, A. *J. Am. Chem. Soc.*, ***1994***, *116*, 12133

1% $(PCy_3)_2Cl_2Ru=CHCH=CPh_2$; PhH , eflux , 6h — 86%

Kinoshita, A.; Mori, M. *Synlett*, ***1994***, 1020

2 eq. morpholine , neat ; MS4Å , 30 min — 92%

Fisher, G.B.; Lee, L.; Klettke, F.W. *Synth. Commun.*, ***1994***, *24*, 1541

Zn(Cu) , MeOH , H_2O , rt ; ultrasound — 53%

De Kimpe, N.; Jolie, R.; De Smaele, D. *J. Chem. Soc. Chem. Commun.*, ***1994***, 1221

LDA , THF , -78°C — 63%

Coldham, I.; Collis, A.J.; Mould, R.J.; Rathmell, R.E. *Tetrahedron Lett.*, ***1995***, *36*, 3557

$HNEt_2$, 20% Et_3NHI , THF ; 10% $Pd(dba)_2$, 2 PPh_3 ; 65°C , 18h — 70% (96:4 *E:Z*) + 8%

Besson, L.; Goré, J.; Cazes, B. *Tetrahedron Lett.*, ***1995***, *36*, 3857

OTf

$NaCH(CO_2Me)_2$, 5% $Pd(PPh_3)_4$
LiCl , DMSO , 60°C

Bn N

CO_2Me
CO_2Me

53%

Bn N

Vicart, N.; Cazes, B.; Goré, J. *Tetrahedron Lett.*, ***1995***, *36*, 5015

C_3H_7 Bn ether , -50°C → -10°C , 4h
N
C_3H_7

Ph Ph
Ti
PrOi OiPr

Ph
Ph NHBn
C_3H_7 C_3H_7

82%

Gao, Y.; Harada, K.; Usato, F. *Tetrahedron Lett.*, ***1995***, *36*, 5913

Ph
N
Ph

Bu_3Sn

15% $Yb(OTf)_3$, CH_2Cl_2 , rt

Ph
HN
Ph

56%

Bellucci, C.; Cozzi, P.G.; Umani-Ronchi, A. *Tetrahedron Lett.*, ***1995***, *36*, 7289

Ph NPh

1. Bu_3Sn , Me_3SiCl
2. aq. NH_4F

NHPh
Ph

96%

Wang, D.-K.; Dai, L.-X.; Hou, X.-L. *Tetrahedron Lett.*, ***1995***, *36*, 8649

Ph_2P
MeO
C_6H_{13}

5 PhMgBr , 5% $NiCl_2(PPh_3)_2$
THF , 22°C

Ph_2P
Ph
C_6H_{13}

+

Ph_2P
Ph
C_6H_{13}

(8 : 1) 80%
(5:1 *cis:tranws*)

Didiuk, M.T.; Morken, J.P.; Hoveyda, A.H. *J. Am. Chem. Soc.*, ***1995***, *117*, 7273

$BF_3 \cdot OEt_2$, 210°C, 2h

47%

Anderson, W.K.; Lai, G. *Synthesis*, ***1995***, 1287

5 eq. 1-octene, DMA

5 eq. morpholine, 100°C

10 eq. H_2O, 2d

C_7H_{15} + C_6H_{13}

(85 : 15) 98%

Larock, R.C.; Tu, C. *Tetrahedron*, ***1995***, *51*, 6635

SECTION 357: ESTER - ESTER

Bu_3SnH, AIBN, Tol

reflux, 2h

72%

Kim, S.; Kee, I.S. *Tetrahedron Lett.*, ***1993***, *34*, 4213

1. Ph–CH₂CH₂–CHO, -78°C

$Ti(OiPr)Cl_3$, CH_2Cl_2

2. H_2O

3. *p*-TsOH, toluene

(92 : 8) 85%

Shimada, S.; Tohno, I.; Hashimoto, Y.; Saigo, K. *Chem. Lett.*, ***1993***, 1117

1/50 $CoCl_2$, 42h; DME, O_2, 80°C

(1 : 5) quant.

Li, P.; Alper, H. *Can. J. Chem.*, ***1993***, *71*, 84

acetone, SmI_2, THF; 4% $Fe(dbm)_3$, reflux

CO_2Me CO_2Me CO_2Me

74%

Imamoto, T.; Hatajima, T.; Yoshizawa, T. *Tetrahedron Lett.*, ***1994***, *35*, 7805

MeO OTMS OMe

$Pd(PPh_3)_4$, TlOAc, HF; reflux, 1d

OMe CO_2Me

89%

Sakamoto, T.; Kondo, Y.; Masumoto, K.; Yamanaka, H. *J. Chem. Soc., Perkin Trans. 1.*, ***1994***, 235

PhCHO, $Cp_2Sn(thf)_2$; Tol, 0°C → 25°C

77%

Takano, M.; Kikuchi, S.; Morita, K.; Nishiyama, Y.; Ishii, Y. *J. Org. Chem.*, ***1995***, *60*, 4974

$NaBO_3{\cdot}4\,H_2O$, Ac_2O, PhH; Na_2CO_3, 55°C, 14h

67%

Bhat, S.; Ramesha, A.R.; Chandrasekaran, S. *Synlett*, ***1995***, 329

Also via Dicarboxylic Acids:	Section 312 (Carboxylic Acids - Carboxylic Acids)
Hydroxy-esters	Section 327 (Alcohol - Ester)
Diols	Section 323 (Alcohol - Alcohol)

SECTION 358: ESTER - ETHER, EPOXIDE, THIOETHER

Bu_3SnH, AIBN, PhH
reflux, 6h, syringe pump

CO_2Et Br CO_2Et 95%

Lee, E.; Tae, J.S.; Lee, C.; Park, C.M. *Tetrahedron Lett.*, ***1993**, 34*, 4831

$SiMe_3$ Ph OH

1. CsF, MeCN
2. MeO_2C CO_2Me

MeO_2C CO_2Me Ph 92%

Hojo, M.; Ohkuma, M.; Ishibashi, N.; Hosomi, A. *Tetrahedron Lett.*, ***1993**, 34*, 5943

SEt OTMS

$(TfO)_2Sn$(bis-amine)
CH_2Cl_2, 9h, -78°C
PhCHO

O OTMS EtS Ph

86% (93:7 *syn:anti*)

Kobayashi, S.; Uchiro, H.; Shiina, I.; Mukaiyama, T. *Tetrahedron*, ***1993**, 49*, 1761

CO_2Et CHO

1.3 Bu_3SnH
0.25 AIBN
0.03M PhH
reflux, 8h

CO_2Et OH + O

(46 : 54) 91%

Lee, E.; Tae, J.S.; Chong, Y.H.; Park, Y.C.; Yun, M.; Kim, S. *Tetrahedron Lett.*, ***1994**, 35*, 129

CHO

OTMS SEt, 40% SnO
20% $Sn(OTf)_2$
chiral diamine
slow addition (3h)

OTMS O SEt

61% (62% ee)

Kobayashi, S.; Kawasuji, T.; Mori, N. *Chem. Lett.*, ***1994***, 217

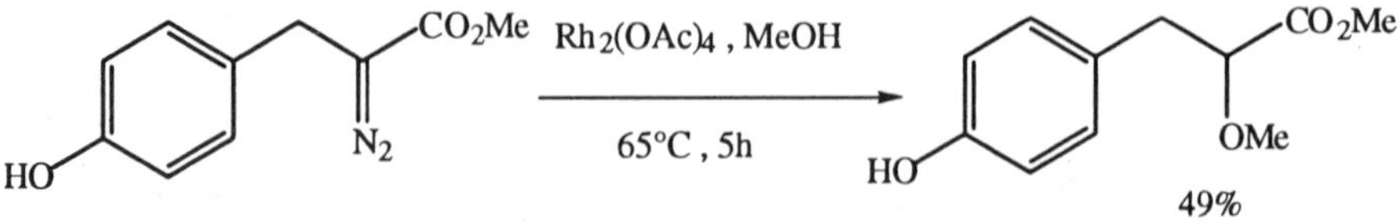
CO2Me
N2
HO
Rh2(OAc)4 , MeOH
65°C , 5h
CO2Me
OMe
HO
49%

Cox, G.G.; Haigh, D.; Hindley, R.M.; Miller, D.J.; Moody, C.J. *Tetrahedron Lett.*, ***1994***, *35*, 3139

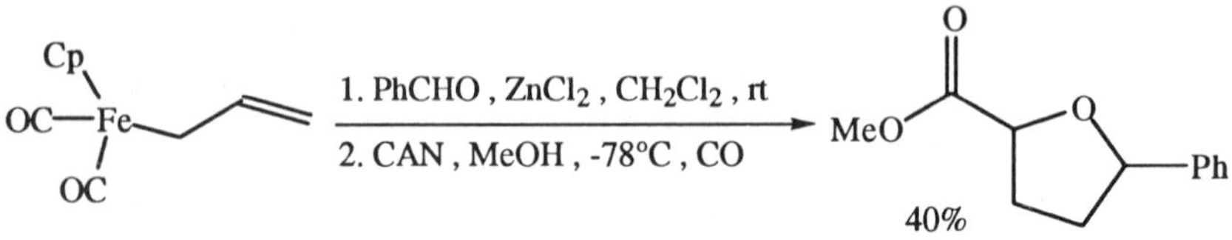
Cp
OC—Fe
OC
1. PhCHO , ZnCl2 , CH2Cl2 , rt
2. CAN , MeOH , -78°C , CO
O
MeO
O
Ph
40%

Jiang, S.; Turos, E. *Tetrahedron Lett.*, ***1994***, *35*, 7889

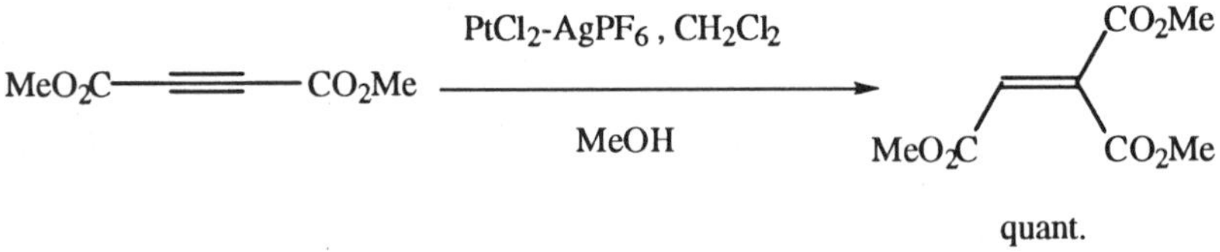
MeO2C
CO2Me
PtCl2-AgPF6 , CH2Cl2
MeOH
CO2Me
MeO2C
CO2Me
quant.

Kataoka, Y.; Matsumoto, O.; Ohashi, M.; Yamagata, T.; Tani, K. *Chem. Lett.*, ***1994***, 1283

SECTION 359: ESTER - HALIDE, SULFONATE

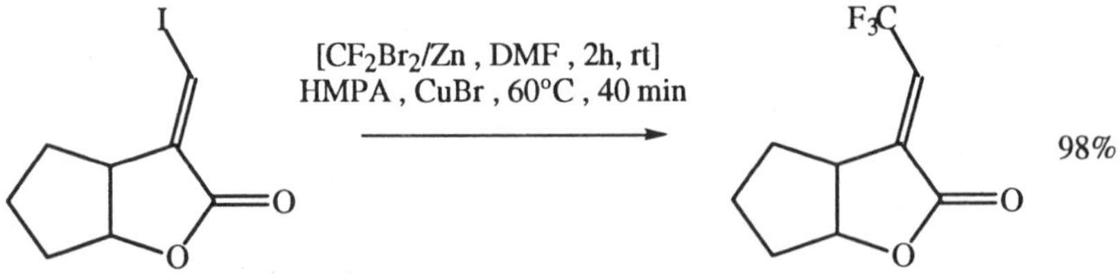
I
O
O
[CF2Br2/Zn , DMF , 2h, rt]
HMPA , CuBr , 60°C , 40 min
F3C
O
O
98%

Mawson, S.D.; Weavers, R.T. *Tetrahedron Lett.*, ***1993***, *34*, 3139

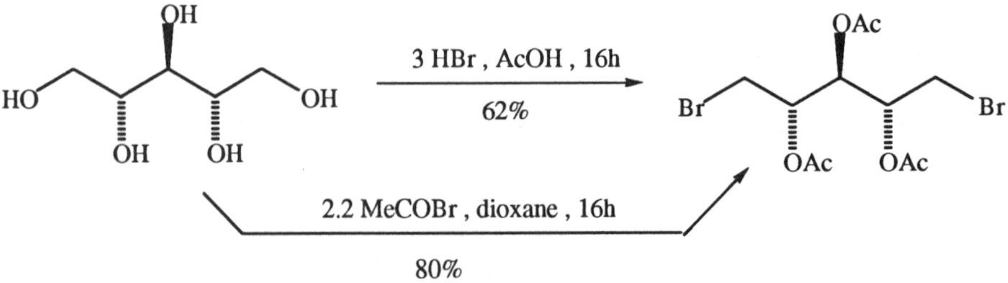
OH
HO
OH
OH
OH
3 HBr , AcOH , 16h
62%
OAc
Br
Br
OAc
OAc
2.2 MeCOBr , dioxane , 16h
80%

El Anzi, A.; Benazza, M.; Fréchou, C.; Demailly, G. *Tetrahedron Lett.*, ***1993***, *34*, 3741

CH_2Cl_2 , 18h

bis-(sym-collidine) iodine(I) hexaflurophosphate

76%

Simonot, B.; Rousseau, G. *J. Org. Chem.*, ***1993***, *58*, 4

1. $Ti(OTf)_4$, CH_2Cl_2
2. I_2 , CH_2Cl_2

89%

Kitagawa, O.; Inoue, T.; Hirano, K.; Takuchi, T. *J. Org. Chem.*, ***1993***, *58*, 3106

KI , sodium persulfate , $NaHCO_3$

H_2O

96%

Royer, A.C.; Mebane, R.C.; Swafford, A.M. *Synlett*, ***1993***, 899

BzCl , 0.4 SmI_3 , MeCN , rt

87%

Yu, Y.; Zhang, Y.; Ling, R. *Synth. Commun.*, ***1993***, *23*, 1973

1. $HC{\equiv}CCO_2H$, NCI , CH_2Cl_2
2. $(BzO)_2$, PhH , eflux

78x44%

Haaima, G.; Lunch, M.-J.; Routledge, A.; Weavers, R.T. *Tetrahedron*, ***1993***, *49*, 4229

ICH_2CO_2Et , DMSO
BEt_3 , rt (open to air)

14 eq.

54%

Bachiocchi, E.; Muraglia, E. *Tetrahedron Lett.*, ***1994***, *35*, 2763

Bn_2N … CO_2H — 1. mcpba; 2. I_2, $NaHCO_3$ → lactone ($ONBn_2$, I) 50%

Reetz, M.T.; Lauterbach, E.H. *Heterocycles*, ***1993***, *35*, 627

Me … — CF_3I, aq. MeCN; $Na_2S_2O_4$-$NaHCO_3$ → Me, I, CF_3 80% (95:5 *E:Z*)

Lu, X.; Wang, Z.; Ji, J. *Tetrahedron Lett.*, ***1994***, *35*, 613

AcCl, 5% $ZnCl_2$; rt, 1h → Cl … OAc + AcO … Cl (3 : 97) 95%

Mimero, P.; Saluzzo, C.; Amouroux, R. *Tetrahedron Lett.*, ***1994***, *35*, 1553

1. Cl-N, N-Cl, O, O, O, N, Cl; DMF; 2. LiOMe, MeOH → Cl, Cl, CO_2Me 96%

Bellesia, F.; Boni, M.; Ghelfi, F.; Pagnoni, U.M. *Tetrahedron Lett.*, ***1994***, *35*, 2961

CF_2Br, MeO_2C, MeO_2C — Bu_3SnH, AIBN, PhH; reflux, 1-15 h addition → F, F, MeO_2C, MeO_2C 51%

Buttle, L.A.; Motherwell, W.B. *Tetrahedron Lett.*, ***1994***, *35*, 3995

BuO–CH₂–thiirane → (PhCOCl; TBAB, 90°C, 6h) → BuO–CH₂–CH(SCOPh)–CH₂Cl

97%

Kameyama, A.; Kiyota, M.; Nishikubo, T. *Tetrahedron Lett.*, ***1994***, *35*, 4571

Ph-epoxide → (AcCl, Cp_2YCl; CH_2Cl_2, 13h) → AcO–CH₂–CH(Cl)–Ph + AcO–CH(Ph)–CH₂Cl

(88 : 12) 95%

Qian, C.; Zhu, D. *Synth. Commun.*, ***1994***, *24*, 2203

cyclohexene → (I_2, $Pb(OAc)_2$•3 H_2O, AcOH; rt, 6h) → trans-2-iodocyclohexyl acetate (I, OAc)

85%

Bedekar, A.V.; Nair, K.B.; Soman, R. *Synth. Commun.*, ***1994***, *24*, 2299

Br-substituted α-methylene lactone → ($PhMe_3N^+$ Br_3^-) → Br-substituted lactone with CH₂Br

78%

Collado, I.G.; Galán, R.H.; Massanet, G.M.; Alonso, M.S. *Tetrahedron*, ***1994***, *50*, 6433

hex-5-enyl dichloroacetate (O, Cl, Cl) → (0.3 Cu(bpy)Cl, DCE; reflux, 18h) → dichloro lactone (O, O, Cl, Cl)

57% (70:30)

Pirrung, F.O.H.; Hiemstra, H.; Speckamp, W.N.; Kaptein, B.; Schoemaker, H.E. *Tetrahedron*, ***1994***, *50*, 12415

O Ph Cl , Zn , ether 25°C , 18h O Ph O Cl

70%

Bhar, S.; Ranu, B.C. *J. Org. Chem.*, ***1995***, *60*, 745

SECTION 360: ESTER - KETONE

OH $C_{11}H_{23}$ CO_2Me

1. Br O Br 55%

Py , ether

2. TMSCl , ether 53%

Zn , reflux

O O $C_{11}H_{23}$ O

Landi Jr., J.J.; Garafalo, L.M.; Ramig, K. *Tetrahedron Lett.*, ***1993***, *34*, 277

N_2 CO_2Me O Ph

2% $Rh_2[O_2CCH(Bn)NPhth]_4$

CH_2Cl_2 , 0°C

O CO_2Me Ph 96% (46%ee , R)

Hashimoto, S.; Watanabe, N.; Sato, T.; Shiro, M.; Ikegami, S. *Tetrahedron Lett.*, ***1993***, *34*, 5109

NNHPh CO_2Et

1.7 PhI(OH)OTs , $CDCl_3$

rt , 2h

O CO_2Et 98%

Barton, D.H.R.; Jaszberenyi, J.Cs.; Shinada, T. *Tetrahedron Lett.*, ***1993***, *34*, 7191

O CO_2Et

PhCHO , $EtNO_2$

$AlCl_3$, 100°C , 5h

O CO_2Et Ph 88%

Sartori, G.; Bigi, E.; Maggi, R.; Bernardi, G.L. *Tetrahedron Lett.*, ***1993***, *34*, 7339

OEt

Ph_3P

CO_2Et

1. PhC(O)CH$_2$Br , Cs_2CO_3 , CH_2Cl_2

48h , 30°C

2. aq. HCl , $CHCl_3$

O

CO_2Et

Ph

92x99%

Hatanaka, M.; Himeda, Y.; Imashiro,; Tanaka, Y.; Ueda, I. *J. Org. Chem.*, ***1994***, *59*, 111

O

1. $(COCl)_2$, hν

2. MeOH

O

60%

CO_2Me

Bashir-Hashemi, A.; Hardee, J.R.; Gelber, N.; Qi, L.; Axenrod, T. *J. Org. Chem.*, ***1994***, *59*, 2131

O

MeO

CO_2Me

KH , THF , 18-c-6 , rt

O

MeO

MeO_2C

60%

Habi, A.; Gravel, D. *Tetrahedron Lett.*, ***1994***, *35*, 4315

OAc

C_7H_{15}

$SnBu_3$

PhC(O)Cl , 4% $Pd(PPh_3)_2Cl_2$

8% CuCN , 75°C , 18h

OAc

C_7H_{15}

Ph

74% O

Ye, J.; Bhatt, R.K.; Falck, J.R. *J. Am. Chem. Soc.*, ***1994***, *116*, 1

Ph

N

O

O

CO_2t-Bu

⊖ ⊖

O O

Ph

CO_2t-Bu

96%

Lygo, B. *Tetrahedron Lett.*, ***1994***, *35*, 5073

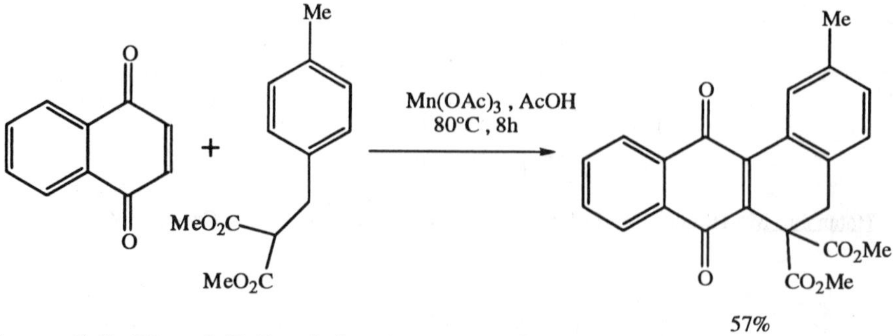

Me

O

+

$Mn(OAc)_3$, AcOH
80°C , 8h

MeO_2C

O

MeO_2C

Me

O

O

CO_2Me

CO_2Me

57%

Chuang, C.-P.; Wang, S.-F. *Tetrahedron Lett.*, ***1994***, *35*, 4365

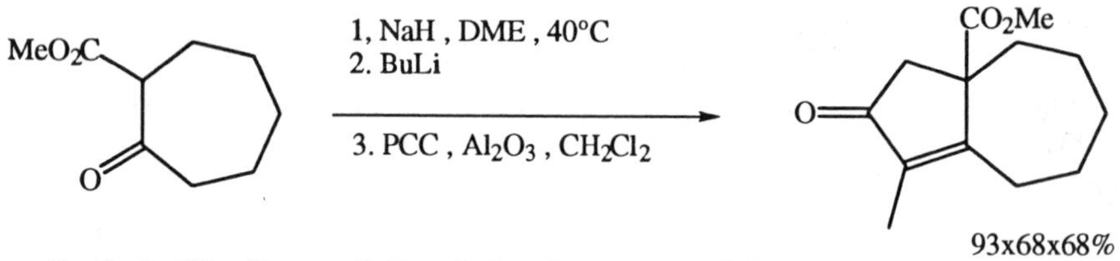

MeO_2C

O

1, NaH , DME , 40°C
2. BuLi

3. PCC , Al_2O_3 , CH_2Cl_2

CO_2Me

O

93x68x68%

Piers, E.; Cook, K.L.; Rogers, C. *Tetrahedron Lett.*, ***1994***, *35*, 8573

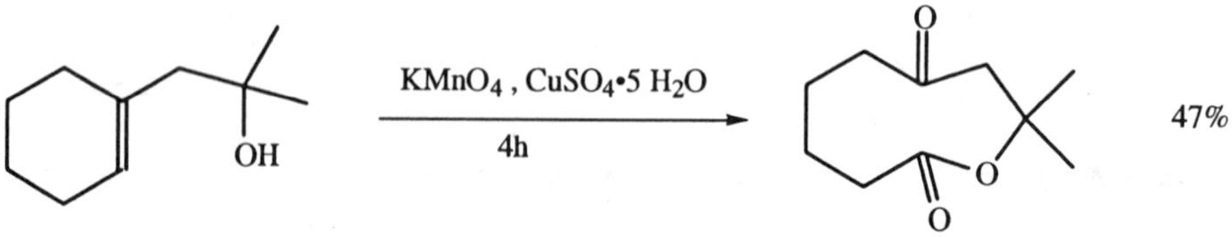

OH

$KMnO_4$, $CuSO_4$•5 H_2O

4h

O

O

O

47%

Das, J.; Chandrasekaran, S. *Tetrahedron*, ***1994***, *50*, 11709

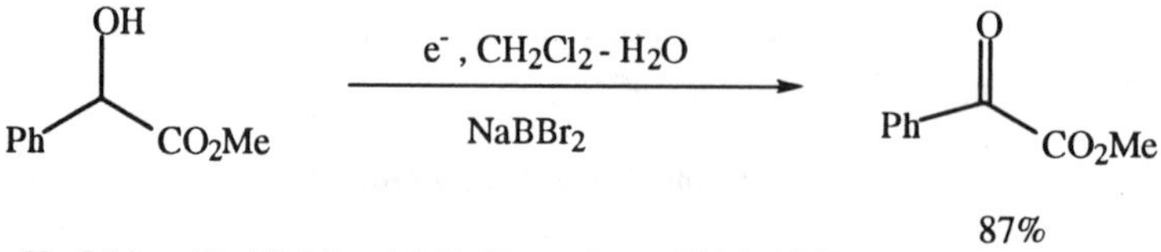

OH

Ph CO_2Me

e^- , CH_2Cl_2 - H_2O

$NaBBr_2$

O

Ph CO_2Me

87%

Maekawa, H.; Ishino, Y.; Nishiguchi, I. *Chem. Lett.*, ***1994***, 1017

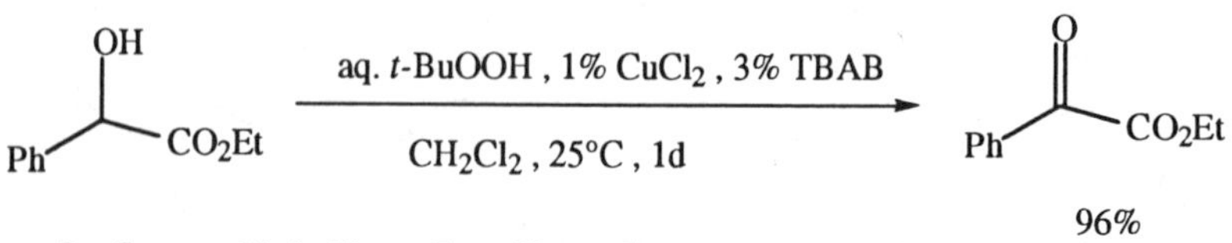

OH

Ph CO_2Et

aq. *t*-BuOOH , 1% $CuCl_2$, 3% TBAB

CH_2Cl_2 , 25°C , 1d

O

Ph CO_2Et

96%

Feldberg, L.; Sasson, Y. *J. Chem. Soc. Chem. Commun.*, ***1994***, 1807

$(Me_3Si)_3SiH$, AIBN

PhH , 80°C , 0.01M

68%

Ryu, I.; Nagahara, K.; Yamazaki, H.; Tsunoi, S.; Sonoda, N. *Synlett*, ***1994***, 643

Bu_3SnH , AIBN
PhH , 80°C

CO_2Me + CO_2Me OH O

(9 : 1) 75%

Enholm, E.J.; Kinter, K.S. *J. Org. Chem.*, ***1995***, *60*, 4850

Ph OCO_2Me e^- , DMF/$LiClO_4$ Ph O O H

67%

Barba, F.; Quintanilla, M.G.; Montero, G. *J. Org. Chem.*, ***1995***, *60*, 5658

OH OEt $KMnO_4$, pH 7 OH O OEt O

97%

Tatlock, J.H. *J. Org. Chem.*, ***1995***, *60*, 6221

O CO_2t-Bu

1. $CuBr_2$, PhI(OH)OTs , 0°C
2. DMAP , dimethyldioxirane

CH_2Cl_2 , 0°C

O CO_2t-Bu O

1st step is quant

Coats, S.J.; Wasserman, H.H. *Tetrahedron Lett.*, ***1995***, *36*, 7735

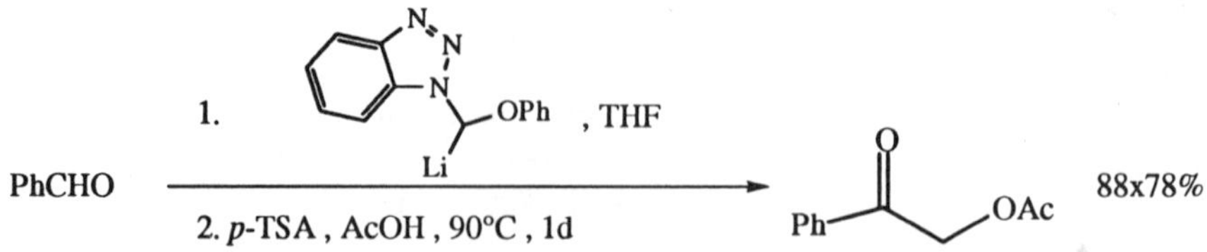

1. (1-(lithio(phenoxy)methyl)benzotriazole), THF

PhCHO → PhC(O)CH₂OAc 88x78%

2. p-TSA , AcOH , 90°C , 1d

Katritzky, A.R.; Yang, Z.; Moutou, J.-L. *Tetrahedron Lett.*, ***1995***, *36*, 841

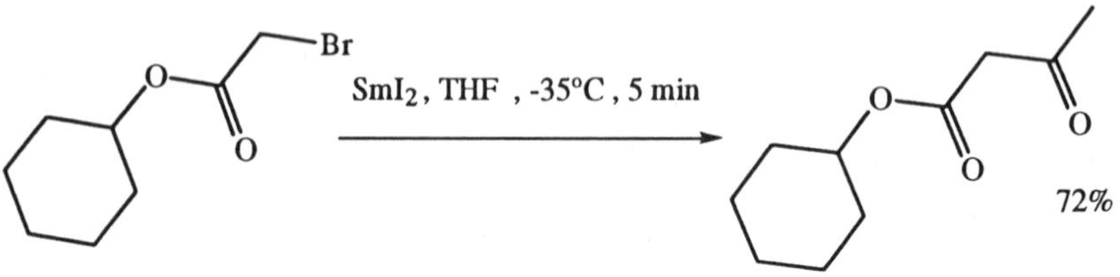

Br

SmI_2, THF , -35°C , 5 min

O O O O

72%

Park, H.S.; Lee, I.S.; Kim, Y.H. *Tetrahedron Lett.*, ***1995***, *36*, 1673

REVIEW:

"Alternate Preparations Of α-Keto Esters From Acid Chlorides," Katritzky, A.R.; Wang, Z.; Wells, A.P. *Org. Prep. Proceed. Int.*, ***1995***, *27*, 457

Also via Ketoacids	Section 320 (Carboxylic Acid - Ketone)
Hydroxyketones	Section 330 (Alcohol - Ketone)

SECTION 361: ESTER - NITRILE

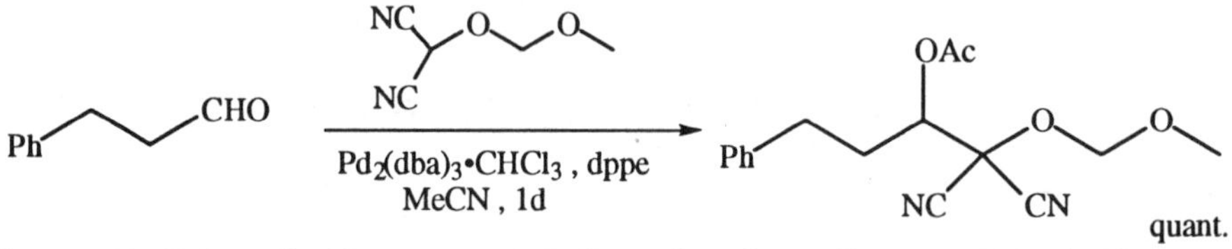

NC NC O O

Ph CHO

$Pd_2(dba)_3{\bullet}CHCl_3$, dppe

MeCN , 1d

OAc Ph O O NC CN

quant.

Nemoto, H.; Kubota, Y.; Yamamoto, Y. *J. Chem. Soc. Chem. Commun.*, ***1994***, 1665

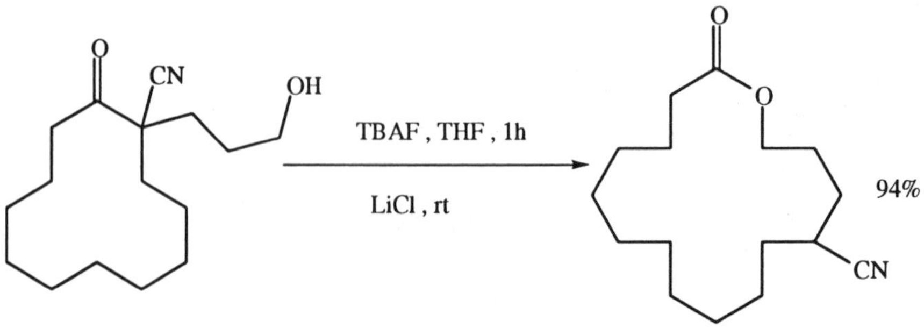

O CN OH

TBAF , THF , 1h

LiCl , rt

O O CN

94%

Stojanova, D.S.; Milenkov, B.; Hesse, M. *Helv. Chim. Acta*, ***1993***, *76*, 2303

SECTION 362: ESTER - ALKENE

This section contains syntheses of enol esters and esters of unsaturated acids as well as ester molecules bearing a remote alkenyl unit.

CO$_2$H, Bu — 1. Br$_2$, NaHCO$_3$; 2. AgNO$_3$, AcOH → O, O, Bu — 66 x 50%

Black, T.H.; Huang, J. *Tetrahedron Lett.*, ***1993***, *34*, 1411

CO$_2$H — 1. K$_2$CO$_3$, 5 kbar; 2. 5% Pd(OAc)$_2$, 10% P(2-furyl)$_3$; 3. HC≡CCMe$_2$OAc, DMSO, 20°C, 14h → O, O — 61%

Bouyssi, D.; Gore, J.; Balme, G.; Louis, D.; Wallach, J. *Tetrahedron Lett.*, ***1993***, *34*, 3129

MeO, O, Me, OTf — CO, K$_2$CO$_3$; Pd(OAc)$_2$(dppp) → Me, MeO, O, O — 90%

Ciattini, P.G.; Mastropietro, G.; Morera, E.; Ortar, G. *Tetrahedron Lett.*, ***1993***, *34*, 3763

Br, CO$_2$H — Ph-C≡CH, NEt$_3$, MeCN, rt; PdCl$_2$(PPh$_3$)$_2$, CuI, 10h → O, O, Ph — 80% (>97:3 *Z:E*)

Lu, X.; Huang, X.; Ma, S. *Tetrahedron Lett.*, ***1993***, *34*, 5963

OSiMe$_3$, Me$_3$SiO — Me$_3$SiCH=C=O, PhH; reflux, 3d → O, O, Me$_3$Si — 51%

Ito, T.; Aoyama, T.; Shioiri, T. *Tetrahedron Lett.*, ***1993***, *34*, 6583

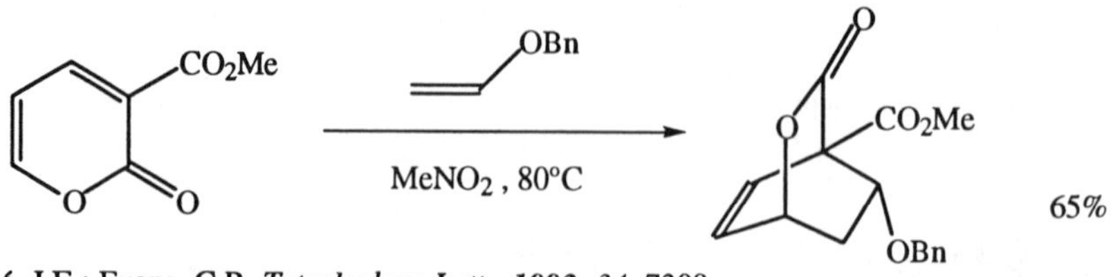

CO_2Me OBn $MeNO_2$, 80°C O O O CO_2Me OBn 65%

Markó, I.E.; Evans, G.R. *Tetrahedron Lett.*, ***1993***, *34*, 7309

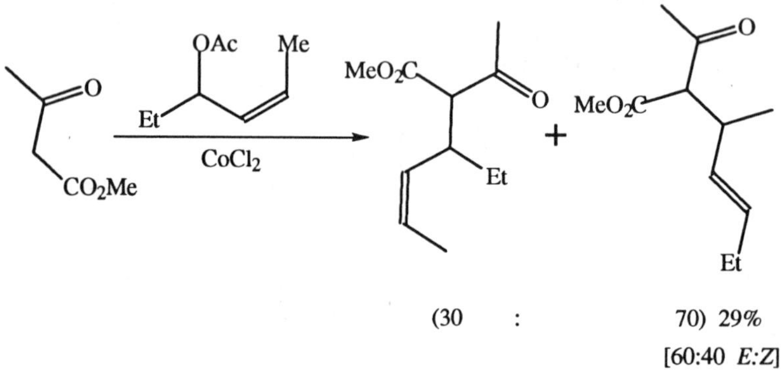

OAc Me O Et CO_2Me $CoCl_2$ MeO_2C O Et + MeO_2C O Et

(30 : 70) 29%

[60:40 *E:Z*]

Bhatia, B.; Reddy, M.M.; Iqbal, J. *Tetrahedron Lett.*, ***1993***, *34*, 6301

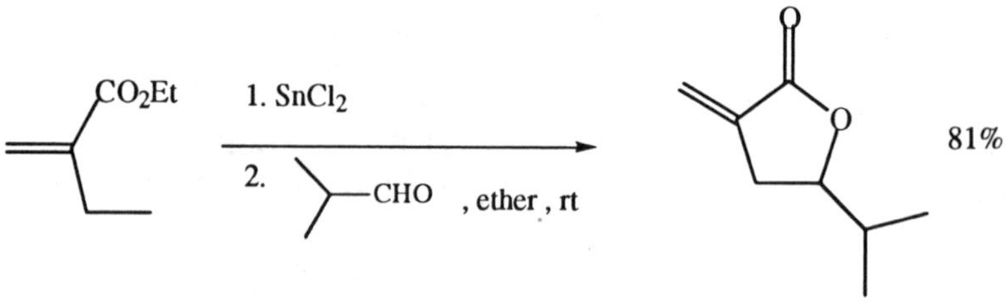

CO_2Et 1. $SnCl_2$ 2. CHO , ether , rt O O 81%

Fouquet, E.; Gabriel, A.; Maillard, B.; Pereyre, M. *Tetrahedron Lett.*, ***1993***, *34*, 7749

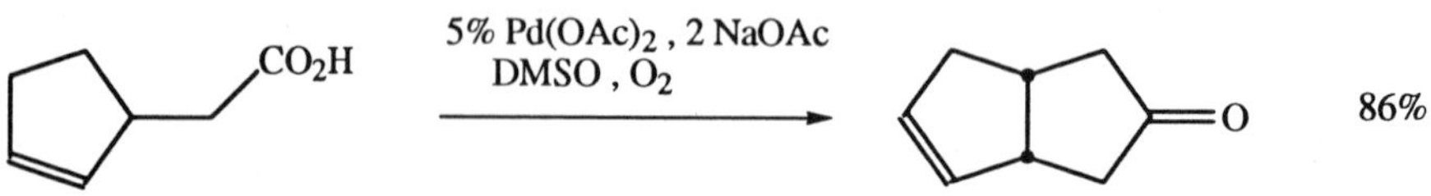

CO_2H 5% $Pd(OAc)_2$, 2 NaOAc DMSO , O_2 O 86%

Larock, R.C.; Hightower, T.R. *J. Org. Chem.*, ***1993***, *58*, 5298

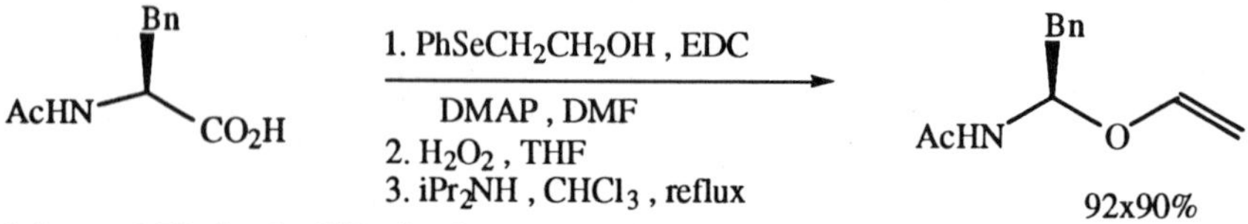

Bn AcHN CO_2H 1. $PhSeCH_2CH_2OH$, EDC DMAP , DMF 2. H_2O_2 , THF 3. iPr_2NH , $CHCl_3$, reflux Bn AcHN O 92x90%

Weinhouse, M.I.; Janda, K.D. *Synthesis*, ***1993***, 81

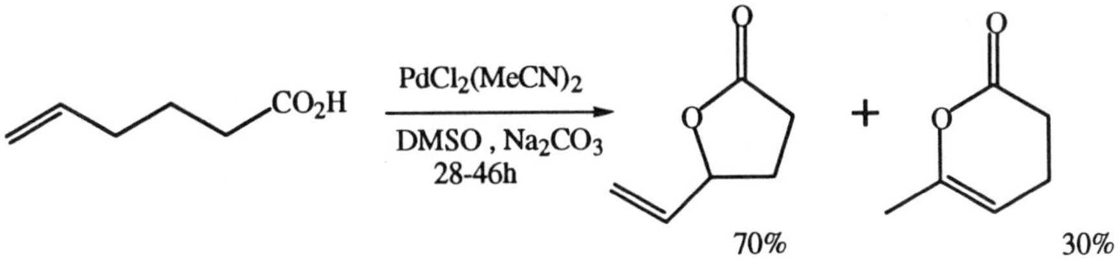

CO_2H → $PdCl_2(MeCN)_2$, DMSO, Na_2CO_3, 28-46h → 70% + 30%

Annby, U.; Stenkula, M.; Andersson, C.-M. *Tetrahedron Lett.*, ***1993***, *34*, 8545

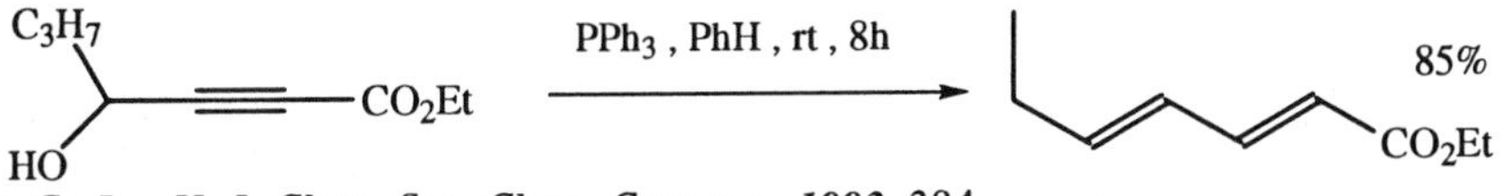

C_3H_7, HO, CO_2Et → PPh_3, PhH, rt, 8h → CO_2Et 85%

Guo, C.; Lu, X. *J. Chem. Soc. Chem. Commun.*, ***1993***, 394

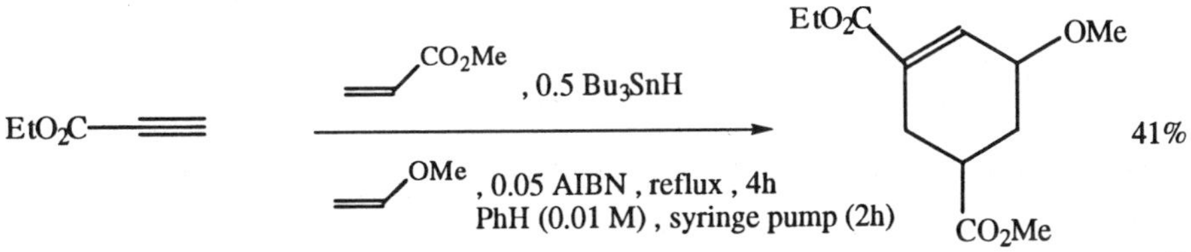

EtO_2C → CO_2Me, 0.5 Bu_3SnH; OMe, 0.05 AIBN, reflux, 4h; PhH (0.01 M), syringe pump (2h) → EtO_2C, OMe, CO_2Me 41%

Lee, E.; Hur, C.U.; Rhee, Y.H.; Park, Y.C.; Kim, S.Y. *J. Chem. Soc. Chem. Commun.*, ***1993***, 1466

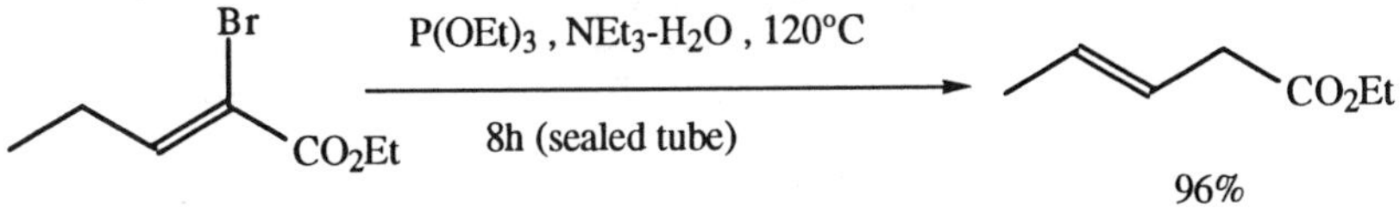

Br, CO_2Et → $P(OEt)_3$, NEt_3-H_2O, 120°C, 8h (sealed tube) → CO_2Et 96%

Hirao, T.; Hirano, K.; Ohshiro, Y. *Bull. Chem. Soc. Jpn.*, ***1993***, *66*, 2781

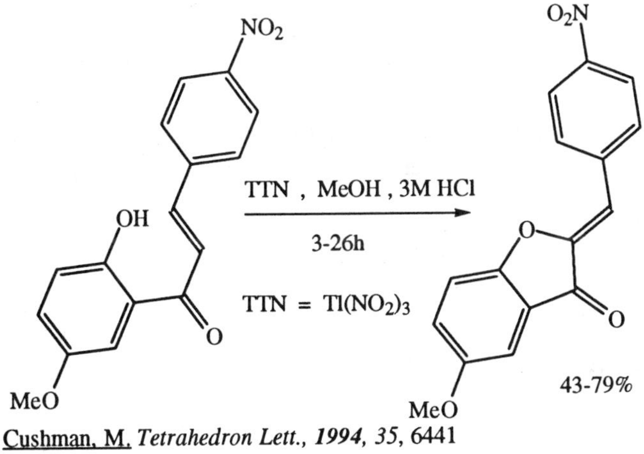

NO_2, OH, O, MeO → TTN, MeOH, 3M HCl, 3-26h; TTN = $Tl(NO_2)_3$ → O_2N, O, O, MeO 43-79%

Thakkar, K.; Cushman, M. *Tetrahedron Lett.*, ***1994***, *35*, 6441

OCO_2Me ... CO_2Me → $Pd(OAc)_2$, dppp , PhMe , MeOH / reflux → CO_2Me, CO_2Me 88%

Mandai, T.; Tsujiguchi, Y.; Tsuji, J.; Saito, S. *Tetrahedron Lett.*, ***1994***, *35*, 5701

Bu, HO → $[Pd(MeCN)_2(PPh_3)_2]\ [BF_4]_2$ / CO , THF , 50°C (10 Kg cm^{-2}) → Bu, O, O + Bu, CO_2H

(17 : 83) 92%

Matsushita, K.; Komori, T.; Oi, S.; Inoue, Y. *Tetrahedron Lett.*, ***1994***, *35*, 5889

O → 1. NaH , TMEDA , THF , reflux 2. inverse addition to PhOCOCl → O, O, OPh 43%

Harwood, L.M.; Houminer, Y.; Manage, A.; Seeman, J.I. *Tetrahedron Lett.*, ***1994***, *35*, 8027

PhCHO → CO_2Me / DABCO , microwave , 10 min → Ph, OH, CO_2Me 34%

Kundu, M.K.; Mukherjee, S.B.; Balu, N.; Padmakumar, R.; Bhat, S.V. *Synlett*, ***1994***, 444

PhCHO → $BrCH_2CO_2Me$, Bu_3As , 10% Cd / 100°C , 20h → Ph, CO_2Me 87% (*E* only)

Zheng, J.; Shen, Y. *Synth. Commun.*, ***1994***, *24*, 2069

N_2 CO_2Me ; Rh_2(S-TBSP)$_4$; MeO_2C

TBSP = *p*-(*t*-butyl)phenylsulfonyl prolinate 76% (63% ee)

Davies, H.M.L.; Peng, Z.-Q.; Houser, J.H. *Tetrahedron Lett.*, ***1994***, *35*, 8939

EtO_2C EtO_2C Br ; O , 3.5% $Pd(OAc)_2$; 8% PPh_3 , 1.2 Ag_2CO_3 , MeCN 90°C , 45 min ; EtO_2C EtO_2C O

80%

Meyer, F.E.; Ang, K.H.; Stenig, A.G.; de Meijere, A. *Synlett*, ***1994***, 191

Ph Ph OH ; OAc , CO , THF , autoclave ; $RuCl_2(PPh_3)_3$-K_2CO_3 , 200°C , 15h ; Ph Ph O O

77%

Kondo, T.; Kodoi, K.; Mitsudo, T.-u.; Watanabe, Y. *J. Chem. Soc. Chem. Commun.*, ***1994***, 755

1. Rieke Mg* , THF , rt
2. ethylene oxide , THF , -78°C
3. CO_2 , 0°C → rt
4. H_3O^+ , 110C

O O 69%

Rieke, R.D.; Sell, M.S.; Xiong, H. *J. Org. Chem.*, ***1995***, *60*, 5143

MeO_2C $SnBu_3$ SPh ; $Cu(NO_3)_2$, THF , rt ; SPh CO_2Me CO_2Me SPh 65%

Beddoes, R.L.; Cheeseright, T.; Wang, J.; Quayle, P. *Tetrahedron Lett.*, ***1995***, *36*, 283

PhCHO → ($Ph_3As=C(CO_2Me)SePh$, $CHCl_3$, 4h) → $PhHC=C(CO_2Me)SePh$ 96% (90:10, *Z*:*E*)

Huang, Z.-Z.; Huang, X.; Huang, Y.-Z. *Tetrahedron Lett.*, ***1995***, *36*, 425

C_3H_7 C_3H_7 Ph O I → (5% $Cl_2Pd(PPh_3)_2$. 4 NEt_3, 20 atm CO, PhH, 110°C, 10h) → C_3H_7 C_3H_7 Ph O O 99%

Copéret, C.; Sugihara, T.; Wu, G.; Shimoyama, I.; Negishi, E. *J. Am. Chem. Soc.*, ***1995***, *117*, 3422

Cl O → (1. NEt_3, CH_2Cl_2, -20°C; 2. BnOH) → O OBn 75%

Cardillo, G.; De Simone, A.; Mingardi, A.; Tomasini, C. *Synlett*, ***1995***, 1131

O O C_5H_{11} → ($CuCl_2$, LiCl, MeCN, $PdCl_2(PhCN)_2$) → Cl Cl O O C_5H_{11} + Cl Cl O O C_5H_{11} (>97 : 3) 95%

Zhu, G.; Lu, X. *Tetrahedron Asymmetry*, ***1995***, *6*, 345

PhO Cl O → (allyl $SiMe_3$, toluene, $AlCl_3$, 20°C, 2h) → PhO O 82%

Mayr, H.; Gabriel, A.O.; Schumacher, R. *Liebigs Ann. Chem.*, ***1995***, 1583

$BuHC{=}C(OSiMe_3)(OPr)$ —[$MeO_2C{-}{\equiv}{-}CO_2Me$; $ZrCl_4$, CCl_4, 0°C → rt]→ product (CO_2Me, PrO_2C, =CHBu, MeO_2C) 80%

Mitani, M.; Sudoh, T.; Koyama, K. *Bull. Chem. Soc. Jpn.*, ***1995***, *68*, 1683

Related Methods:	Section 60A (Protection of Aldehydes).
	Section 180A (Protection of Ketones).
Also via Acetylenic Esters:	Section 306 (Alkyne - Ester).
Alkenyl Acids:	Section 322 (Carboxylic Acid - Alkene).
β-Hydroxy-esters:	Section 327 (Alcohol - Ester).

SECTION 363: ETHER, EPOXIDE, THIOETHER - ETHER, EPOXIDE, THIOETHER

See Section 60A (Protection of Aldehydes) and Section 180A (Protection of Ketones) for reactions involving formation of Acetals and Ketals.

1. DDQ, MeOH, CH_2Cl_2, rt, 1d; 2. aq. $NaHCO_3$ — OMe — 78%

Xu, Y.-C.; Lebeau, E.; Gillard, J.W.; Attardo, G. *Tetrahedron Lett.*, ***1993***, *34*, 3841

OH — iPrOH, 20% $CuCl_2$, 5% $Pd(NO)_2Cl(MeCN)_2$, 2h — O, O — 50% isolated

Meulemans, T.M.; Kiers, N.H.; Feringa, B.L.; van Leeuwen, P.W.N.M. *Tetrahedron Lett.*, ***1994***, *35*, 455

O, HO, OH — HO–C6H4–OMe; DEAD, PPh_3, CH_2Cl_2, 0°C — O, O, HO, OMe — 80% (α-only)

Sobti, A.; Sulikowski, G.A. *Tetrahedron Lett.*, ***1994***, *35*, 3661

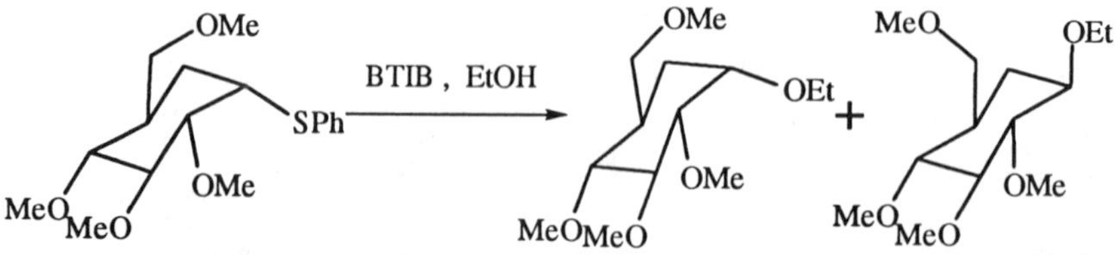

BTIB , EtOH

(α:β = 1:15) 81%

BTIB = *bis*-(trifluoroacetoxy)iodobenzene

Sun, L.; Li, P.; Zhao, K. *Tetrahedron Lett.*, ***1994***, *35*, 7147

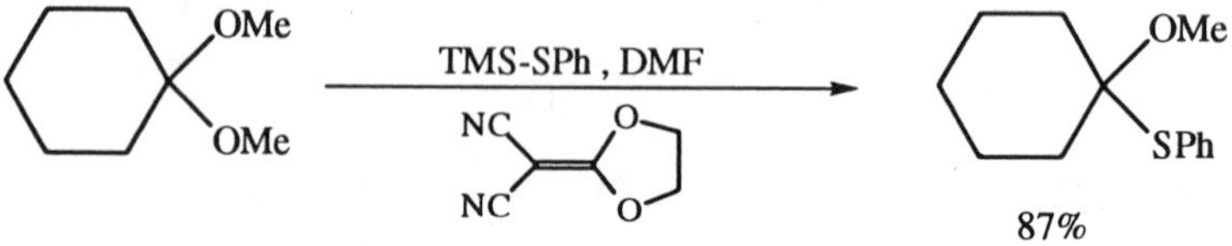

TMS-SPh , DMF

87%

Miura, T.; Masaki, Y. *Tetrahedron Lett.*, ***1994***, *35*, 7961

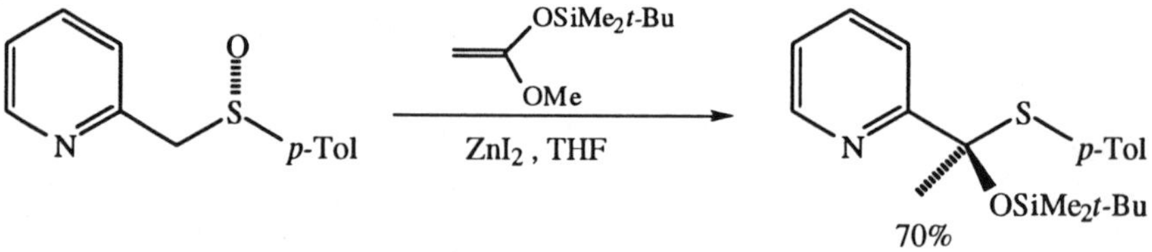

ZnI_2 , THF

70%
(>99% e)

Kita, Y.; Shibata, N.; Fukui, S.; Fujita, S. *Tetrahedron Lett.*, ***1994***, *35*, 9733

SECTION 364: ETHER, EPOXIDE, THIOETHER - HALIDE, SULFONATE

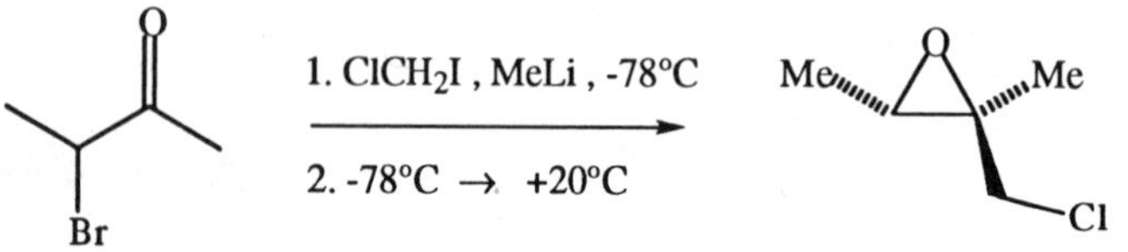

1. $ClCH_2I$, MeLi , -78°C
2. -78°C → +20°C

62%

Barluenga, J.; Llavona, L.; Bernad, P.L.; Concellón, J.M. *Tetrahedron Lett.*, ***1993***, *34*, 3173

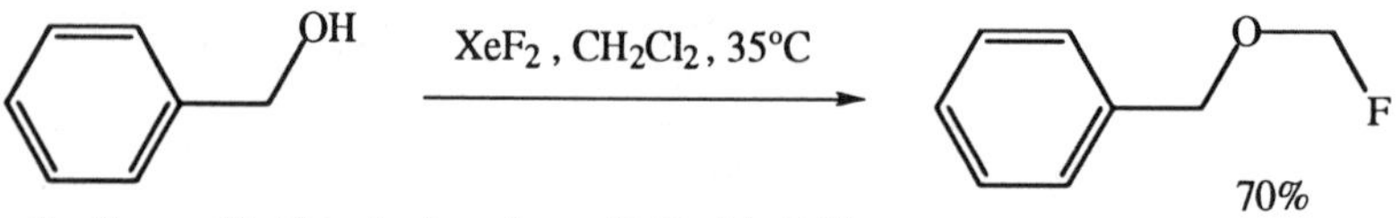

XeF_2 , CH_2Cl_2 , 35°C

70%

Stavber, S.; Zupan, M. *Tetrahedron Lett.*, ***1993***, *34*, 4355

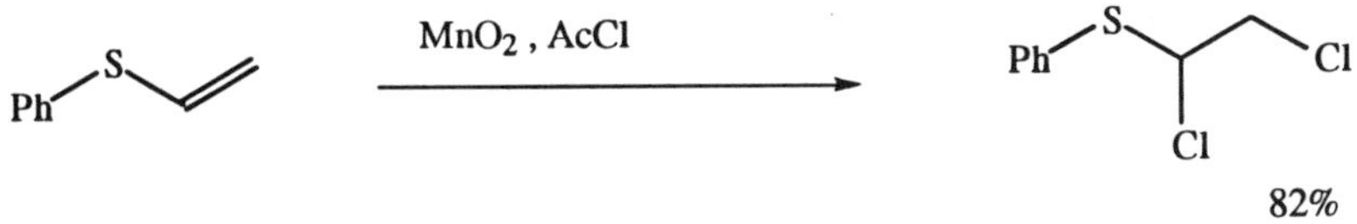

MnO_2 , AcCl

82%

Bellesia, F.; Ghelfi, F.; Pagnoni, U.M.; Pinetti, A. *Gazz. Chim. Ital.*, ***1993***, *123*, 289

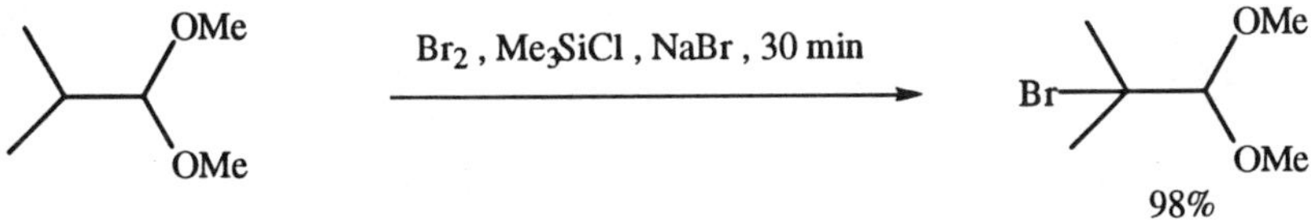

Br_2 , Me_3SiCl , NaBr , 30 min

98%

Bellesia, F.; Boni, M.; Ghelfi, F.; Pagnoni, U.M. *Gazz. Chim. Ital.*, ***1993***, *123*, 629

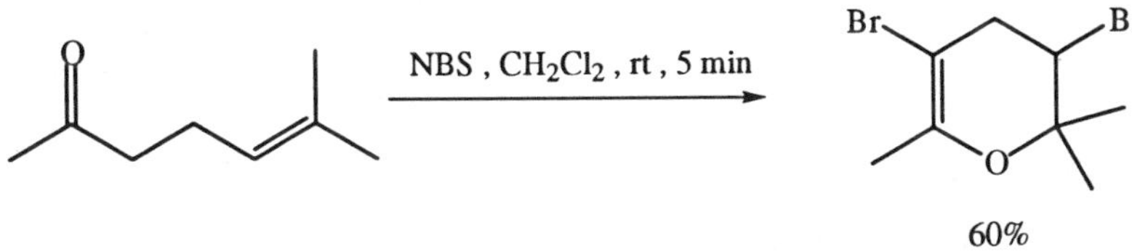

NBS , CH_2Cl_2 , rt , 5 min

60%

Antonioletti, R.; Magnanti, S.; Screttri, A. *Tetrahedron Lett.*, ***1994***, *35*, 2619

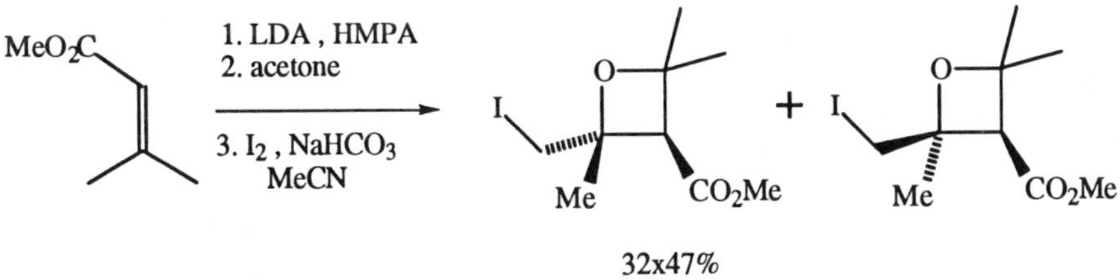

1. LDA , HMPA
2. acetone
3. I_2 , $NaHCO_3$ MeCN

32x47%

Galatsis, P.; Parks, D.J. *Tetrahedron Lett.*, ***1994***, *35*, 6611

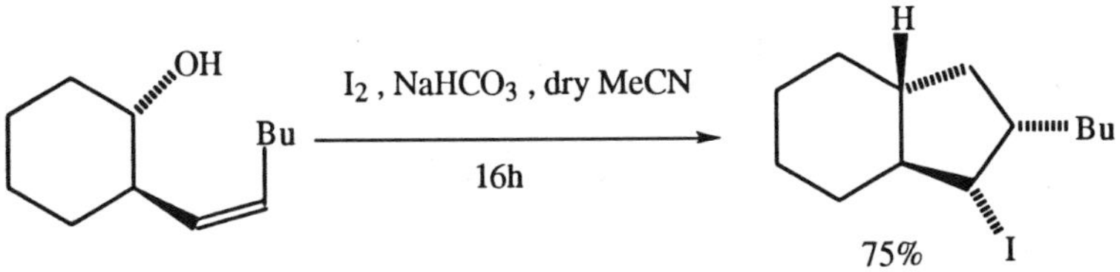

I_2 , $NaHCO_3$, dry MeCN
16h

75%

Barks, J.M.; Knight, D.W.; Weingarten, G.G. *J. Chem. Soc. Chem. Commun.*, ***1994***, 719

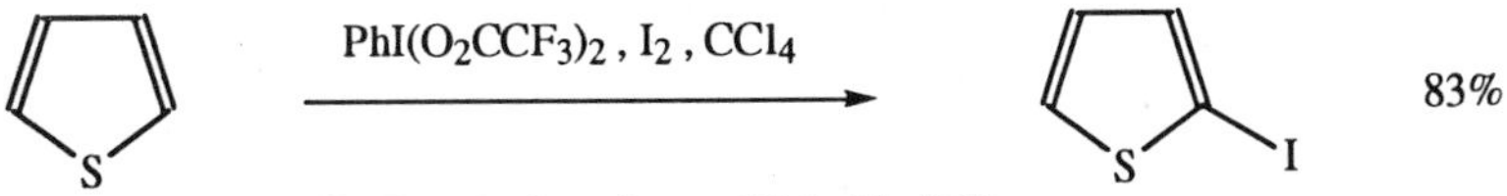

$PhI(O_2CCF_3)_2$, I_2 , CCl_4

83%

D'Auria, M.; Mauriello, G. *Tetrahedron Lett.*, ***1995***, *36*, 4883

C_3H_7 ; Cl ; Cl ; $(CH_2O)_n/CH_2Cl_2$, rt ; $AlCl_3$, 1d ; C_3H_7 ; C_3H_7 ; O ; O ; +

(4 : 1) 60%

Metzger, J.O.; Biermann, U. *Bull. Soc. Chim. Belg.*, ***1994***, *103*, 393

OH ; $(collidine)_2I^+ PF_6^-$; CH_2Cl_2 ; I ; O

95%

Brunel, Y.; Rousseau, G. *Synlett*, ***1995***, 323

I_2 , CAN , MeOH ; rt , 2h ; OMe ; I ; + ; OMe ; I

(2.6 : 1) 88%

Horiuchi, C.A.; Hosokawa, H.; Kanamori, M.; Muramatsu, Y.; Ochiai, K.; Takahashi, E. *Chem. Lett.*, ***1995***, 13

SECTION 365: ETHER, EPOXIDE, THIOETHER - KETONE

$OSiMe_3$; MeO ; OH ; BF_3•OEt_2, $MeNO_2$, -25°C ; 30 min ; O ; OMe

51%

(7:3 *syn:anti*)

Duhamel, P.; Guillemont, J.; Poirier, J.-M. *Tetrahedron Lett.*, ***1993***, *34*, 4197

O ; N_2 ; O ; 2% $Cu(hfacac)_2$, CH_2Cl_2 ; reflux ; O ; O

40%

hfacac = hexafluoroacetyl acetonate

Clark, J.S.; Krowiak, S.A.; Street, L.J. *Tetrahedron Lett.*, ***1993***, *34*, 4385

Me O O OAc hν Me O O OAc 93% (9:1 2Me:3Me)

Ciufolini, M.A.; Rivera-Fortin, M.A.; Byrne, N.E. *Tetrahedron Lett.*, ***1993***, *34*, 3505

O SPh Ph_2t-BuO , Me_2AlCl CH_2Cl_2 , -45°C O SPh Ph_2t-BuO Me

(96:4) where 4 = other isomers

Horiguchi, Y.; Suehiro, I.; Sasaki, A.; Kuwajima, I. *Tetrahedron Lett.*, ***1993***, *34*, 6077

OTMS MeO OMe 5% $TMSN(SO_2F)_2$ CH_2Cl_2 , -78°C O OMe

Trehan, A.; Vij, A.; Walia, M.; Kaur, G.; Verma, R.D.; Trehan, S. *Tetrahedron Lett.*, ***1993***, *34*, 7335

O O N—SePh O p-TsOH , H_2O , CH_2Cl_2 20°C , 12h O SePh 70%

Cossy, J.; Furet, N. *Tetrahedron Lett.*, ***1993***, *34*, 7755

$[RhCl(C_2H_4)_2]_2$, $SnCl_2$
acetone , Ar , 80°C
sealed tube , 10h

Ipatkin, V.V.; Kovalev, I.P.; Ignatenko, A.V.; Nikishin, G.I. *Tetrahedron Lett.*, ***1993***, *34*, 7971

OTMS OTMS
OMe
$TrSbCl_6$, CH_2Cl_2 ,-78°C
CHO
CO_2Me
(>200:1) 76%

Molander, G.A.; Cameron, K.O. *J. Org. Chem.*, ***1993***, *58*, 5931

$Al(OC_6F_5)_3$, toluene
0°C , 2h
C_5H_{11}
>99% (86:14 *S*:*R*)

Ishihara, K.; Hanaki, N.; Yamamoto, H. *Synlett*, ***1993***, 127

1. KH
2. $ZnCl_2$
3. MeC(Cl)SPh
SPh
63%

Groth, U.; Huhn, T.; Richter, N. *Liebigs Ann. Chem.*, ***1993***, 49

MeCHO , PhSH , CH_2Cl_2
SiO_2 , rt
OH SPh
93%

Fuchs, K.; Paquette, L.A. *J. Org. Chem.*, ***1994***, *59*, 528

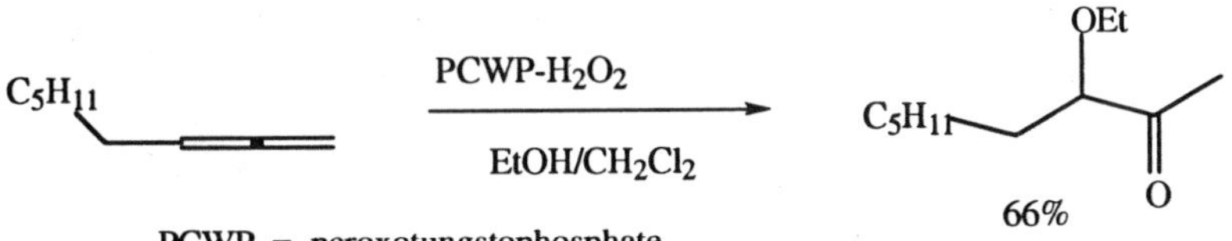

C_5H_{11} (allene) → [$PCWP-H_2O_2$, $EtOH/CH_2Cl_2$] → OEt, C_5H_{11}, O — 66%

PCWP = peroxotungstophosphate

Sakaguchi, S.; Watase, S.; Katayama, Y.; Sakata, Y.; Nishiyama, Y.; Ishii, Y. *J. Org. Chem.*, ***1994***, *59*, 5681

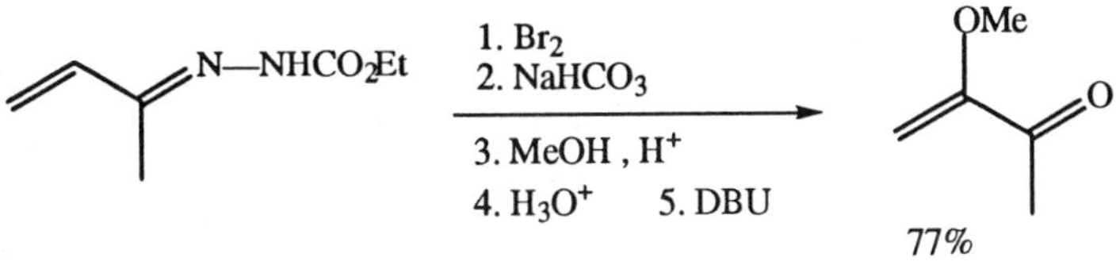

N—$NHCO_2Et$ → [1. Br_2 2. $NaHCO_3$ 3. MeOH , H^+ 4. H_3O^+ 5. DBU] → OMe, O — 77%

Feuerer, A.; Severin, T. *J. Org. Chem.*, ***1994***, *59*, 6026

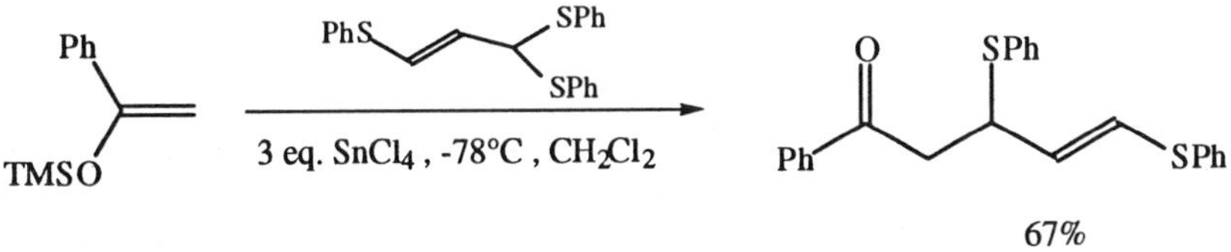

Ph, TMSO → [PhS, SPh, SPh; 3 eq. $SnCl_4$, -78°C , CH_2Cl_2] → O, SPh, Ph, SPh — 67%

Hunter, R.; Michael, J.P.; Walter, D.S. *Tetrahedron Lett.*, ***1994***, *35*, 5481

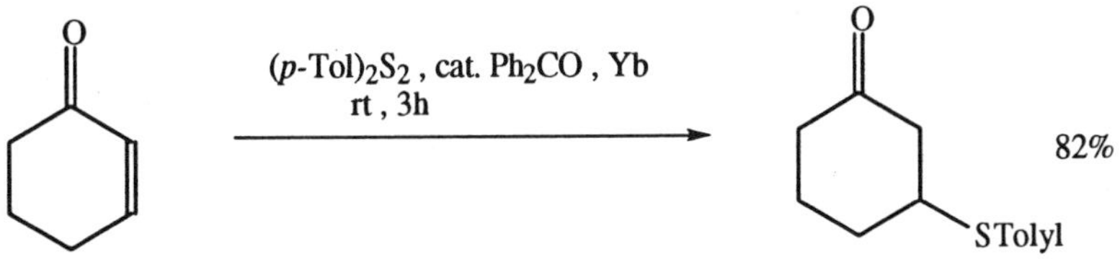

O → [(*p*-Tol)$_2S_2$, cat. Ph_2CO , Yb, rt , 3h] → O, STolyl — 82%

Taniguchi, Y.; Maruo, M.; Takaki, K.; Fujiwara, Y. *Tetrahedron Lett.*, ***1994***, *35*, 7789

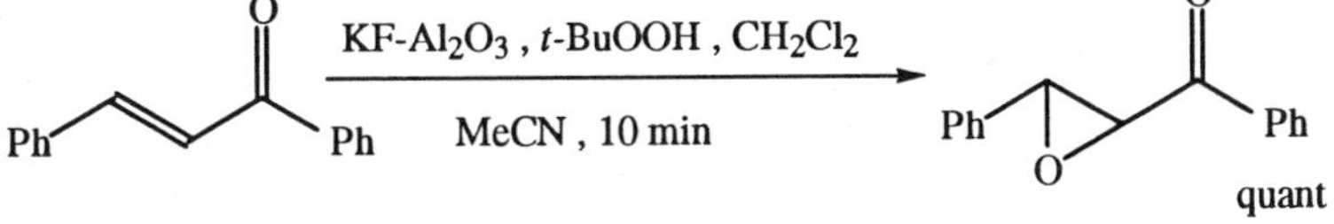

Ph, O, Ph → [KF-Al_2O_3 , *t*-BuOOH , CH_2Cl_2, MeCN , 10 min] → O, Ph, O, Ph — quant

Yadav, V.K.; Kapoor, K.K. *Tetrahedron Lett.*, ***1994***, *35*, 9481

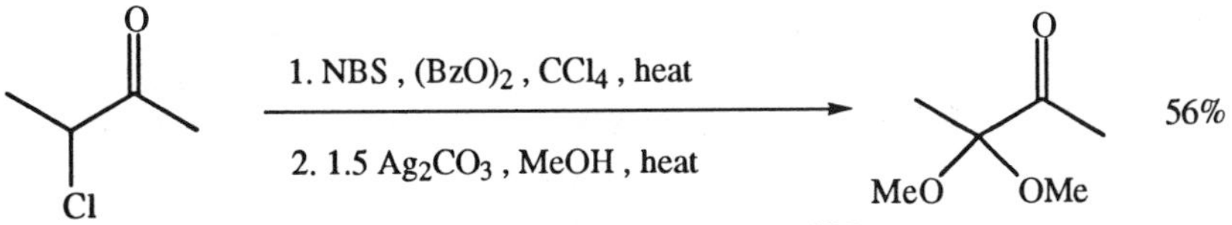

O, Cl → [1. NBS , $(BzO)_2$, CCl_4 , heat 2. 1.5 Ag_2CO_3 , MeOH , heat] → O, MeO, OMe — 56%

De Kimpe, N.; Stanoeva, E.; Boeykens, M. *Synthesis*, ***1994***, 427

Et_2N ... OTMs → 1. $EtCH(OMe)_2$, $TiCL_4$, CH_2Cl_2, -78°C, 1h; 2. SiO_2 → Et, OMe, O quant.

Hojo, M.; Nagoyoshi, M.; Fujii, A.; Yanagi, T.; Ishibashi, N.; Miura, K.; Hosomi, A. *Chem. Lett.*, ***1994***, 719

C_5H_{11}, O, OH → 2 (OAc), CH_2Cl_2; 2 $snCl_2$, 2 NCS, -10°C → C_5H_{11}, O, O

80% (40:60 *syn:anti*)

Masuyama, Y.; Kobayashi, Y.; Kurusu, Y. *J. Chem. Soc. Chem. Commun.*, ***1994***, 1123

Ph, O, CO_2Me, PPh_3 → 3 eq. (dimethyldioxirane); 25°C, 1h → Ph, O, CO_2Me, O

quant.

Wasserman, H.H.; Baldino, C.M.; Coats, S.J. *J. Org. Chem.*, ***1995***, *60*, 8231

O → $NaBO_3$, $(C_6H_{13})_4N\ HSO_4$; CH_2Cl_2, rt, 2h → O, O 95%

Straub, T.S. *Tetrahedron Lett.*, ***1995***, *36*, 663

HO → PhI, 10% $Pd(PPh_3)_4$, DMF; K_2CO_3, 60°C, 18h → O, O, Ph 63% (23:77 *cis:trans*)

Walkup, R.D.; Guan, L.; Kim, Y.S.; Kim, S.W. *Tetrahedron Lett.*, ***1995***, *36*, 3805

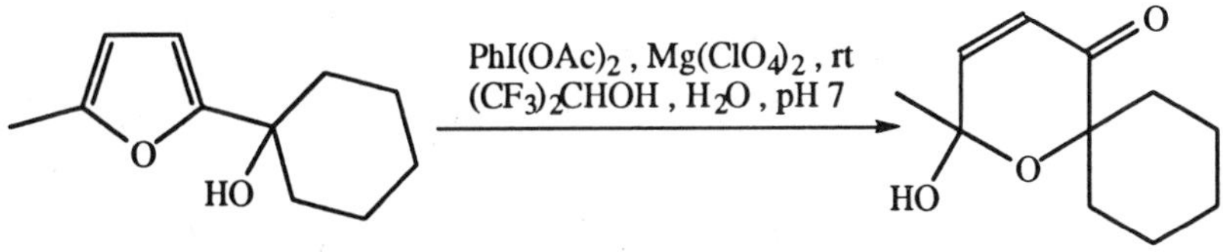

$PhI(OAc)_2$, $Mg(ClO_4)_2$, rt
$(CF_3)_2CHOH$, H_2O , pH 7

HO HO O O O

99%

De Mico, A.; Magarita, R.; Piancatelli, G. *Tetrahedron Lett.*, ***1995***, *36*, 3553

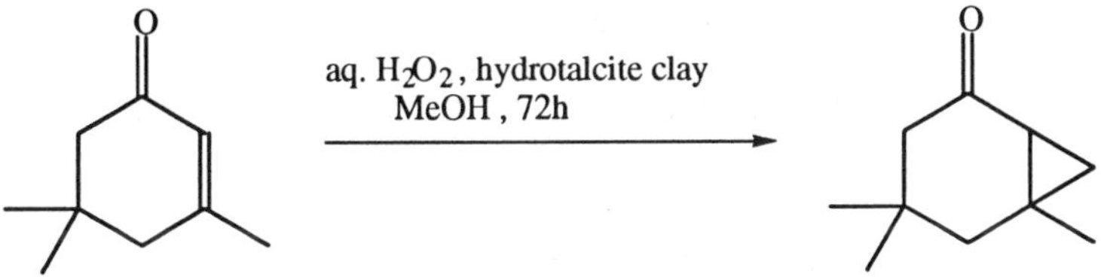

aq. H_2O_2 , hydrotalcite clay
MeOH , 72h

O O

66%

Cativiela, C.; Figueras, F.; Fraile, J.M.; García, J.I.; Mayoral, J.A. *Tetrahedron Lett.*, ***1995***, *36*, 4125

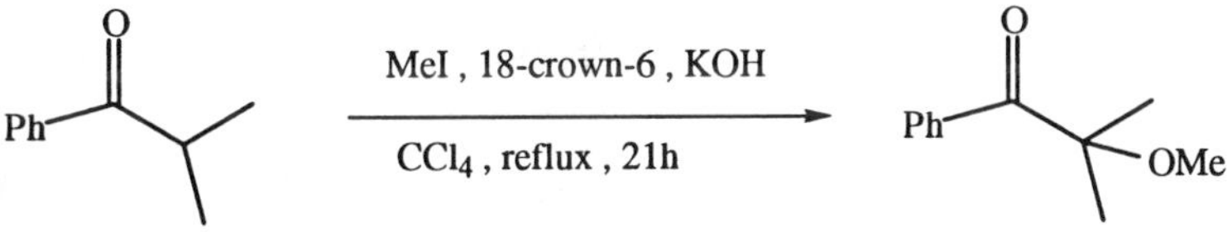

MeI , 18-crown-6 , KOH
CCl_4 , reflux , 21h

Ph Ph OMe O O

66%

Abele, E.; Rubina, K.; Shymanska, M.; Kukevics, E. *Synth. Commun.*, ***1995***, *25*, 1371

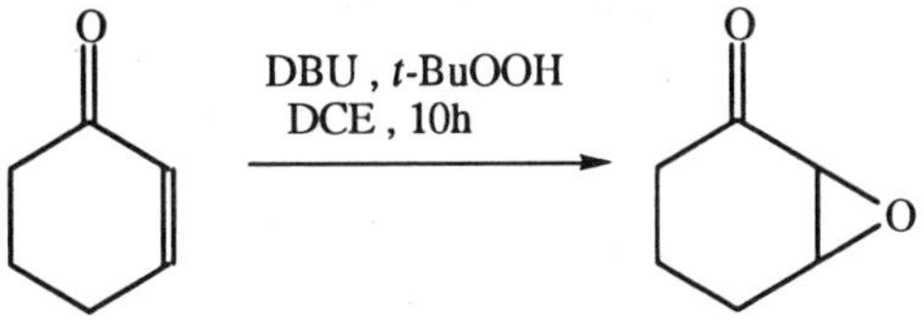

DBU , *t*-BuOOH
DCE , 10h

O O O

92%

Yadav, V.K.; Kapoor, K.K. *Tetrahedron*, ***1995***, *51*, 8573

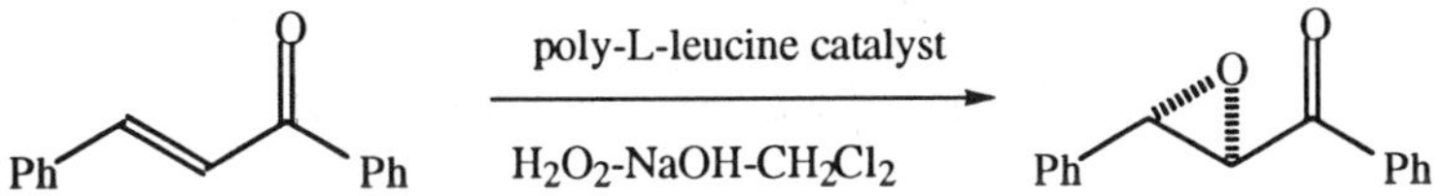

poly-L-leucine catalyst
H_2O_2-NaOH-CH_2Cl_2

Ph Ph Ph Ph O O O

85% (93% ee)

Sánchez, M.E.L.; Roberts, S.M. *J. Chem. Soc., Perkin Trans. 1.*, ***1995***, 1467

1. PhI(OH)OTs
2. PhOH , K_2CO_2 , dry acetone

Ph OPh

67%

Prakash, O.; Saini, N.; Sharma, P.K. *J. Indian Chem. Soc.*, ***1995***, *72*, 129

SECTION 366: ETHER, EPOXIDE, THIOETHER - NITRILE

PhCHO → $[HC(Py)_3W(NO)_2(CO)][SbF_6)_2$ / TMS-CN , rt , 1h , $MeNO_2$ → OTMS Ph CN 85%

Faller, J.W.; Gundersen, L-L. *Tetrahedron Lett.*, ***1993***, *34*, 2275

PhCHO → Me_3SiCH_2CN , KF , 90 sec. / microwave → Ph CN TMSO 70%

Latouche, R.; Texier-Boullet, F.; Hamelin, J. *Bull. Soc. Chim. Fr.*, ***1993***, *130*, 535

PhCHO → TMSCN , MeCN , 20°C , 12h → OTMS Ph CN

a chemoselective reaction since ketones react very slowly

95%

Manju, K.; Trehan, S. *J. Chem. Soc., Perkin Trans. 1.*, ***1995***, 2383

SECTION 367: ETHER, EPOXIDE, THIOETHER - ALKENE

Enol ethers are found in this section as well as alkenyl ethers.

Bu OH → 1. Jones oxidation 2. 10% H_2SO_4 / acetone , 5h reflux → O Bu O 82%

Majetich, G.; Zhang, Y.; Dreyer, G. *Tetrahedron Lett.*, ***1993***, *34*, 449

O → $PhMe_2SiH$, hexane , rt / $(PPh_3)_4RhH$ → $OSiMe_2Ph$

84%

Chan, T.H.; Zheng, G.Z. *Tetrahedron Lett.*, ***1993***, *34*, 3095

TfO ... OMe

$Pd(PPh_3)_4$, K_2CO_3 , DMF
60°C , 1.5h

66%

OMe

Arcadi, A.; Cacchi, S.; Larock, R.C.; Marinelli, F. *Tetrahedron Lett.*, ***1993***, *34*, 2813

OMe (xs) , rt

PhCHO

0.5% $Yb(fod)_3$

OMe

O OMe

Ph

addition of K_2CO_3 prevents formation of this

Deaton, M.V.; Ciufolini, M.A. *Tetrahedron Lett.*, ***1993***, *34*, 2409

TMS

OTBS

HO C_6H_{13}

1. iBuMgBr , cat. Cp_2TiCl_2
2. PhCN

TMS

Ph O C_6H_{13}

OTBS

70%

Tani, K.; Sato, Y.; Okamoto, S.; Sato, F. *Tetrahedron Lett.*, ***1993***, *34*, 4977

O H NHBoc + OMe OTBDMS

1. 0.5M $LiClO_4$, ether
2. 1N HCl , THF
15 min , 0°C

O H O NHBoc 79%

(8:1 *threo:erythro*)

Grieco, P.A.; Moher, E.D. *Tetrahedron Lett.*, ***1993***, *34*, 5567

OEt OEt

1. Me_2S , TMSOTf , -78°C
2. PhSLi

OEt SPh 91%

Kim, S.; Park, J.H.; Lee, J.M. *Tetrahedron Lett.*, ***1993***, *34*, 5769

Me3Si, OH, Bn; OMe, OMe, 0.3 p-TsOH; neat, rt; Bn; 71%

Mohr, P. *Tetrahedron Lett.*, ***1993***, *34*, 6251

OH; 0.1 $Pd(OAc)_2$, CuCl; O_2, DMF, 24h; 5 eq. CH_2=CHPh; O; Ph; 85%

Semmelhack, M.F.; Epa, W.R. *Tetrahedron Lett.*, ***1993***, *34*, 7205

Ph_3P; Ph; O; F_3C; SEt; NaH, rt, 2h; Ph; SEt; CF_3; H; 62% (94:6 *Z:E*)

Bégué, J.-P.; Bonnet-Delpon, D.; M'Bida, A. *Tetrahedron Lett.*, ***1993***, *34*, 7753

OMe; OMe; TMSOf, iPr_2NEt; -20°C → rt, 4h; OMe; 94%

Gassman, P.G.; Burns, S.J.; Pfister, K.B. *J. Org. Chem.*, ***1993***, *58*, 1449

SMe; 1. catecholborane, 3% $PdCl_2$/dppf; 2. Ph, Br, reflux; aq. NaOH, 2h; SMe; Ph; 85%

Gridnev, I.D.; Miyaura, N.; Suzuki, A. *J. Org. Chem.*, ***1993***, *58*, 5351

$C_6H_{13}CHO$ — 1. Me_3Si / Cl, $BF_3 \cdot OEt_2$; 2. DBU/$LiClO_4$/MeCN, 60°C → C_6H_{13}, O — 73x55%

D'Aniello, F.; Mattii, D.; Taddei, M. *Synlett*, ***1993***, 119

OH, CHO — $Ph_3\overset{\oplus}{P}CH_2CH_2OH$, K_2CO_3; PhCl, heat → O — 37%

Billeret, D.; Blondeau, D.; Sliwa, H. *Synthesis*, ***1993***, 881

MeCN — 1. MeOH, HCl_{gas}, ether 0°C, 5d; 2. NaOMe, heat → OMe, OMe — 52%

Argade, A.B.; Joglekar, B.R. *Synth. Commun.*, ***1993***, *23*, 1979

O — $Mo(CO)_5$, Bu, OMe, PhH, 100°C; sealed tube → Bu, O, OMe — 62%

Harvey, D.F.; Neil, D.A. *Tetrahedron*, ***1993***, *49*, 2145

HO, OH — $Pd_2(dba)_3 \cdot CHCl_3$, PBu_3, 130°C; Nafion-H perfluorinated resin sulfonic acid, 5h → O — 71%

Ji, J.; Lu, X. *J. Chem. Soc. Chem. Commun.*, ***1993***, 764

NaOCl, Mn(salen) cat.; ether, 4-Ph-pyridine-N-oxide → O — 45% (64% ee)

Chang, S.; Heid, R.M.; Jacobsen, E.W. *Tetrahedron Lett.*, ***1994***, *35*, 669

Ph, SH — Ph-C≡CH, PhH, 90°C; AIBN, sealed tube 2h → Ph, S (65) + Ph, SMe, Ph (23%)

Benati, L.; Capella, L.; Montevecchi, P.C.; Spagnolo, P. *J. Org. Chem.*, ***1994***, *59*, 2818

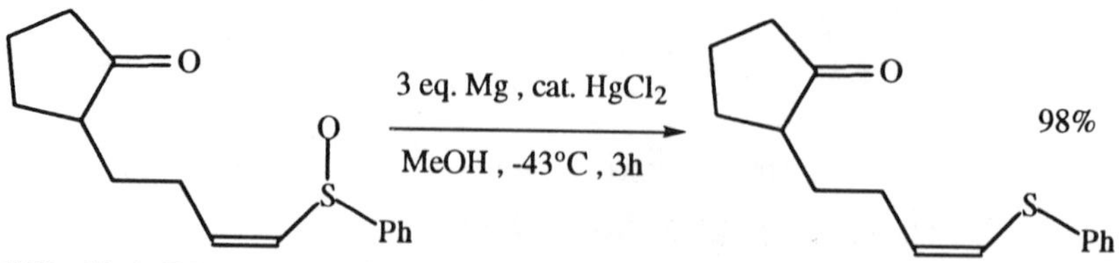

3 eq. Mg , cat. $HgCl_2$ / MeOH , -43°C , 3h — 98%

Lee, G.H.; Choi, E.B.; Lee, E.; Pak, C.S. *Tetrahedron Lett.*, ***1994***, *35*, 2195

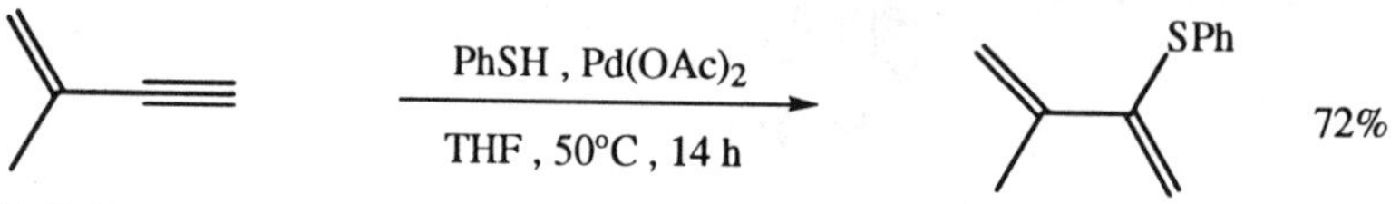

PhSH , $Pd(OAc)_2$ / THF , 50°C , 14 h — SPh — 72%

Bäckvall, J.-E.; Ericsson, A. *J. Org. Chem.*, ***1994***, *59*, 5850

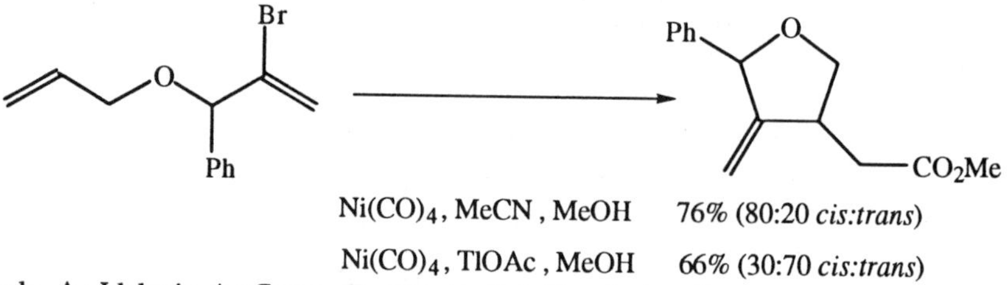

Br, O, Ph → Ph, O, CO_2Me

$Ni(CO)_4$, MeCN , MeOH 76% (80:20 *cis:trans*)

$Ni(CO)_4$, TlOAc , MeOH 66% (30:70 *cis:trans*)

Delgado, A.; Llebaria, A.; Camps, F.; Moretó, J.M. *Tetrahedron Lett.*, ***1994***, *35*, 4011

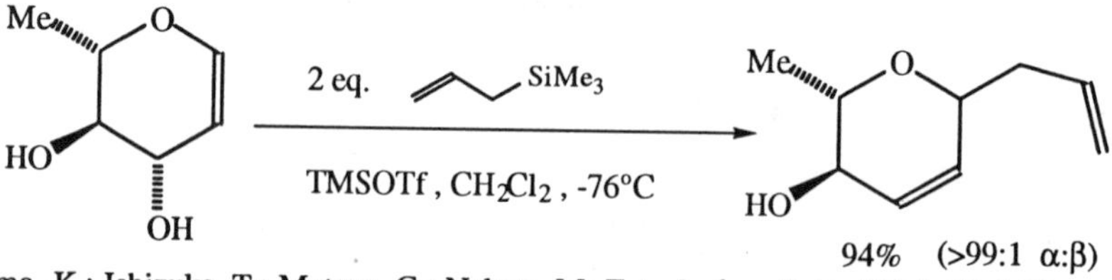

Me, O, HO, OH — 2 eq. allyl-$SiMe_3$ / TMSOTf , CH_2Cl_2 , -76°C — Me, O, HO

94% (>99:1 α:β)

Toshima, K.; Ishizuka, T.; Matsuo, G.; Nakata, M. *Tetrahedron Lett.*, ***1994***, *35*, 5673

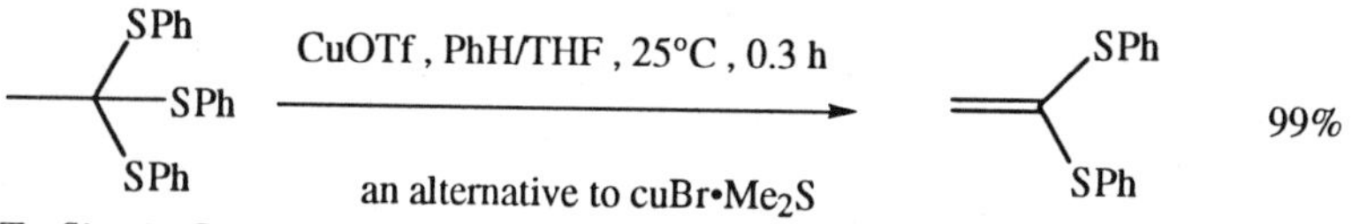

SPh, SPh, SPh — CuOTf , PhH/THF , 25°C , 0.3 h / an alternative to $cuBr{\bullet}Me_2S$ — SPh, SPh — 99%

Cohen, T.; Shook, C.; Thiruvazhi, M. *Tetrahedron Lett.*, ***1994***, *35*, 6041

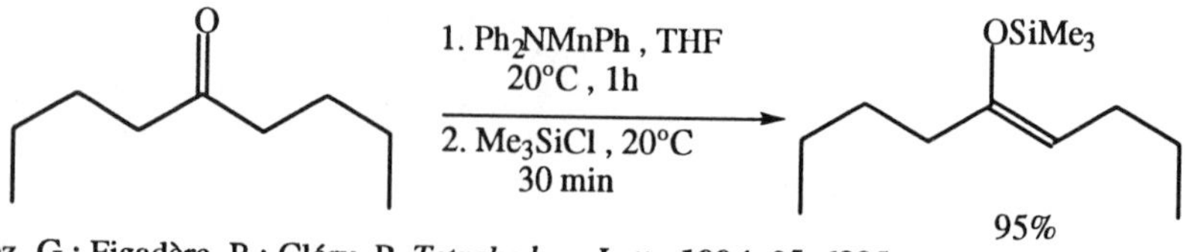

O — 1. Ph_2NMnPh , THF 20°C , 1h / 2. Me_3SiCl , 20°C 30 min — $OSiMe_3$ — 95%

Cahiez, G.; Figadère, B.; Cléry, P. *Tetrahedron Lett.*, ***1994***, *35*, 6295

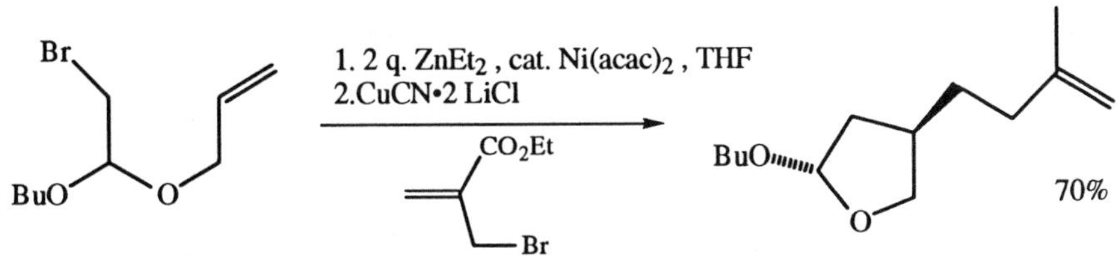

1. 2 q. $ZnEt_2$, cat. $Ni(acac)_2$, THF
2. $CuCN{\cdot}2\,LiCl$

CO_2Et ; Br

BuO ; O ; Br ; BuO ; O

70%

Vaupel, A.; Knochel, P. *Tetrahedron Lett.*, ***1994***, *35*, 8349

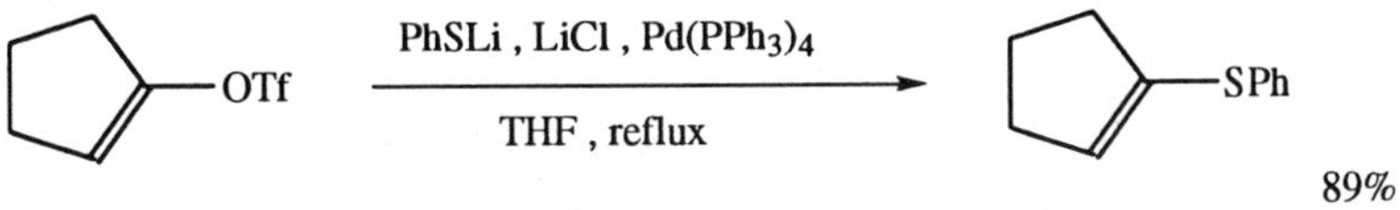

OTf ; $PhSLi$, LiCl , $Pd(PPh_3)_4$; THF , reflux ; SPh

89%

Martínez, A.G.; Barcina, J.O.; Cerezo, A. de F.; Subramanian, L.R. *Synlett*, ***1994***, 561

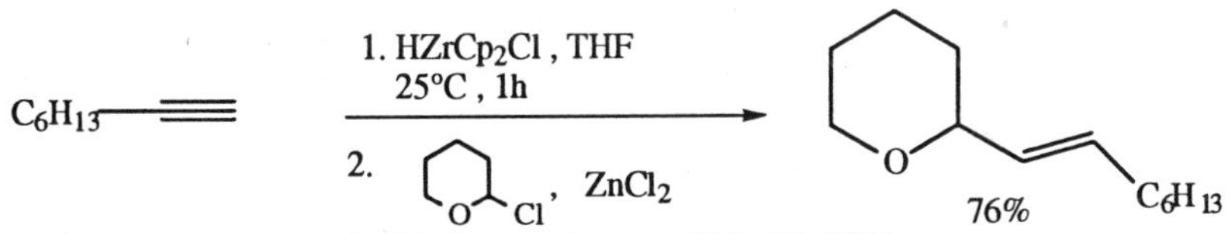

C_6H_{13}

1. $HZrCp_2Cl$, THF
25°C , 1h

2. O Cl , $ZnCl_2$

O ; C_6H_{13}

76%

Pereira, S.; Zheng, B.; Srebnik, M. *J. Org. Chem.*, ***1995***, *60*, 6260

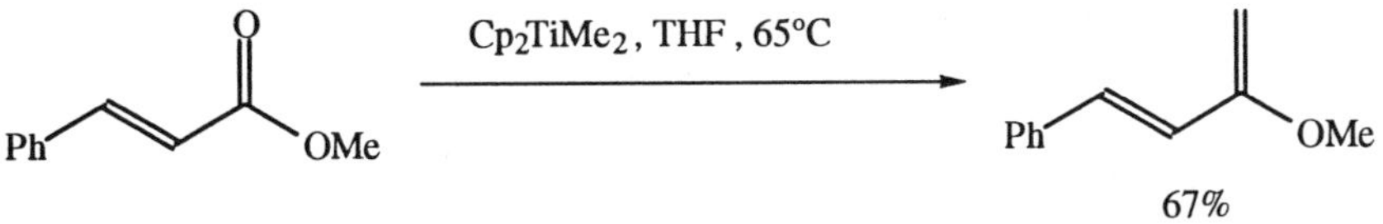

O ; Ph ; OMe ; Cp_2TiMe_2 , THF , 65°C ; Ph ; OMe

67%

Petasis, N.A.; Lu, S.-P. *Tetrahedron Lett.*, ***1995***, *36*, 2393

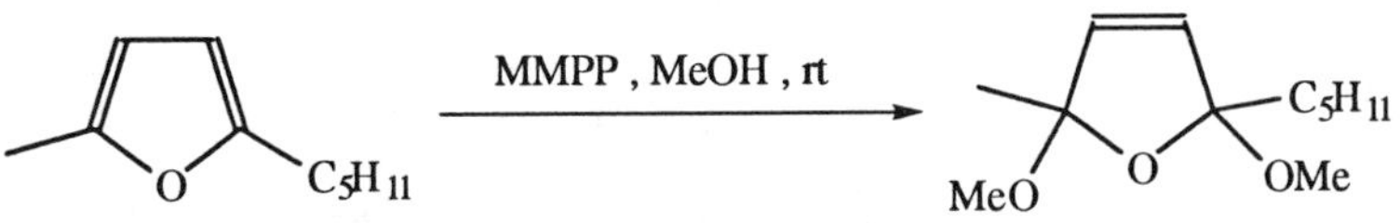

O ; C_5H_{11} ; MMPP , MeOH , rt ; C_5H_{11} ; MeO ; O ; OMe

MMPP = monoperoxyphthalate hexahydrate 91%

D'Annibale, A.; Scettri, A. *Tetrahedron Lett.*, ***1995***, *36*, 4659

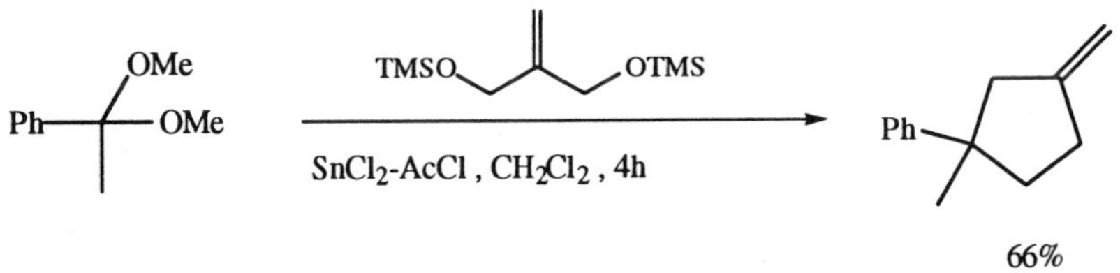

OMe ; Ph ; OMe ; TMSO ; OTMS ; $SnCl_2$-AcCl , CH_2Cl_2 , 4h ; Ph

66%

Oriyama, T.; Ishiwata, A.; Sano, T.; Matsuda, T.; Takahashi, M.; Koga, G. *Tetrahedron Lett.*, ***1995***, *36*, 5581

$W(OAr)(OPr)(CHCMe_3)Cl(OEt)_2$ quant.

Leconte, M.; Pagano, S.; Mutch, A.; Lefebvre, F.; Basset, J.M. *Bull. Soc. Chim. Fr.*, ***1995***, *132*, 1069

Related Methods: Section 180A (Protection of Ketones)

SECTION 368: HALIDE, SULFONATE - HALIDE, SULFONATE

Halocyclopropanations are found in Section 74F (Alkyls from Alkenes).

Ph — $Cl_3C\text{-}CCl_3$, toluene , 3h; $RuCl_2(PPh_3)_3$, 120°C — Cl, Cl, Ph 41%

Sakai, K.; Sugimoto, K.; Shigeizumi, S.; Kondo, K. *Tetrahedron Lett.*, ***1994***, *35*, 737

=N-OH — 1. Cl_2 , $NF_3{\bullet}OEt_2$; 2. HCl — Cl, Cl 82%

Tordeux, M.; Boumizane, K.; Wakselman, C. *J. Org. Chem.*, ***1993***, *58*, 1939

=NOH — $NO^+BF_4^-$, PPHF — F, F 80%

PPHF = pyridinium polyhydrogen fluoride

York, C.; Prakash, G.K.S.; Wang, Q.; Olah, G.A. *Synlett*, ***1994***, 425

$C_{10}NMe_3MnO_4$, Me_3SiBr; CH_2Cl_2 , 0°C — Br, Br

Hazra, B.G.; Chordia, M.D.; Bahule, B.B.; Pore, V.S.; Basu, S. *J. Chem. Soc., Perkin Trans. 1.*, ***1994***, 1667

Ph — $KICl_2$, H_2O — Ph, I, Cl + Ph, Cl, I

(6 : 1) 94%

Zefirov, N.S.; Sereda, G.A.; Sosonuk, S.E.; Zyk, N.V.; Likhomanova, T.I. *Synthesis*, ***1995***, 1359

SECTION 369: HALIDE, SULFONATE - KETONE

$Cl_2C{=}C{=}O$, hexane, rt — 92%

de Faria, A.R.; Matos, C.R.; Correia, C.R.D. *Tetrahedron Lett.*, ***1993***, *34*, 27

NBS, aq. MeCN, cat. HCl; rt, 6h — 81%

Morton, H.E.; Leanna, M.R. *Tetrahedron Lett.*, ***1993***, *34*, 4481

1. $ClCH_2I$, MeLi, -78°C
2. Me_3SiCl, -78°C → rt

83%

Barluenga, J.; Pedregal, B.; Concellón, J.M. *Tetrahedron Lett.*, ***1993***, *34*, 4563

HTIB, I_2, MeCN, rt; reflux, 20h — 72%

HTIB = hydroxy *p*-tosyloxyiodo benzene

Bovonsombat, P.; McNelis, E. *Tetrahedron*, ***1993***, *49*, 1525

1. pyridinium polyhydrogen fluoride CH_2Cl_2
2. $PhI(O_2CCF_3)_2$
3. K_2CO_3

61%

Karam, O.; Jacquesy, J.-C.; Jouannetaud, M.-P. *Tetrahedron Lett.*, ***1994***, *35*, 2541

1. PhCHO, 5% ($BiCl_3$•1.5 $ZnCl_2$)
2. Me_3SiCl

>88%

Le Roux, C.; Gaspard-Iloughmane, H.; Dubac, J. *J. Org. Chem.*, ***1994***, *59*, 2238

1. 2 LDA, THF-HMPA, -78°C
2. $C_8H_{17}CHO$
3. FVP (508°C)

Ph, S, O, F → C_8H_{17}, O, F 82x48%

Reutrakul, V.; Kruahong, T.; Pohmakotr, M. *Tetrahedron Lett.*, ***1994***, *35*, 4853

PDC, I_2, CH_2Cl_2
rt, overnight

O → O, I

49% conversion

Bovonsombat, P.; Angara, G.J.; McNelis, E. *Tetrahedron Lett.*, ***1994***, *35*, 6787

OH → NBS, *t*-BuOH, 30°C 1.5h → O, Br 54%

Cossy, J.; Furet, N. *Tetrahedron Lett.*, ***1995***, *36*, 3691

C_6H_{13}, B, O, O

1. $HZrCp_2Cl$
2. AcCl, 15% $CuBr{\bullet}SMe_2$
3. NBS

Br, C_6H_{13}, O 93%

Zheng, B.; Srebnik, M. *Tetrahedron Lett.*, ***1995***, *36*, 5665

poly(4-methyl-5-vinylthiozolium) hydrotribromide
1.5h

Ph, O → Ph, O, Br quant.

Babadjamian, A.; Kessat, A. *Synth. Commun.*, ***1995***, *25*, 2203

TMSO, H

2.2 eq. $FeCl_3$, DMF
-40°C

O, Cl 51%

Booker-Milburn, K.I.; Thompson, D.F. *J. Chem. Soc., Perkin Trans. 1.*, ***1995***, 2315

DMSO , $(COCl)_2$, MeOH

NEt_3 , CH_2Cl_2 , -60°C

O

Cl

93%

Raina, S.; Singh, V.K. *Tetrahedron*, ***1995***, *51*, 2467

O O

F_2/N_2-HCO_2H , 10-15°C , 2h

O O

F

70%

Chambers, R.D.; Greenhall, M.P.; Hutchinson, J. *J. Chem. Soc. Chem. Commun.*, ***1995***, 21

SECTION 370: HALIDE, SULFONATE - NITRILE

NO ADDITIONAL EXAMPLES

SECTION 371: HALIDE, SULFONATE - ALKENE

O

1. LDA
2. $ClP(O)(OEt)_2$
3. TMS-I

I

83%

Lee, K.; Wiemer, D.F. *Tetrahedron Lett.*, ***1993***, *34*, 2433

$OSiMe_2t$-Bu

Bu Br

Br

LDA , ether

-78°C → rt

$OSiMe_2t$-Bu

Bu Br

93%

Shinokubo, H.; Oshima, K.; Utimoto, K. *Tetrahedron Lett.*, ***1993***, *34*, 4985

OH

C_7H_{15}

1. Cl_3CCN
2. I_2 , K_2CO_3 , ether

76% x 88%

CCl_3 O N I H C_7H_{15}

+

CCl_3 O N I C_7H_{15} H

(33 : 1)

Friesen, R.W.; Giroux, A.; Cook, K.L. *Tetrahedron Lett.*, ***1993***, *34*, 5983

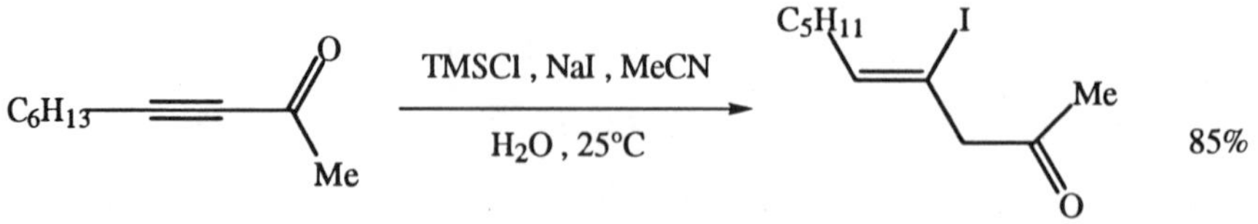

C_6H_{13}–C≡C–C(O)Me → (TMSCl, NaI, MeCN; H_2O, 25°C) → C_5H_{11}, I, Me, O — 85%

Luo, F.-T.; Kumar, K.A.; Hsieh, L.-C.; Wang, R.-T. *Tetrahedron Lett.*, ***1994***, *35*, 2553

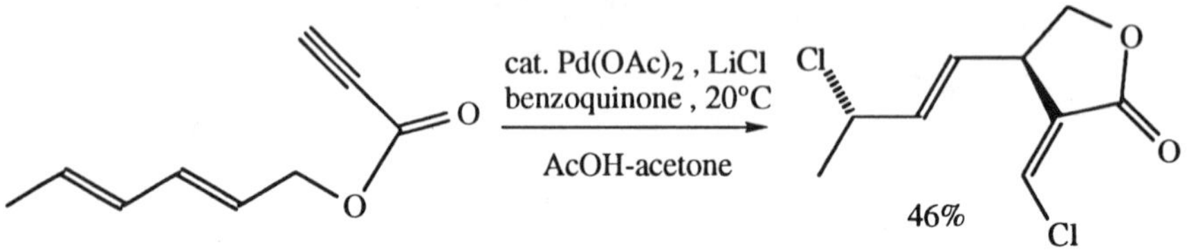

cat. $Pd(OAc)_2$, LiCl, benzoquinone, 20°C; AcOH-acetone → Cl, O, O, Cl — 46%

Bäckvall, J.-E.; Nilsson, Y.I.M.; Andersson, P.G.; Gatti, R.G.P.; Wu, J. *Tetrahedron Lett.*, ***1994***, *35*, 5713

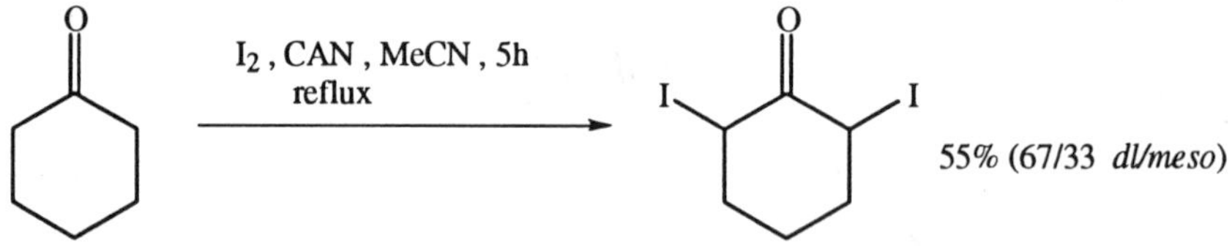

I_2, CAN, MeCN, 5h, reflux → 55% (67/33 *dl/meso*)

Horiuchi, C.A.; Takahashi, E. *Bull. Chem. Soc. Jpn.*, ***1994***, *67*, 271

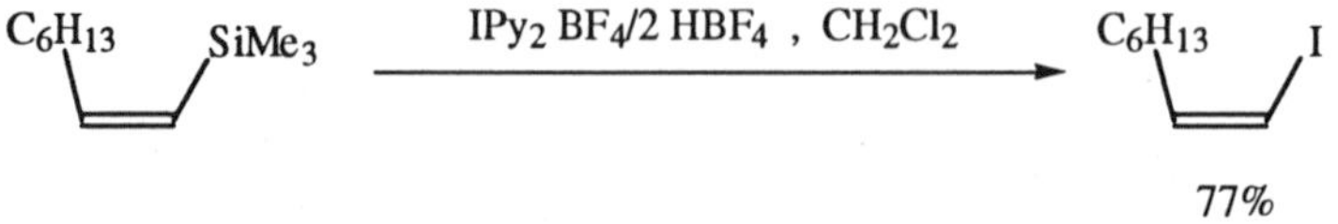

C_6H_{13}, $SiMe_3$ → ($IPy_2\,BF_4/2\,HBF_4$, CH_2Cl_2) → C_6H_{13}, I — 77%

Barluenga, J.; Alvarez-García, L.J.; Gonzálex, J.M. *Tetrahedron Lett.*, ***1995***, *36*, 2153

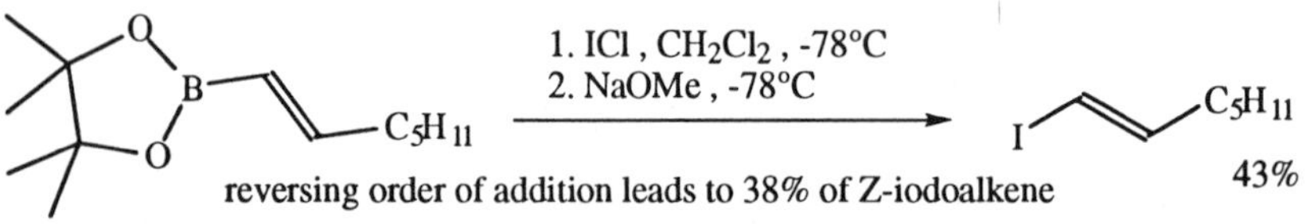

1. ICl, CH_2Cl_2, -78°C
2. NaOMe, -78°C

C_5H_{11} → I, C_5H_{11} — 43%

reversing order of addition leads to 38% of Z-iodoalkene

Stewart, S.K.; Whiting, A. *Tetrahedron Lett.*, ***1995***, *36*, 3929

SECTION 372: KETONE - KETONE

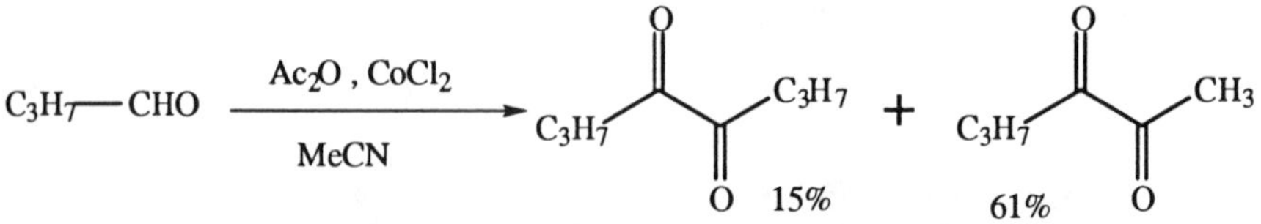

C_3H_7—CHO → (Ac_2O, $CoCl_2$; MeCN) → C_3H_7C(O)C(O)C_3H_7 15% + C_3H_7C(O)C(O)CH_3 61%

Bhatia, B.; Punniyamurthy, T.; Iqbal, J. *J. Org. Chem.*, ***1993***, *58*, 5518

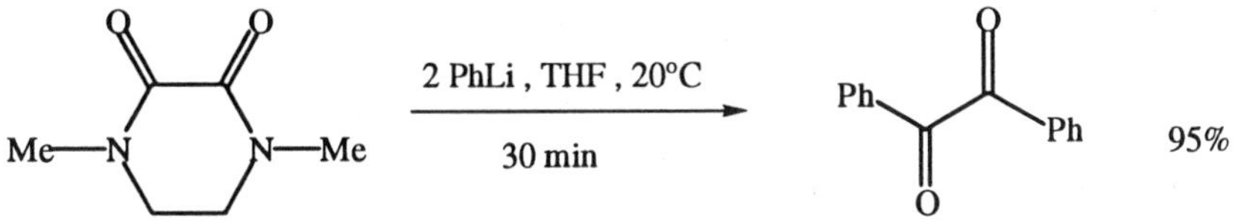
Me
N
N
Me
2 PhLi , THF , 20°C
30 min
Ph
Ph
95%

Mueller-Westerhoff, U.T.; Zhou, M. *Tetrahedron Lett.*, ***1993***, *34*, 571

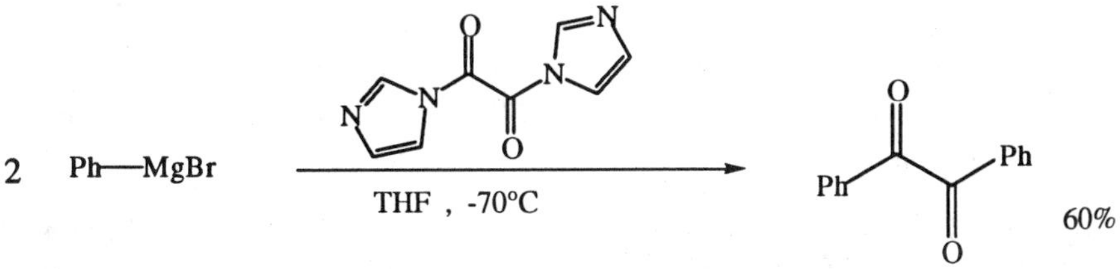
2 Ph—MgBr
THF , -70°C
Ph
Ph
60%

Mitchell, R.H.; Iyer, V.S. *Tetrahedron Lett.*, ***1993***, *34*, 3683

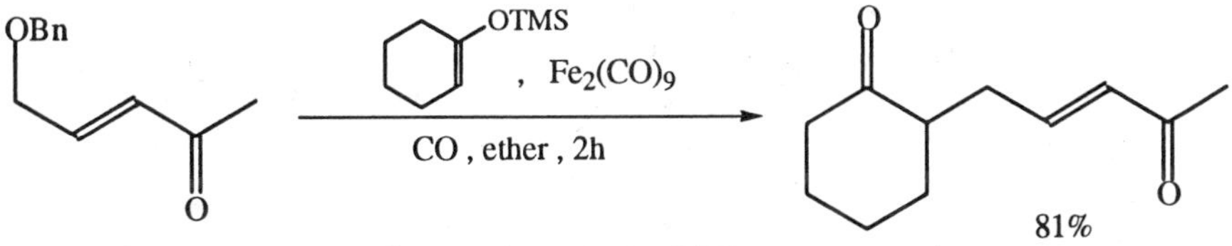
OBn
OTMS
, $Fe_2(CO)_9$
CO , ether , 2h
81%

Zhou, T.; Green, J.R. *Tetrahedron Lett.*, ***1993***, *34*, 4497

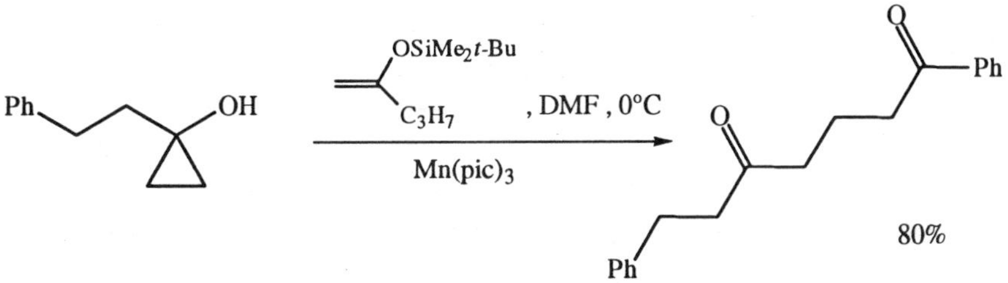
Ph
OH
$OSiMe_2t$-Bu
C_3H_7
, DMF , 0°C
$Mn(pic)_3$
Ph
Ph
80%

Iwasawa, N.; Hayakawa, S.; Funahashi, M.; Isobe, K.; Narasaka, K. *Bull. Chem. Soc. Jpn.*, ***1993***, *66*, 819

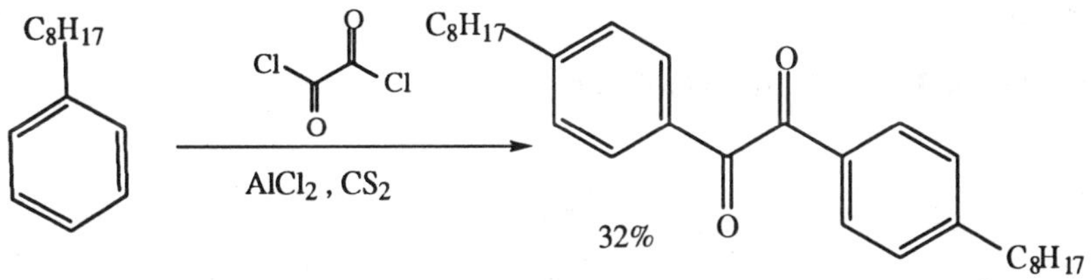
C_8H_{17}
Cl
Cl
$AlCl_2$, CS_2
C_8H_{17}
C_8H_{17}
32%

Mohr, B.; Enkelmann, V.; Wegner, G. *J. Org. Chem.*, ***1994***, *59*, 635

OTMS
OTMS
$BF_3 \cdot OEt_2$
61%

Jenkins, T.J.; Burnell, D.J. *J. Org. Chem.*, ***1994***, *59*, 1485

Yb-TMS-Br
THF, reflux
68%

Taniguchi, Y.; Nakahashi, M.; Kuno, T.; Tsuno, M.; Makioka, Y.; Takaki, K.; Fujiwara, Y. *Tetrahedron Lett.*, ***1994***, *35*, 4111

OAc
Ph
$CoCl_2$
Ph
Ph
83%

Maikap, G.C.; Reddy, M.M.; Mukhopadhyay, M.; Bhatia, B.; Iqbal, J. *Tetrahedron*, ***1994***, *50*, 9145

OH
Bu
PCWP, DCE, reflux, 3h
Bu
OH PCP = $[p\text{-}C_5H_5N(CH_2)_{15}Me]_3\ \{PO_4[W(O)(O_2)_2]_4\}$ O 93%

Iwahama, T.; Sakaguchi, S.; Nishiyama, Y.; Ishii, Y. *Tetrahedron Lett.*, ***1995***, *36*, 1523

REVIEW:

"α-Diones from Cyclic Oxamides and Organolithium Reagents: A New, General and Environmentally Beneficial Synthetic Method," Mueller-Westerhoff, U.T.; Zhou, M. *Synlett*, ***1994***, 975

SECTION 373: KETONE - NITRILE

Ph
Ph CN , DCE, 70°C
20% $Pd(OAc)_2\text{-}PPh_3\text{-}dppb$
65h
Ph
NC
Ph
O
74%

Nozaki, K.; Sato, N.; Takaya, H. *J. Org. Chem.*, ***1994***, *59*, 2679

SECTION 374: KETONE - ALKENE

For the oxidation of allylic alcohols to alkene ketones, see Section 168 (Ketones from Alcohols and Phenols)

For the oxidation of allylic methylene groups ($C=C-CH_2 \rightarrow C=C-C=O$), see Section 170 (Ketones from Alkyls and Methylenes).

For the alkylation of alkene ketones, also see Section 177 (Ketones from Ketones) and for conjugate alkylations see Section 74E (Alkyls form Alkenes).

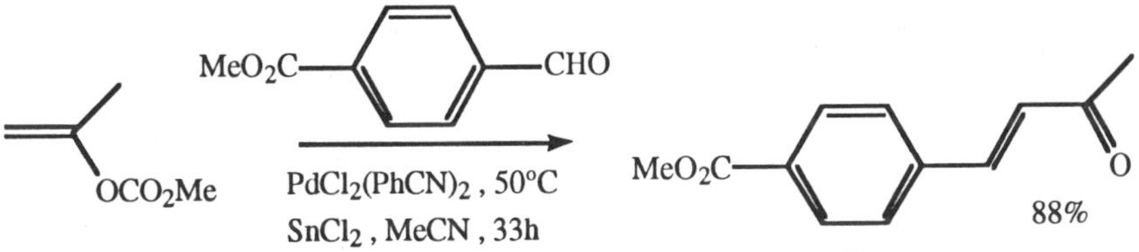

MeO_2C–C_6H_4–CHO

OCO_2Me $PdCl_2(PhCN)_2$, 50°C $SnCl_2$, MeCN, 33h MeO_2C O 88%

Masuyama, Y.; Sakai, T.; Kurusu, Y. *Tetrahedron Lett.*, ***1993***, *34*, 653

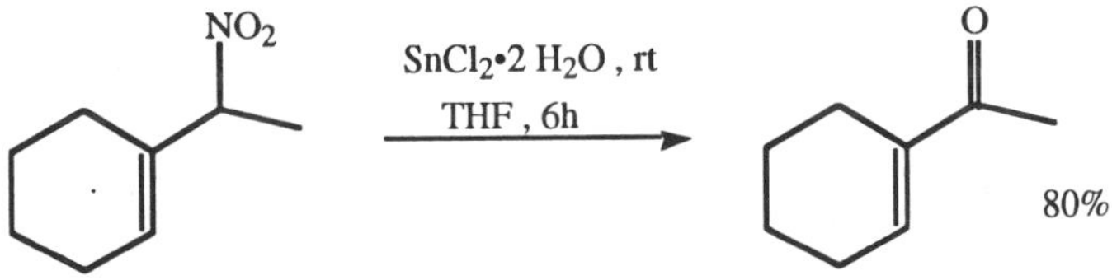

NO_2 $SnCl_2$•2 H_2O, rt THF, 6h O 80%

Das, N.B.; Sarma, J.C.; Sharma, R.P.; Bordoloi, M. *Tetrahedron Lett.*, ***1993***, *34*, 869

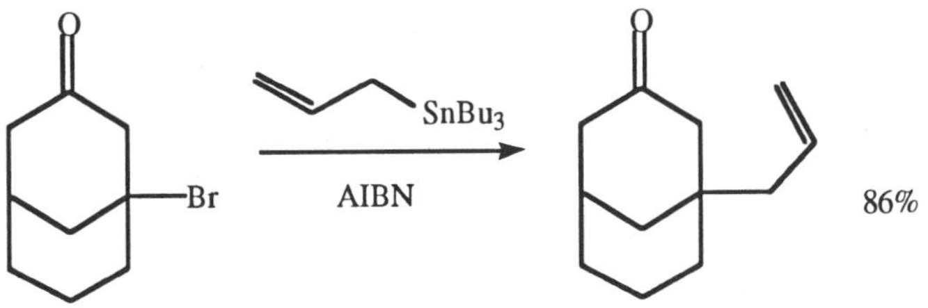

O SnBu$_3$ O Br AIBN 86%

Kraus, G.A.; Andersh, B.; Su, Q.; Shi, J. *Tetrahedron Lett.*, ***1993***, *34*, 1741

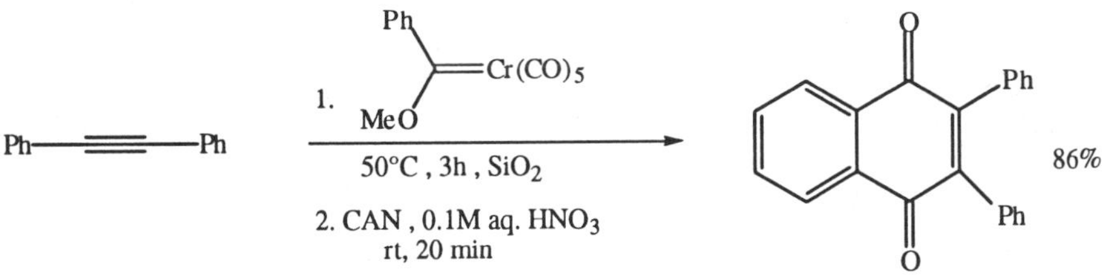

Ph Ph=Cr(CO)$_5$ 1. MeO O Ph Ph 50°C, 3h, SiO_2 86% 2. CAN, 0.1M aq. HNO_3 rt, 20 min Ph O

Harrity, J.P.A.; Kerr, W.J.; Middlemiss, D. *Tetrahedron Lett.*, ***1993***, *34*, 2995

1. PhH , $Cu(OAc)_2{\bullet}2\ H_2O$, $Pb(OAc)_4$, 2d
2. ethylene glycol , CH_2Cl_2

CO_2Et → CO_2Et 78°C

Schultz, A.G.; Holoboski, M.A. *Tetrahedron Lett.*, ***1993***, *34*, 3021

MTO , CH_2Cl_2 , rt
SiO_2/Al_2O_3

51%

MTO = methyltrioxorhenium

Junga, H.; Blechert, S. *Tetrahedron Lett.*, ***1993***, *34*, 3731

Ph, O → Ph, O, O

$Mo(CO)_6$, DMSO , Toluene
100°C , 12h

65%

Jeong, N.; Lee, S.J.; Lee, B.Y.; Chung, Y.K. *Tetrahedron Lett.*, ***1993***, *34*, 4027

O, Ph, Br → O, Ph, NO_2

O_2N–C_6H_4–CHO
SmI_3 , THF , rt , 2h

94%
(mostly *E*)

Yu, Y.; Lin, R.; Zhang, Y. *Tetrahedron Lett.*, ***1993***, *34*, 4547

TBDMSO, O, O → TBDMSO, O, O, O, N–Me, O

$CHCl_3$, rt , 8h
N–Me (N-methylmaleimide)

93%

Konopelski, J.P.; Kasar, R.A. *Tetrahedron Lett.*, ***1993***, *34*, 4587

Bu_3SnH , AIBN , 60°C

syringe pump (3h) , 0.02M

60%

Rawal, V.H.; Zhong, H.M. *Tetrahedron Lett.*, ***1993***, *34*, 5197

Bu_3Sn ... OEt, OEt

0.01 $PdCl_2(MeCN)_2$, 20°C
DMF , 2h

75%

Parrain, J.-L.; Beaudet, I.; Duchêne, A.; Watrelot, S.; Quintard, J.-P. *Tetrahedron Lett.*, ***1993***, *34*, 5445

F_3C ... $=\overset{\oplus}{N}Me_2$ $CF_3CO_2^{\ominus}$

CH_2Cl_2 , reflux , 96h

88%

Ales, C.; Janousek, Z.; Viehe, H.G. *Tetrahedron Lett.*, ***1993***, *34*, 5711

1. BuMgBr , 10% CuI , TMSCl
HMPA , THF , -78°C
2. CH_2I_2 , 2 eq. Et_2Zn
3. 2.2 eq. $FeCl_3$, DMF , 0°C
4. NaOAc , MeOH

99x93x54%

Booker-Milburn, K.I.; Thompson, D.F. *Tetrahedron Lett.*, ***1993***, *34*, 7291

Bu_3SnH , AIBN
PhH
slow addition

(7:3 *Z:E*) 81%

fast addition

(2:1 *Z:E*) 70%

Sha, C.-K.; Shen, C.-Y.; Jean, T.-S.; Chiu, R.-T.; Tseng, W.-H. *Tetrahedron Lett.*, ***1993***, *34*, 7641

Rh(II) octanoate , 25°C
CH_2Cl_2

97%

Padwa, A.; Kassir, J.M.; Semones, M.A.; Weingarten, M.D. *Tetrahedron Lett.*, ***1993***, *34*, 7853

I–C6H4–NO_2 , DMF

2.5 $Pd_2(dba)_3$, 50% CuI

$SnBu_3$

85%

Liebeskind, L.S.; Riesinger, S.W. *J. Org. Chem.*, ***1993***, *58*, 408

$Pd_2(dba)_3$•$CHCl_3$
PPh_3 , LiCl

THF , H_2O
70°C

$PhCO_2$
Me

(73 : 27) 67%

Ihle, N.C.; Heathcock, C.H. *J. Org. Chem.*, ***1993***, *58*, 560

EtO_2C—≡—Bu

$(EtO_2C\sim\sim)_2Zn$

CuBr•SMe_2 , HMPA
TMSCl

CO_2Et
Bu

65%

Crimmins, M.T.; Nantermet, P.G.; Trotter, B.W.; Vallin, I.M.; Watson, P.S.; McKerlie, L.A.; Reinhold, T.L.; Cheung, A.W.; Stetson, K.A.; Dedopoulou, D.; Gray, J.L. *J. Org. Chem.*, ***1993***, *58*, 1038

1. KH, (2-pyridyl)S(O)OMe

2. 2 eq. $CuSO_4$, 110°C

53%

Trost, B.M.; Parquette, J.R. *J. Org. Chem.*, ***1993***, *58*, 1579

$Co_2(CO)_8$, MeCN

rt → 75°C

91%

Hoye, T.R.; Suriano, J.A. *J. Org. Chem.*, ***1993***, *58*, 1659

2 eq. PhCHO, $ZrOCl_2$•8 H_2O

250°C

73%

Yuki, T.; Hashimoto, M.; Nishiyama, Y.; Ishii, Y. *J. Org. Chem.*, ***1993***, *58*, 4497

Ph—≡—Ph

5% $Pd(OAc)_2$, 4 NaOAc

100°C, DMF, Bu_4NCl

84%

Larock, R.C.; Doty, M.J.; Cacchi, S. *J. Org. Chem.*, ***1993***, *58*, 4579

$W(CO)_6$, THF, CO, sealed tube

hν, 65°C → 110°C

57%

Hoye, T.R.; Suriano, J.A. *J. Am. Chem. Soc.*, ***1993***, *115*, 1154

OTMS

PhC≡CH , $SnCl_4$-NBu_3

O

Ph

85% (20:1 *E:Z*)

Yamaguchi, M.; Hayashi, A.; Hirama, M. *J. Am. Chem. Soc.*, ***1993***, *115*, 3362

O

Ph

5% $RhCl(PPh_3)_3$, toluene , 120°C

Ph

O

84%

Huffman, M.A.; Liebeskind, L.S. *J. Am. Chem. Soc.*, ***1993***, *115*, 4895

Ph

O

Me_3SiCN

$Cp_2Ti(PMe_3)_2$

Ph

O

O

80%

Berk, S.C.; Grossman, R.B.; Buchwald, S.L. *J. Am. Chem. Soc.*, ***1993***, *115*, 4912

MeO_2CO

CO , 10% $Pd(OAc)_2$-dppp
PhH-MeCN-MeOH

CO_2Me

O

81%

Mandai, T.; Tsuji, J.; Tsujiguchi, Y.; Saito, S. *J. Am. Chem. Soc.*, ***1993***, *115*, 5865

O

CO_2Et

OMe

OMe

$TiCl_4$, CH_2Cl_2 , -40°C , 1h

O

CO_2Et

78%

Funk, R.L.; Fitzgerald, J.F.; Olmstead, T.A.; Para, K.S.; Wos, J.A. *J. Am. Chem. Soc.*, ***1993***, *115*, 8849

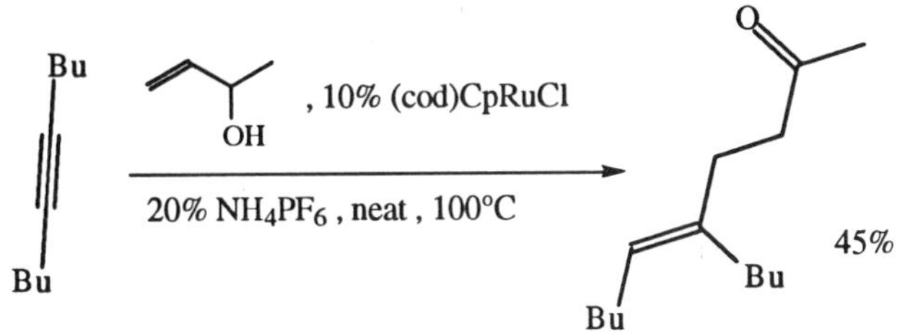

Bu, Bu; , 10% (cod)CpRuCl; OH; 20% NH_4PF_6 , neat , 100°C; Bu, Bu; 45%

Trost, B.M.; Martinez, J.A.; Kulawiec, R.J.; Indolese, A.F. *J. Am. Chem. Soc.*, ***1993***, *115*, 10402

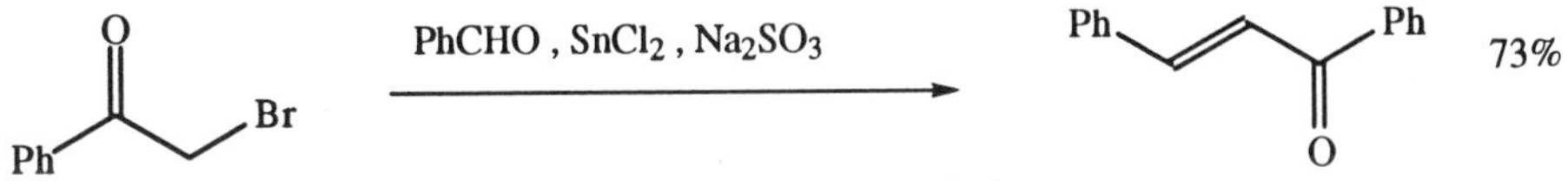

PhCHO , $SnCl_2$, Na_2SO_3; Ph, Br; Ph, Ph; 73%

Lin, R.; Yu, Y.; Zhang, Y. *Synth. Commun.*, ***1993***, *23*, 271

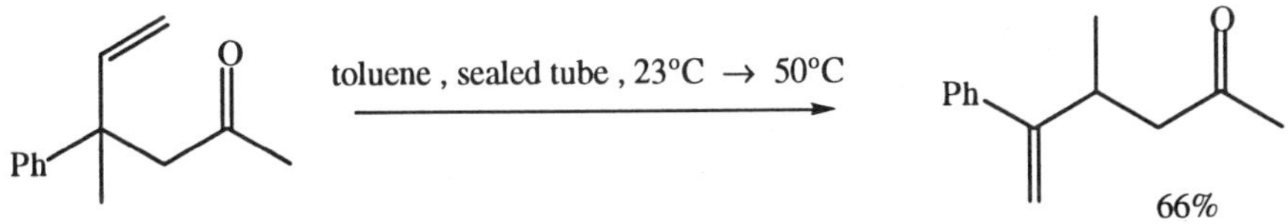

toluene , sealed tube , 23°C → 50°C; Ph; Ph; 66%

Srikrishna, A.; Krishnan, K.; Van Kateswarlu, S. *J. Chem. Soc. Chem. Commun.*, ***1993***, 143

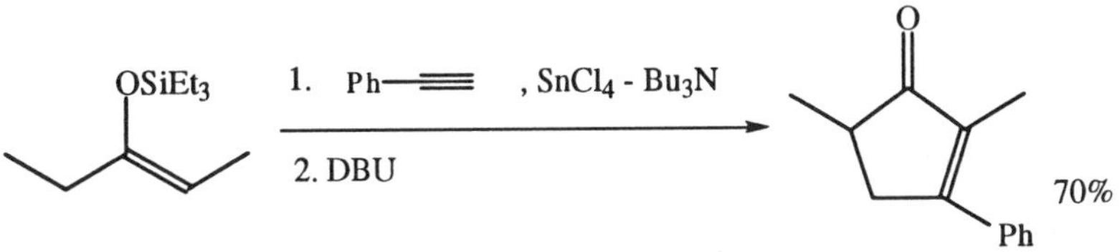

$OSiEt_3$; 1. Ph—≡ , $SnCl_4$ - Bu_3N; 2. DBU; Ph; 70%

Yamaguchi, M.; Sehata, M.; Hayashi, A.; Hirama, M. *J. Chem. Soc. Chem. Commun.*, ***1993***, 1708

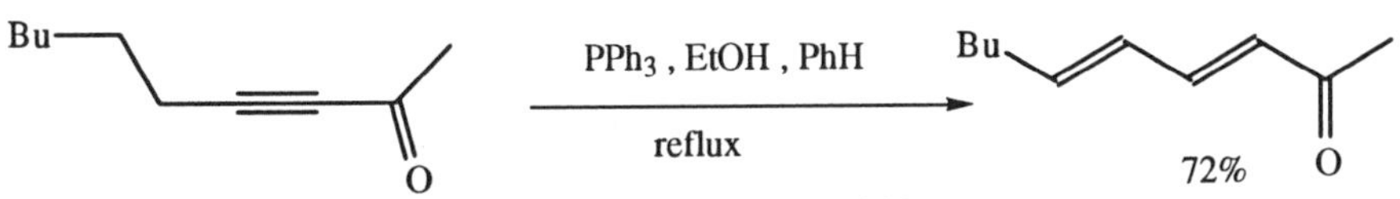

Bu; PPh_3 , EtOH , PhH; reflux; Bu; 72%

Guo, C.; Lu, X. *J. Chem. Soc., Perkin Trans. 1.*, ***1993***, 1921

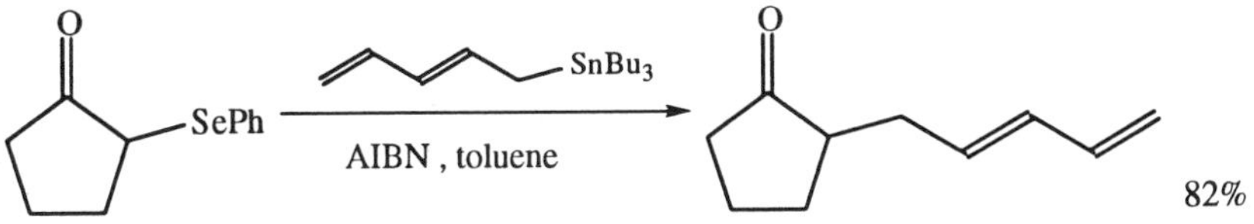

SePh; $SnBu_3$; AIBN , toluene; 82%

Watanabe, Y.; Yoneda, T.; Okumura, T.; Ueno, Y.; Toru, T. *Bull. Chem. Soc. Jpn.*, ***1993***, *66*, 3030

0.01 mol dm^{-3}

toluene , reflux

53%

Makatani, K.; Takada, K.; Odagaki, Y.; Isoe, S. *J. Chem. Soc. Chem. Commun.*, ***1993***, 556

1. $Bu_3SnOSnBu_3$
2. PhCHO , HMPA
3. H_2O , ($-CO_2$)

62%

Shibata, I.; Nishio, M.; Baba, A.; Matsuda, H. *Chem. Lett.*, ***1993***, 1219

1. Me_3SiI , $CHCl_3$, rt , 30 min
2. $PdCl_2(MeCN)_2$, 9h

$SiPh_3$ $SnBu_3$ $SiPh_3$

64%

Degl'Innocenti, A.; Capperucci, A.; Bartoletti, L.; Mordini, A.; Reginato, G. *Tetrahedron Lett.*, ***1994***, *35*, 2081

SeO_2H , PhH

	OH	O
1 eq. , rt , 14h	35%	36%
2 eq. , 20°C , 14h	20%	75%
2 eq. , reflux , 2h	-	90%

Barton, D.H.R.; Wang, T.-L. *Tetrahedron Lett.*, ***1994***, *35*, 5149

Li / OEt

90%

enolate free α-alkoxy vinyllithium reagent - improved prep & procedure

Shimano, M.; Meyers, A.I. *Tetrahedron Lett.*, ***1994***, *35*, 7727

O
C_6H_{13}
Me
I

TMSCl, MeCN, rt, 3h

C_5H_{11}
I
O

92%

Luo, F.-T.; Hsieh, L.-C. *Tetrahedron Lett.*, ***1994***, *35*, 9585

EtO_2C
EtO_2C

CO, 120°C, 3% $Co_2(CO)_8$

10% $P(OPh)_3$, DME, 1d

EtO_2C
EtO_2C
O

82%

Jeong, N.; Hwang, S.H.; Lee, Y.; Chung, Y.K. *J. Am. Chem. Soc.*, ***1994***, *116*, 3159

Ph
O

$Cp_2Ti(PMe_3)_2$, Schlenk tube

Me_3SiCN, PhH

Ph
O
O

80%

Berk, S.C.; Grossman, R.B.; Buchwald, S.L. *J. Am. Chem. Soc.*, ***1994***, *116*, 8593

O
NHMe
Me
Me

1. PhMgCl, $CeCl_3$, -78°C, THF

2. 10% AcOH

Me
Me
Ph
O

85%

Bartoli, G.; Cimarelli, C.; Marcantoni, E.; Palmieri, G.; Petrini, M. *J. Chem. Soc. Chem. Commun.*, ***1994***, 715

O
CO_2Me

1. KH
2. $ZnCl_2$
3. MeCH(Cl)SPh
4. DABCO

O

90x88%

Arnecke, R.; Groth, U.; Köhler, T. *Liebigs Ann. Chem.*, ***1994***, 891

MsTO
CHEt
C_3H_7

$TiCl_4$, $LiAlH_4$, CCl_4

O
C_3H_7
Cl
Et

63% (Z)

Mitani, M.; Kabayashi, Y. *Bull. Chem. Soc. Jpn.*, ***1994***, *67*, 284

1. allyl iodide, rt, 3 min
SmI_2-HMPA
2. MeOH, rt, 3 min
3. 0.1 N aq. HCl, rt
3 min

Ph Ph Ph O O +

(9 : 1) 84%

Miyoshi, N.; Takeuchi, S.; Ohgo, Y. *Bull. Chem. Soc. Jpn.*, ***1994***, *67*, 445

OAc

PhCHO, Ni(cod)$_2$PPh$_3$, Zn
dioxane, 60°C, 4d

Ph O 52%

Masuyama, Y.; Sakai, T.; Kato, T.; Kurusu, Y. *Bull. Chem. Soc. Jpn.*, ***1994***, *67*, 2265

O

2 eq. Mn(OAc)$_3$, Cu(OAc)$_2$
AcOH, 80°C

O

52%

Snider, B.B.; Cole, B.M. *J. Org. Chem.*, ***1995***, *60*, 5376

t-BuMe$_2$SiH, Rh$_4$(CO)$_{12}$
CO, PhH, reflux

t-BuMe$_2$Si O + *t*-BuMe$_2$Si O

63% 14%

Matsuda, I.; Ishibashi, H.; Ii, N. *Tetrahedron Lett.*, ***1995***, *36*, 241

X

O Co(CO)$_4$
NaH

O X

X = NHTs

X = NTs, 69%

X can also be OH, CHNO$_2$, C(CO$_2$Me)$_2$

Bates, R.W.; Devi, T.R. *Tetrahedron Lett.*, ***1995***, *36*, 509

OSiiPr$_3$ — Ph

2.25 CAN , DMF , 0°C , 1.75h

Ph — O — 83%

Evans, P.A.; Longmire, J.M.; Modi, D.P. *Tetrahedron Lett.*, ***1995***, *36*, 3985

OAc — SiMe$_3$

e$^-$, Et$_4$NOTs

O — SiMe$_3$ — 81%

Lin, L.C.; Chueh, L.L.; Tsay, S.-C.; Hwu, J.R. *Tetrahedron Lett.*, ***1995***, *36*,4093

C$_7$C$_{15}$ — HO — CO$_2$Et

(PPh$_3$)$_3$RhCl , Bu$_3$P , toluene

reflux , 12h

C$_7$C$_{15}$ — CO$_2$Et — O — 82%

Saïah, M.K.E.; Pellicciari, R. *Tetrahedron Lett.*, ***1995***, *36*, 4497

t-Bu — O

PhS(O)OMe/KH , THF

t-Bu — O

88%

Resek, J.E.; Meyers, A.I. *Tetrahedron Lett.*, ***1995***, *36*, 7051

C$_5$H$_{11}$ — OMOM — EtO$_2$C

EtO$_2$C ZnI

Me$_3$SiCl , CuCN , HMPA

CO$_2$Et — C$_5$H$_{11}$ — O — OMOM

68%

Crimmins, M.T.; Huang, S.; Guise, L.E.; Lacy, D.B. *Tetrahedron Lett.*, ***1995***, *36*, 7061

Ph — Bt

Bt = benzotriazole

1. BuLi

2. cyclohexanone , ZnBr$_2$, toluene
110°C , 12h

O — Ph

60%

Katritzky, A.R.; Xie, L.; Toader, D.; Serdyuk, L. *J. Am. Chem. Soc.*, ***1995***, *117*, 12015

$iPrNO_2$, DBU , MeCN

rt , 4h

65%

Ballini, R.; Bosica, G. *Tetrahedron*, ***1995***, *51*, 4213

CO_2Me

CO_2Me

1. MeI , $NaCo(CO)_4$, CO

2. K_2CO_3

O MeO_2C CO_2Me

78%

Bates, R.W.; Rama-Devi, T.; Ko, H.-H. *Tetrahedron*, ***1995***, *51*, 12939

8% $MeReO_3$, Cl_3CCF_3

reflux , 5d

80%

Schneider, M.F.; Junga, H.; Blechert, S. *Tetrahedron*, ***1995***, *51*, 13003

Ph——≡——Ph , toluene , 135°C

$Ru(H)_2(CO)(PPh_3)_3$, 1h

Ph Ph O

85% (9:1)

Kakiuchi, F.; Yamamoto, Y.; Chatani, N.; Murai, S. *Chem. Lett.*, ***1995***, 681

SECTION 375: NITRILE - NITRILE

NO ADDITIONAL EXAMPLES

SECTION 376: NITRILE - ALKENE

1. LDA, THF, 3 eq. HMPA, -78°C; 2. MeI, -78°C — 81%, 100% Z

Chang, C.-J.; Fang, J.-M.; Liao, L.-F. *J. Org. Chem.*, ***1993***, *58*, 1754

1. PPh_3, PhH, rt; 2. Ph_2C=C=O, PhH, rt — 51%

Molina, P.; Alajarín, M.; López-Leonardo, C.; Alcántara, J. *Tetrahedron*, ***1993***, *49*, 5153

1. Me_3SiCH_2MgCl, THF, -20°C; 2. PCl_3, THF, 67°C — 36%

Tso, H.-H.; Gilbert, B.A.; Hwu, J.R. *J. Chem. Soc. Chem. Commun.*, ***1993***, 669

2 LiCN, 0.07 $Pd(PPh_3)_4$, 0.07 12-crown-4, PhH, rt — 78%

Piers, E.; Fleming, F.F. *Can. J. Chem.*, ***1993***, *71*, 1867

$Me_2N^{\oplus}$=CCl_2 $Cl^{\ominus}$, CH_2Cl_2; 2 eq. NEt_3, -15°C — 79%

Bouvy, D.; Janousek, Z.; Viehe, H.G. *Bull. Soc. Chim. Belg.*, ***1993***, *102*, 129

MeCN, THF; $Pd_2(dba)_3$•$CHCl_3$ — 60%

Yamamoto, Y.; Al-Masum, M.; Asao, N. *J. Am. Chem. Soc.*, ***1994***, *116*, 6019

O—$SiPh_2Cl$ 1. $Pd_2(dba)_3$•$CHCl_3$, 10h

2. Br , 4% CuI , 2 eq. KF , DMF

$PdCl(PPh_3)_2(CH_2Ph)$, rrt , 4h

CN Me HO 57x90%

Suginome, M.; Kinugasa, H.; Ito, Y. *Tetrahedron Lett.*, ***1994***, *35*, 8635

CHO EtO_2C CN , THF

6% $RuH_2(PPh_3)_4$

CO_2Et 83% CN

Murahashi, S.-I.; Naota, T.; Taki, H.; Mizuno, M.; Takaya, H.; Komiya, S.; Mizuho, Y.; Oyasato, N.; Hiraoka, M.; Hirano, M.; Fukuoka, A. *J. Am. Chem. Soc.*, ***1995***, *117*, 12436

Ph Br KCN , DMF , $NiBr_2(PPh_3)_2$

Zn , PPh_3 , 50°C

Ph CN

99% (95/5 *E/Z*)

Sakakibara, Y.; Enami, H.; Ogawa, H.; Fujimoto, S.; Kato, H.; Kunitake, K.; Sasaki, K.; Sakai, M. *Bull. Chem. Soc. Jpn.*, ***1995***, *68*, 3137

SECTION 377: ALKENE - ALKENE

Cl_3TiO 1. $(Me_2N)_3P=CH_2$ 4 eq. $NaN(SiMe_3)_2$

2. 5 eq. CHO

85%

(7.9:1 *E:Z*)

Reynolds, K.A.; Dopico, P.G.; Sundermann, M.J.; Hughes, K.A.; Finn, M.G. *J. Org. Chem.*, ***1993***, *58*, 1298

$C_{10}H_{21}$—CHO 1. Cp_2ClZr $SnBu_3$

CH_2Cl_2

2. BF_3•OEt_2

$C_{10}H_{21}$

92% (96:4 *E:Z*)

Maeta, H.; Hasegawa, T.; Suzuki, K. *Synlett*, ***1993***, 341

Bu, OTIPS; C_5H_{11}; (cod)RuCpCl; Bu, OTIPS, C_5H_{11} + Bu, OTIPS, C_5H_{11}

(1 : 4) 68%

Trost, B.M.; Indolese, A. *J. Am. Chem. Soc.*, ***1993***, *115*, 4361

PhO, OPh, P, O, O, C_7H_{15}; MgCl; -20°C , 1h; C_7H_{15}

78% (94:6 γ:α)

Yanagisawa, A.; Hibino, H.; Nomura, N.; Yamamoto, H. *J. Am. Chem. Soc.*, ***1993***, *115*, 5879

Br, B, O, O; ZnCl; $Pd(PPh_3)_4$, THF , rt; B, O, O

79%

Mazal, C.; Vaultier, M. *Tetrahedron Lett.*, ***1994***, *35*, 3089

Bu—≡—Bu; 1. Cp_2ZrEt_2 2. 50°C , 1h , OTMS 3. cat. CuCl/2 LiCl , Cl 50°C , 1h; Bu, Bu

87%

Takahashi, T.; Kotora, M.; Kasai, K.; Suzuki, N. *Tetrahedron Lett.*, ***1994***, *35*, 5685

PhS, Br; 1. $C_5H_{11}MgB$, $NiCl_2(dppe)$, THF 2. PhMgBr , $NiCl_2(dppe)$; C_5H_{11}, Ph

52%

Babudri, F.; Fiandanese, V.; Mazzone, L.; Naso, F. *Tetrahedron Lett.*, ***1994***, *35*, 8847

Br

Ph, $Pd(OAc)_2$, 10 kbar, 2d

PPh_3, NEt_3, THF/MeCN, 20°C

Ph

82%

Voigt, K.; Schick, U.; Meyer, F.E.; de Meijere, A. *Synlett*, ***1994***, 189

ArS

1. H_2O_2, AcOH, rt, 14h
2. C_7H_8, $NaHCO_3$

reflux, 3h

79%

MeO_2C

Ar = 4-chlorophenyl

MeO_2C

Schmitz, C.; Harvey, J.N.; Viehe, H.G. *Bull. Soc. Chim. Belg.*, ***1994***, *103*, 105

HO

$(Me_2C{=}CHCH_2)_3In_2Br_3$

DMF, 140°C

OH

OH

+

(65 : 35) 91%

Araki, S.; Imai, A.; Shimizu, K.; Yamada, M.; Mori, A.; Butsugan, Y. *J. Org. Chem.*, ***1995***, *60*, 1841

$Pd(OAc)_2$, TPPTS
aq. MeCN, iPr_2NH

I CO_2Me

Bu

80%

CO_2Me

Bu B O O

Genêt, J.P.; Linquist, A.; Blart, E.; Mouriès, V.; Savignac, M.; Vaultier, M. *Tetrahedron Lett.*, ***1995***, *36*, 1443

OCO_2Et

MeF_2Si C_6H_{13}

$Pd(OAc)_2$, PPh_3, DMF, 28h
rt, sealed tube

Ph

C_6H_{13}

73%

Ph

Matsuhashi, H.; Hatanaka, Y.; Kuroboshi, M.; Hiyama, T. *Tetrahedron Lett.*, ***1995***, *36*, 1539

EtO
O=P—OEt
O
Ph

SmI_2, cat. $Pd(PPh_3)_4$
THF, *t*-BuOH

Ph + Ph

(>99 : <1) 80%

Mikami, K.; Yoshida, A.; Matsumoto, S.; Feng, F.; Matsumoto, Y.; Sugino, A.; Hanamoto, T.; Inanaga, J. *Tetrahedron Lett.*, ***1995***, *36*, 907

OSiMe$_3$
Ph

SiMe$_3$

10% $ZnCl_2$, CH_2Cl_2
25°C, 1.5h

Ph + Ph

(22 : 78) 94%

Yokozawa, T.; Furuhashi, K.; Natsume, H. *Tetrahedron Lett.*, ***1995***, *36*, 5243

C_3H_7—≡

Bu
OH, DMF:H_2O

5% CpRu(cod)Cl, 100°C

Bu
C_3H_7

56%

Trost, B.M.; Indolese, A.F.; Müller, T.J.J.; Trepton, B. *J. Am. Chem. Soc.*, ***1995***, *117*, 615

Bu—≡—Bu

1. Cp_2ZrEt_2
2. CH_2=CHOEt
3. H^+

Bu Bu

50%

Takahashi, T.; Kondakov, D.Y.; Xi, Z.; Suzuki, N. *J. Am. Chem. Soc.*, ***1995***, *117*, 5871

$Me_3Si-C\equiv C-C_6H_{13}$ → (allyl OPh, PMe_3; Cp_2ZrBu_2) → Me_3Si / C_6H_{13} diene

81%

Suzuki, N.; Kondakov, D.Y.; Kageyama, M.; Kotora, M.; Hara, R.; Takahashi, T. *Tetrahedron*, ***1995***, *51*, 4519

10% $RhCl(PPh_3)_3$, PhMe

110°C

80%

Wender, P.A.; Takahashi, H.; Witulski, B. *J. Am. Chem. Soc.*, ***1995***, *117*, 4720

$C_5H_{11}-C\equiv C-C_5H_{11}$

1. $TaCl_5$, Zn, DME, PhH
2. THF
3. allyl $O^- Li^+$
4. aq. NaOH

65%

Takai, K.; Yamada, M.; Odaka, H.; Utimoto, K.; Fujii, T.; Furukawa, I. *Chem. Lett.*, ***1995***, 315

1. $:CCl_2$
2. CsF, DMF

72%

Mitani, M.; Kobayashi, Y.; Koyama, K. *J. Chem. Soc., Perkin Trans. 1.*, ***1995***, 653

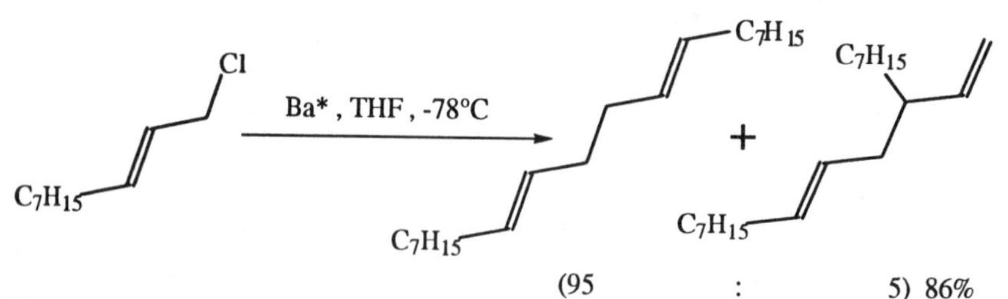

(95 : 5) 86%

Yanagisawa, A.; Hibino, H.; Habaue, S.; Hisada, Y.; Yasue, K.; Yamamoto, H. *Bull. Chem. Soc. Jpn.*, ***1995***, *68*, 1263

SECTION 378: OXIDES - ALKYNES

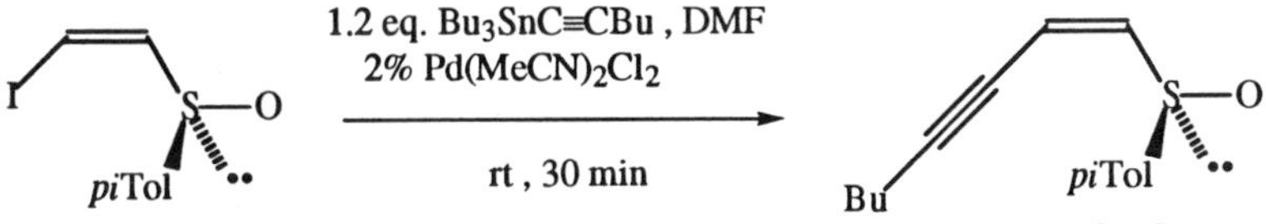

1.2 eq. $Bu_3SnC{\equiv}CBu$, DMF
2% $Pd(MeCN)_2Cl_2$
rt , 30 min

I, S—O, *p*iTol → Bu, S—O, *p*iTol 77%

Paley, R.S.; Lafontaine, J.A.; Ventura, M.P. *Tetrahedron Lett.*, ***1993***, *34*, 3663

SECTION 379: OXIDES - ACID DERIVATIVES

NO ADDITIONAL EXAMPLES

SECTION 380: OXIDES - ALCOHOLS, THIOLS

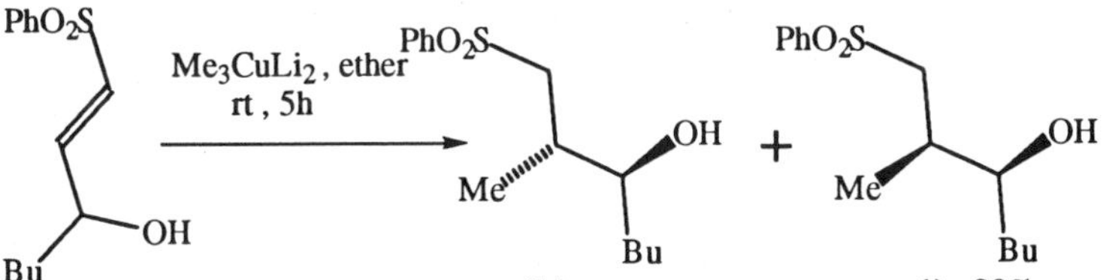

PhO_2S, OH, Bu

Me_3CuLi_2 , ether
rt , 5h

PhO_2S, Me, OH, Bu + PhO_2S, Me, OH, Bu

(11 : 1) 90%

Domínguez, E.; Carretero, J.C. *Tetrahedron Lett.*, ***1993***, *34*, 5803

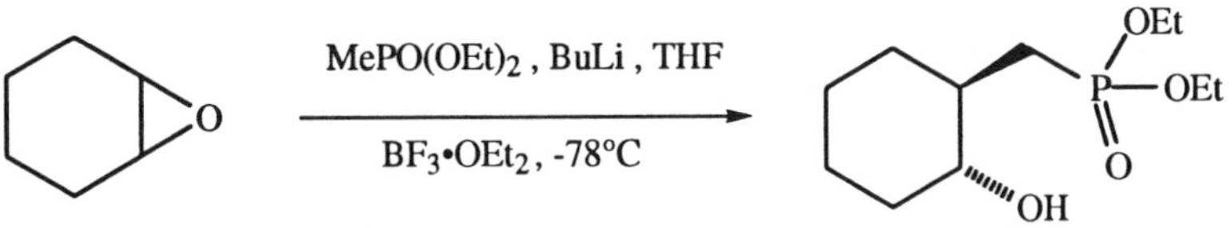

O

$MePO(OEt)_2$, BuLi , THF
$BF_3{\bullet}OEt_2$, -78°C

OEt, P—OEt, O, OH 83%

Li, Z.; Racha, S.; Dan, L.; El-Subbagh, H.; Abushanab, E. *J. Org. Chem.*, ***1993***, *58*, 5779

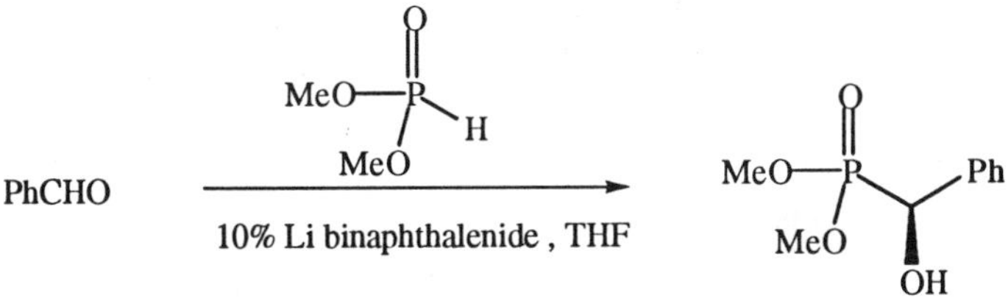

PhCHO

O, MeO—P—H, MeO
10% Li binaphthalenide , THF

O, MeO—P, Ph, MeO, OH 98% (20% ee)

Rath, N.P.; Spilling, C.D. *Tetrahedron Lett.*, ***1994***, *35*, 227

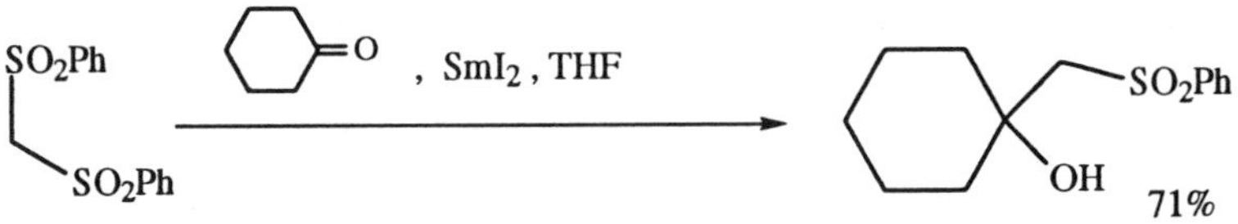

SO_2Ph, SO_2Ph

=O , SmI_2 , THF

SO_2Ph, OH 71%

Chandrasekhar, S.; Yu, J.; Falck, J.R.; Mioskowski, C. *Tetrahedron Lett.*, ***1994***, *35*, 5441

CHO, NPhTh → 3.3 eq. La-Li-(R)-BINOL complex; $MeNO_2$, THF, -40°C, 72h → OH, NO_2, NPhTh 99%

(99:1 *erythro:threo*; 96% ee)

Sasai, H.; Kim, W.-S.; Suzuki, T.; Shibasaki, M. *Tetrahedron Lett.*, ***1994***, *35*, 6123

Cl, O, S, S, O → 1. LHMDS, 0°C; 2. PhCHO, 0°C; 3. HCl, EtOH → HO, Ph, O, S, S, O, Cl + HO, Ph, O, S, S, O, Cl

(8 : 92) 62%

Aggarwal, V.K.; Worrall, J.M.; Adams, H.; Alexander, R. *Tetrahedron Lett.*, ***1994***, *35*, 6167

O, O, P, Ph, Ph → 1. EtLi; 2. $CeCl_3$, THF, -78°C; 3. 10% AcOH → O, OH, P, Ph, Ph, Et

71%

Bartoli, G.; Sambri, L.; Marcantoni, E.; Petrini, M. *Tetrahedron Lett.*, ***1994***, *35*, 8453

OH, OH → H_2O_2, TS-1 → OH, O

TS-1 = metal doped zeolite

quant.

Bovicelli, P.; Lupattelli, P.; Sanetti, A.; Mincione, E. *Tetrahedron Lett.*, ***1994***, *35*, 8477

SO_2Ph, Ph → 1. 2 BuLi, THF; 2. *t*-BuOOH, 15h, -70°C → 20°C → Ph, HO, SO_2Ph, Ph

90% (55:45 *erythro:threo*)

Chemla, F.; Julia, M.; Uguen, D. *Bull. Soc. Chim. Fr.*, ***1994***, *131*, 639

Bakers yeast , 4h

74% (99% ee , S)

Guarna, A.; Occhiato, E.G.; Spinetti, L.M.; Vallecchi, M.E.; Scaarpi, D. *Tetrahedron*, ***1995***, *51*, 1775

SECTION 381: OXIDES - ALDEHYDES

CO/H_2 (600 psi) , CH_2Cl_2

BPh$_3$, dppb , 75•C

Rh(COD)

98%

Totland,K.; Alper, H. *J. Org. Chem.*, ***1993***, *58*, 3326

SECTION 382: OXIDES - AMIDES

$BnNH_2$, NaH , 2.5h

82%

Satoh, T.; Motohashi, S.; Kimura, S.; Tokutake, N.; Yamakawa, K. *Tetrahedron Lett.*, ***1993***, *34*, 4823

1. CAN , MeCN , $NaNO_2$, 1d
2. H_2O

64%

Reddy, M.V.R.; Mehrotra, B.; Vankar, Y.D. *Tetrahedron Lett.*, ***1995***, *36*, 4861

SECTION 383: OXIDES - AMINES

p-TolSO$_2$SePh

AIBN , PhH
reflux

1.3:1 (94%)

Brumwell, J.E.; Simpkins, N.S.; Terrett, N.K. *Tetrahedron Lett.*, ***1993***, *34*, 1215

C_3H_7—≡—P(O)(OEt)(EtO)

95°C , neat , 60h

OEt OEt 71%

C_3H_7

Ruder, S.M.; Norwood, B.K. *Tetrahedron Lett.*, ***1994***, *35*, 3473

X NO_2 NO_2 bakers yeast , 4d X NH_2 NO_2 + X NO_2 NH_2

X = Et (1 : 3) 37%
X = OMe (5 : 1) 36%

Davey, C.L.; Powell, L.W.; Turner, N.J.; Wells, A. *Tetrahedron Lett.*, ***1994***, *35*, 7867

N-Tr H [La , BINOL , Li] , THF

H-P(O)(OMe)(OMe) , 60°V , 18h

NHTr OMe OMe O

47% (69% ee)

Sasai, H.; Arai, S.; Tahara, Y.; Shibasaki, M. *J. Org. Chem.*, ***1995***, *60*, 6656

SECTION 384: OXIDES - ESTERS

O MeO—P MeO Ph

1. 2 eq. BuLi , Ph_3CH , -78°C

2. $ClCO_2Me$, -78°C

O MeO—P MeO Ph CO_2Me

80%

Geirsson, J.K.F.; Njardarson, J.T. *Tetrahedron Lett.*, ***1994***, *35*, 9071

SO_2Ph
EtO_2C
CO_2Et I
1. NaH , THF , 0°C , 30 min
2. DBU , CH_2Cl_2 , 20°C , 1h
SO_2Ph
EtO_2C CO_2Et
73x92%

Short, K.M.; Ziegler Jr., C.B. *Tetrahedron Lett.*, ***1995***, *36*, 355

SECTION 385: OXIDES - ETHERS, EPOXIDES, THIOETHERS

2 Ph OH
1. NaH , THF , rt
2. NO_2 , -40°C
3. 1N HCl
Ph O NO_2
84%

Duffy, J.L.; Kurth, J.A.; Kurth, M.J. *Tetrahedron Lett.*, ***1993***, *34*, 1259

OH
SO_2Tol
LiOO*t*-Bu , THF
rt , 2h
OH
SO_2Tol
O
72%

Bueno, A.B.; Carreño, M.C.; Ruano, J.L.G. *Tetrahedron Lett.*, ***1993***, *34*, 5007

O
SO_2Ph
$Ph_2C{=}CH_2$, $O_2/Mn(OAc)_3$
AcOH , 80°C
SO_2Ph
Ph
Ph O
88%

Qian, C.-Y.; Nishino, H.; Kurosawa, K. *J. Heterocyclic Chem.*, ***1993***, *30*, 209

Ph SO_2Ph
1. BuLi , THF
2. 5 eq. Me_3SiOO*t*-Bu
-40°C , 16h
Ph SO_2Ph
OTMS
92%

Chemla, F.; Julia, M.; Uguen, D. *Bull. Soc. Chim. Fr.*, ***1993***, *130*, 547

OEt
PhSeTos , $Ru(bpy)_3^{2+}$
MeCN , hv
OEt
Tos SePh
75%

Barton, D.H.R.; Csiba, M.A.; Jaszberenyi, J.Cs. *Tetrahedron Lett.*, ***1994***, *35*, 2869

SECTION 386: OXIDES - HALIDES, SULFONATES

Ph–S–CH_3 → ($AgNO_3$, SO_2Cl_2, MeCN) → Ph–S(=O)–CH_2Cl 85%

Kim, Y.H.; Shin, H.H.; Park, Y.J. *Synthesis*, ***1993***, 209

$PhSO_2t$-Bu → (1. BuLi, THF, -78°C; 2. $(PhSO_2)_2NF$) → 2-F-$C_6H_4SO_2t$-Bu 74%

Snieckus, V.; Beaulieu, F.; Mohri, K.; Han, W.; Murphy, C.K.; Davis, F.A. *Tetrahedron Lett.*, ***1994***, *35*, 3465

SECTION 387: OXIDES - KETONES

OTMS → (1. $C(NO_2)_4$, C_5H_{12}, 0.17h, rt; 2. KF, H_2O) → NO_2 95%

Rathore, R.; Lin, Z.; Kochi, J.K. *Tetrahedron Lett.*, ***1993***, *34*, 1859

PhO_2S, Me (epoxide) → (THF, rt, 30 min; $(EtO)_2P(O)H$/NaH) → $(EtO)_2P(O)CH_2C(O)Me$ 70%

Koh, Y.J.; Oh, D.Y. *Tetrahedron Lett.*, ***1993***, *34*, 2147

2-nitrocyclohexanone → (Me_3SiCH_2MgCl, THF, -30°C → 0°C, 1h) → $Me_3SiCH_2C(O)(CH_2)_5NO_2$ 85%

Ballini, R.; Bartoli, G.; Giovannini, R.; Marcantoni, E.; Petrini, M. *Tetrahedron Lett.*, ***1993***, *34*, 3301

Ph(Me)C=N–c-C_6H_{11} → (1. $BnSO_2Cl$, NEt_3, rt; 2. 3N HCl) → $PhC(O)CH_2SO_2Bn$ 79%

Kataoka, T.; Iwama, T. *Synlett*, ***1994***, 1017

NO_2/O_3 , CH_2Cl_2, -10°C — 99% (*o:m* 52:48)

Suzuki, H.; Murashima, T. *J. Chem. Soc., Perkin Trans. 1.*, ***1994***, 903

1. LDA , THF , -90°C; 2. PhO_2S–CH=CH–CO_2Et — 55% (2:1 dr)

Datta, A.; Schmidt, R.R. *Tetrahedron Lett.*, ***1993***, *34*, 4161

$(EtO)_2PONa$, THF , rt , 5h — 60%

Kim, D.Y.; Kong, M.S. *J. Chem. Soc., Perkin Trans. 1.*, ***1994***, 3359

Me_3SiNO_3 , CrO_3 , MeCN , 1d — 61%

Redy, M.V.R.; Kumareswaran, R.; Vankar, Y.D. *Tetrahedron Lett.*, ***1995***, *36*, 7149

SECTION 388: OXIDES - NITRILES

NO ADDITIONAL EXAMPLES

SECTION 389: OXIDES - ALKENES

Ph–CH=CH–$SnBu_3$, THF; 2% $Pd_2(dba)_3 \cdot CHCl_3$, reflux — 87%

Paley, R.S.; de Dios, A.; de la Pradilla, R.F. *Tetrahedron Lett.*, ***1993***, *34*, 2429

SO_2Me $(BzO)_2$, 1d, *t*-BuOH → MeO_2S 55%

Philips, E.D.; Whitham, G.H. *Tetrahedron Lett.*, ***1993***, *34*, 2541

PhO_2S 1. BuLi 2. Me_3SiCl 3. TfOH, Tol, 27°C, 12h → PhO_2S 96 x 92% 93:7 (*E:Z*)

Funk, R.L.; Unstead-Daggett, J.; Brummond, K.M. *Tetrahedron Lett.*, ***1993***, *34*, 2867

Ph $BF_3{\bullet}OEt_2$, rt, CH_2Cl_2 → Ph 75%

Duffy, D.E.; Condit, F.H.; Teleha, C.A.; Wang, C.-L.J.; Calabrese, J.C. *Tetrahedron Lett.*, ***1993***, *34*, 3667

C_8H_{17} PhO_2S IPh BF_4^- $PhSO_2Na$, THF 0•C, 30 min → C_8H_{17} PhO_2S SO_2Ph

Ochiai, M.; Oshima, K.; Masaki, Y.; Kunishima, M.; Tani, S. *Tetrahedron Lett.*, ***1993***, *34*, 4829

OH 5% $Pd(OAc)_2$, 5% dppp $(PhSO_2)_2CH_2$, THF 70°C → $CH(SO_2Ph)_2$ HO H 82% + $(PhO_2S)_2HC$ HO H 7%

Trost, B.M.; Zhi, L.; Imi, K. *Tetrahedron Lett.*, ***1994***, *35*, 1361

5% $Pd(PPh_3)_4$, NEt_3
5 eq. AgN_3 , MeCN
reflux , 3.5h

Br SO$_2$Ph O SO$_2$Ph O 97%

Jin, Z.; Fuchs, P.L. *Tetrahedron Lett.*, ***1993***, *34*, 5205

$P(OEt)_2$ $P(OEt)_2$

1. PhCHO , NaH , THF
2. *t*-BuOK , DMSO

$P(OEt)_2$ O

95x85%

Kiddle, J.J.; Babler, J.H. *J. Org. Chem.*, ***1993***, *58*, 3572

OEt SO$_2$Ph

1. 2 BuLi , 2 HMPA , THF
-78°C → rt
2. aq. NH_4Cl

OEt SO$_2$Ph + OEt SO$_2$Ph

(75 : 25) 53%

Funk, R.L.; Bolton, G.L.; Brummond, K.M.; Ellestad, K.E.; Stallman, J.B. *J. Am. Chem. Soc.*, ***1993***, *115*, 7023

H Me Me H

SO_3

Me Me O_2S—O no yield

Schonk, R.M.; Bakker, B.H.; Cerfontain, H. *Recl. Trav. Chim. Pays-Bas*, ***1993***, *112*, 201

$TolO_2S$

p-TsNa , aq. AcOH
100°C

68% SO_2Tol

Harvey, I.W.; Whitham, G.H. *J. Chem. Soc., Perkin Trans. 1.*, ***1993***, 185, 191

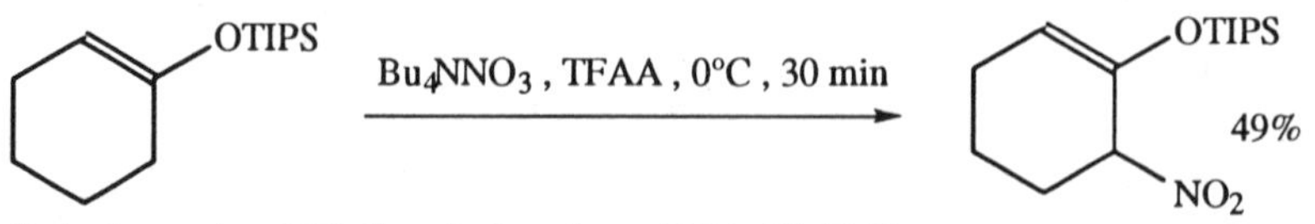
OTIPS
Bu4NNO3 , TFAA , 0°C , 30 min
OTIPS
49%
NO2

Evans, P.A.; Longmire, J.M. *Tetrahedron Lett.*, ***1994***, *35*, 8345

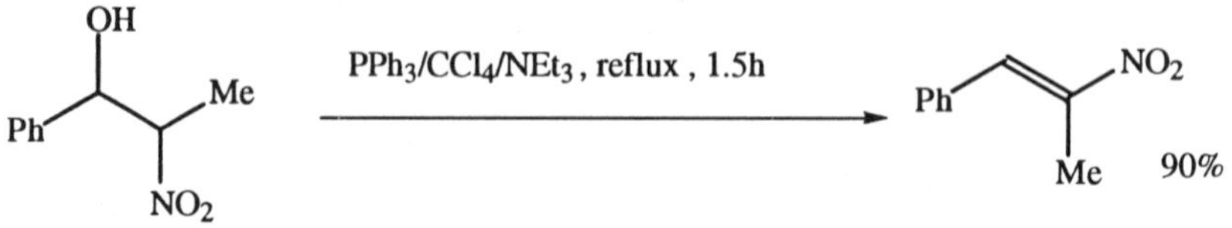
OH
Me
Ph
NO2
PPh3/CCl4/NEt3 , reflux , 1.5h
Ph
NO2
Me
90%

Saikia, A.K.; Barua, N.C.; Sharma, R.P.; Ghosh, A.C. *Synthesis*, ***1994***, 685

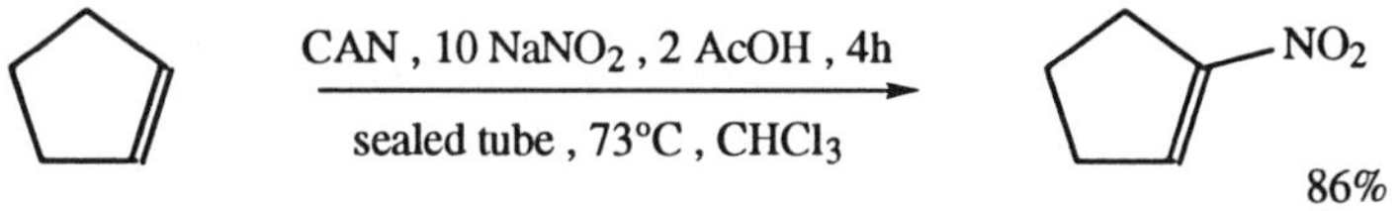
CAN , 10 NaNO2 , 2 AcOH , 4h
sealed tube , 73°C , CHCl3
NO2
86%

Huu, J.R.; Chen, K.-L.; Ananthan, S. *J. Chem. Soc. Chem. Commun.*, ***1994***, 1425

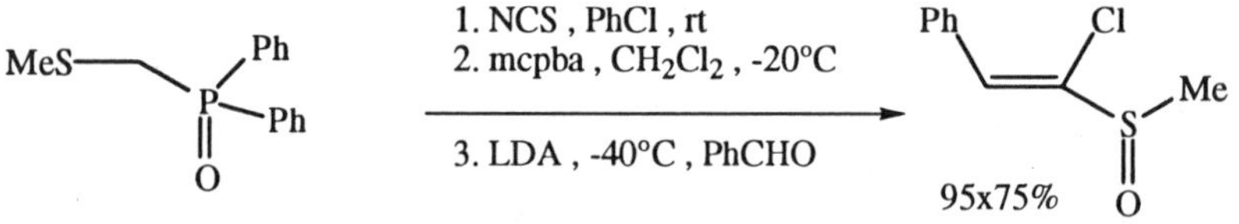
MeS
Ph
P
Ph
O
1. NCS , PhCl , rt
2. mcpba , CH2Cl2 , -20°C
3. LDA , -40°C , PhCHO
Ph
Cl
Me
S
O
95x75%

Otten, P.A.; Davies, H.M.; van der Gen, A. *Tetrahedron Lett.*, ***1995***, *36*, 781

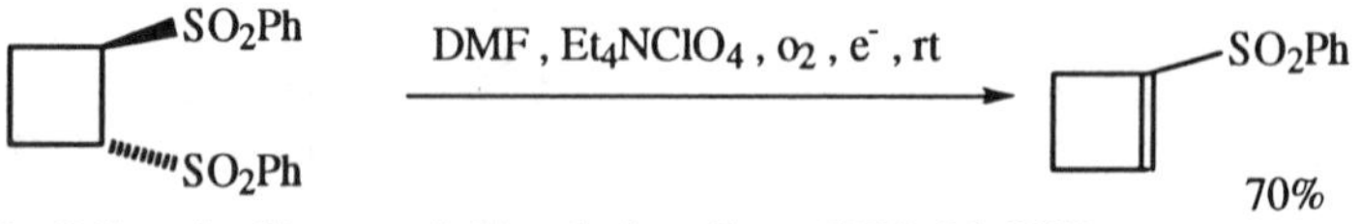
SO2Ph
SO2Ph
DMF , Et4NClO4 , o2 , e- , rt
SO2Ph
70%

Delaunay, J.; Orliac, A.; Simonet, J. *Tetrahedron Lett.*, ***1995***, *36*, 2083

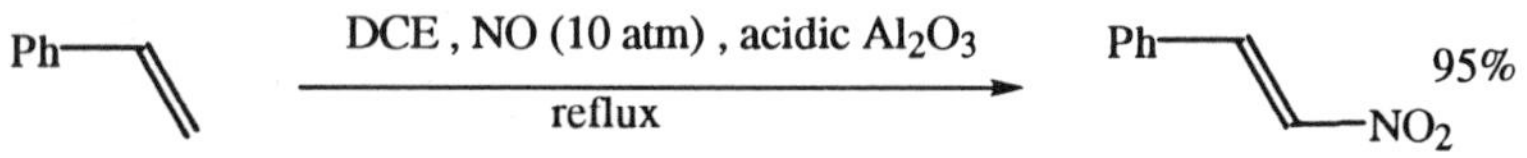
Ph
DCE , NO (10 atm) , acidic Al2O3
reflux
Ph
95%
NO2

Mukaiyama, T.; Hata, E.; Yamada, T. *Chem. Lett.*, ***1995***, 505

SECTION 390: OXIDES - OXIDES

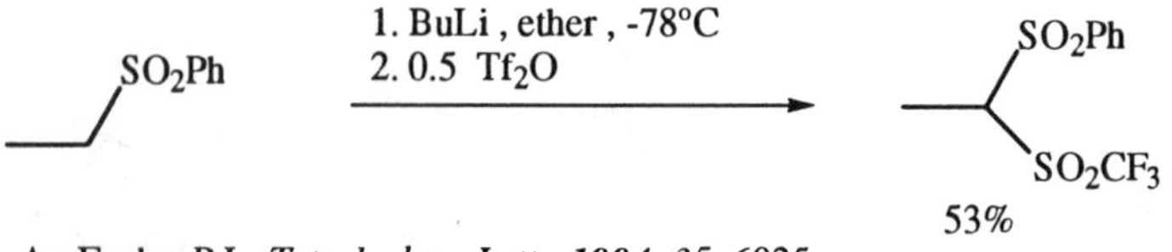
SO2Ph
1. BuLi , ether , -78°C
2. 0.5 Tf2O
SO2Ph
SO2CF3
53%

Mahadevan, A.; Fuchs, P.L. *Tetrahedron Lett.*, ***1994***, *35*, 6025